S. Hrg. 108–504

DETECTION OF LEAD IN THE DC DRINKING WATER SYSTEM

HEARING

BEFORE THE

SUBCOMMITTEE ON FISHERIES, WILDLIFE, AND WATER

OF THE

COMMITTEE ON ENVIRONMENT AND PUBLIC WORKS UNITED STATES SENATE

ONE HUNDRED EIGHTH CONGRESS

SECOND SESSION

APRIL 7, 2004

Printed for the use of the Committee on Environment and Public Works

U.S. GOVERNMENT PRINTING OFFICE

94–604 PDF WASHINGTON : 2006

For sale by the Superintendent of Documents, U.S. Government Printing Office
Internet: bookstore.gpo.gov Phone: toll free (866) 512–1800; DC area (202) 512–1800
Fax: (202) 512–2250 Mail: Stop SSOP, Washington, DC 20402–0001

CONTENTS

Page

APRIL 7, 2004

OPENING STATEMENTS

WITNESSES

ADDITIONAL MATERIAL

DETECTION OF LEAD IN THE DC DRINKING WATER SYSTEM

WEDNESDAY, APRIL 7, 2004

U.S. SENATE,
COMMITTEE ON ENVIRONMENT AND PUBLIC WORKS,
SUBCOMMITTEE ON FISHERIES, WILDLIFE, AND WATER,
Washington, DC.

The subcommittee met, pursuant to notice, at 2:40 p.m. in room 406, Senate Dirksen Building, Hon. Michael D. Crapo (chairman of the subcommittee) presiding.

Present: Senators Crapo, Warner, Clinton, Jeffords [ex officio], and Inhofe [ex officio].

OPENING STATEMENT OF HON. JOHN W. WARNER, U.S. SENATOR FROM THE COMMONWEALTH OF VIRGINIA

Senator WARNER [assuming the chair]. The hearing will come to order.

We are in the process of voting and I will start the hearing simply by giving my statement.

I feel very strongly about this subject and have spent a good deal of time on it and would like to express a few thoughts.

I thank all for bringing this hearing together. It is an issue that directly impacts my constituents in Virginia. I must say I work very closely with the Nation's capital and the governmental authorities there as I have through these 26 years that I have been privileged to be a Senator. Therefore, this is a hearing that affects a good deal of my interests and my career.

My constituents, particularly in Arlington County and the city of Falls Church, because they are the primary customers of the Washington Aqueduct System along with the District of Columbia. The facts of this situation as they have unfolded over the past 2 months are really very disturbing. It is even more disturbing, however, that we and the public became aware of this ongoing problem only after reports in the local media. Every one of the government officials sitting before us on the first panel, the EPA, the Corps, the Water and Sewer Authority, had some measure of knowledge that testing showed some level of lead. That level we will hear more about today and that that water was used for drinking.

The levels we understand here on the committee exceeded the Federal action levels. The rest, we know there was no immediate action taken even though that knowledge was in the hands of responsible government officials. We will have the opportunity today to give a full explanation of that.

We must start correcting the problem. We will have time to address the past but in short, the Corps must determine if a better treatment regime will reduce the leaching of lead from service lines. The Water and Sewer Authority must take immediate steps to provide filters to residents who are served by the over 37,000 service lines that are "undetermined." Those are residents in the category where WASA does not know if they have had lead service lines. If water sampling of some of these residents with "undetermined" service lines reveal lead contamination above the 15 ppb action level, all of these residents, in my judgment, must be provided with water filters. If WASA does not provide the filters for those with undetermined service lines, EPA must exercise its emergency authority to ensure that this occurs because of the imminent public health threat.

I also call on EPA to examine the need to set an enforceable maximum contaminant level, MCL, for lead in drinking water instead of the current 15 ppb. Such an approach may be the only recourse to protect public health and ensure that all necessary steps are taken to reduce lead contamination in drinking water. In this situation, it does not appear that the additional regulatory requirements that should have been implemented when sampling showed high lead levels were enforced by either the EPA or WASA, but you will be given that opportunity today to set the record in your own perspective.

The first order of business that must be taken by responsible agencies before us today is restoring the public trust. I underline that. You have a long way to go and can start with your commitment to provide water filters to all persons served by undetermined service lines. You must also look to ways to finance the full replacement of lead service lines all the way up to the home, not just that portion of the lead service line that is owned by WASA.

I say to my colleagues, I look forward to working with the leadership of this committee to see that we do the responsible thing here in the Congress.

As you see, the second vote has been called and I must go over and make that vote. The committee will stand in recess until the Chairman appears.

[The prepared statement of Senator Warner follows:]

STATEMENT OF HON. JOHN WARNER, U.S. SENATOR FROM THE COMMONWEALTH OF VIRGINIA

Mr. Chairman, thank you for conducting this important hearing this morning. It is an issue that directly impacts my constituents in Arlington County and the city of Falls Church because they are the primary customers of the Washington Aqueduct system along with the District of Columbia.

The facts of this situation, as they have unfolded over the past 2 months are very disturbing. It is even more disturbing, however, that we and the public became aware of this ongoing problem only after reports in the *Washington Post*.

Every one of the government officials sitting before us on the first panel—the EPA, the Corps, and the Water and Sewer Authority—knew that testing showed lead levels in the drinking water were exceeding the Federal action levels. No one took action. No one properly notified the public. And, it seems that you are still finger pointing at each other as to who's to blame.

We must start correcting the problem. In the short-term, the Corps must determine if a better treatment regime will reduce the leaching of lead from service lines. The Water and Sewer Authority must take immediate steps to provide filters to residents who are served by the over 37,000 service lines that are "undetermined." Those are residents in a category where WASA does not know if they have lead service lines. Yet, water sampling of some of these residences with "undetermined" service lines reveal lead contamination above the 15 ppb action level. All of these residences must be provided with water filters.

If WASA does not provide filters for those with "undetermined" service lines, EPA must exercise its emergency authority to ensure that this occurs because of the imminent public health threat. I also call on EPA to examine the need to set an enforceable maximum contaminant level (MCL) for lead in drinking water instead of the current 15 parts per billion action level. Such an approach may be the only recourse to protect public health and ensure that all necessary steps are taken to reduce lead contamination in drinking water. In this situation, it does not appear that the additional regulatory requirements that should have been implemented when sampling showed high lead levels were enforced by either EPA or WASA.

The first order of business that must be taken by the responsible agencies appearing before us today is restoring the public trust. You've got a long way to go. It can start with your commitment to provide water filters to all persons served by "undetermined" service lines. You must also look to ways to financing the full replacement of lead service lines all the way up to the home, not just that portion of the lead service line that is owned by WASA.

Mr. Chairman, I look forward to working with you on the specific challenges facing this region. I also share your concerns that this could be a public health problem confronting any city with lead service lines.

[Recess.]

OPENING STATEMENT OF HON. MICHAEL D. CRAPO, U.S. SENATOR FROM THE STATE OF IDAHO

Senator CRAPO. This hearing will come to order.

Ladies and gentlemen, it is my understanding Senator Warner already convened the hearing so he could make his statement and go vote. Those of us who are here have been on the floor and have voted and we will now officially convene the Senate Committee on Environment and Public Works' hearing of the Subcommittee on Fisheries, Wildlife and Water.

This is an oversight hearing on the detection of lead in the DC drinking water system, focusing on the needed improvements in the public communications and the status of short- and long-term solutions.

Today's hearing will review the detection of lead in DC drinking water, specifically on needed improvements in the communication and the status of immediate actions and long-term solutions. Mayor Anthony Williams of the District of Columbia and Council Member Carol Schwartz were among those who requested that we hold this hearing. I appreciate their efforts and look forward to working with them and others to address the immediate risks of this situation.

Mayor Williams joined me yesterday in a meeting with city residents and Council Member Schwartz has been very helpful and would have come with us but for a regularly scheduled legislative session. Council Member Schwartz has also forwarded to my attention the letter that she and Mayor Williams wrote to the Appropriations Committee last week which, without objection, will be entered in the record.

[The referenced document follows:]

April 1, 2004.

Hon. PETE V. DOMENICI, *Chairman,*
Subcommittee on Energy and Water Development,
Committee on Appropriations,
127 Dirksen Senate Office Building,
Washington, DC.

Hon. HARRY REID, *Ranking Member,*
Subcommittee on Energy and Water Development,
Committee on Appropriations,
156 Dirksen Senate Office Building,
Washington, DC.

DEAR CHAIRMAN DOMENICI AND RANKING MEMBER REID: We are writing you in support of Congresswoman Eleanor Holmes Norton's written request to you for $12,145,000 in response to the elevated lead levels in the District of Columbia's drinking water. The total projected cost to the District for FY '04 is $25,824,101.

As you are aware, the Government of the District of Columbia and the District of Columbia Water and Sewer Authority (WASA) have expended considerable effort and resources to deal with the presence of elevated levels of lead in the drinking water of some residences in the District. The lead appears to be entering the water through corrosion of lead service lines that connect water mains to residences. Although the lead service lines have been in place in most cases for more than fifty years, the elevation of lead levels in the water is a very recent phenomenon.

The District is requesting this Federal support because this drinking water crisis was apparently created by Federal action: specifically by the actions of the United States Army Corps of Engineers and the United States Environmental Protection Agency. The apparent cause for this recent rise in lead levels is a change in the treatment chemistry initiated by the United States Army Corps of Engineers' Washington Aqueduct, the provider of the District's water, and an agency over which your Subcommittee has jurisdiction, and approved by the United States Environmental Protection Agency (EPA), the regulator of the District's water. EPA regulates the District's water because, unlike 49 of the States, the District does not have primacy for regulation in this area, despite multiple requests for such in the past 25 years. Thus, the responsibility and funding for regulation of the District's drinking water is delegated by EPA to its regional office in Philadelphia, rather than to the District.

The total costs that the District is projected to incur for fiscal year 2004 are $25,824,101. We are requesting funding from the Energy and Water Development Subcommittee in the amount of $12,145,000 to help replace lead service lines that are currently part of the District's drinking water infrastructure.

As it now stands, the significant expenditures associated with addressing the lead problem will have to be borne by the District's taxpayers and WASA's ratepayers, which is inherently unfair.

By this letter, we are requesting full reimbursement to both the District and to WASA for these cost.

The regulatory decisions of EPA appear to have generated these costs, and the resources to address them reside within EPA. It would be wholly inappropriate and unjust for the people of the District to bear these costs. Even had the actions of EPA not been the cause of this problem, the structural imbalance the District faces due to its unique situation relative to the Federal Government leaves it with insufficient resources to support its basic needs, let alone extraordinary demands such as have been created by the lead in water problem.

We are working with Congresswoman Norton to advance this critical issue. On behalf of the people of the District of Columbia, we respectfully request a prompt and favorably reply.

Sincerely,

ANTHONY A. WILLIAMS,
Mayor.

CAROL SCHWARTZ,
Councilmember, At-Large,
Chair, Committee on Public Works and the Environment.

Senator CRAPO. Many members of this committee also advocated for this hearing. First, let us recognize the obvious. Clean water is everyone's need and everyone's priority, even though we may sometimes take it for granted. Second, we must appreciate this subject is both complex and emotional. We must proceed accordingly without covering the facts with hard feelings and without disregarding hard feelings with factual arguments.

There is a lot of work to do, some technical and some digging up service lines. In order to do these jobs correctly, we need clear heads, clear messages and clear agreements. We need to fix this problem and we must fix it now. An important fact already in evidence is that lead is toxic but historically, it was used for plumbing and as an ingredient in paint and automotive fuel. Because plumbing, paint chips, dust and exhaust fumes surround most Americans, lead is very troubling.

We have made progress by phasing out leaded gasoline and more slowly rehabilitating lead painted homes. Lead in plumbing represents an enormous part of the Nation's need to replace and rehabilitate its water system. Health risks of lead are generally widely accepted and a recent study may add new concerns. Lead poisoning delays physical and mental development in children and in adults, causes increases in blood pressure and after long-term exposure, damages kidneys.

Another important fact is that many people were surprised in January of this year when they read in the newspaper that lead levels were high, in some cases very high in many homes in Washington, DC. The fact that people were surprised means that to communicate effectively from now on, we must communicate differently from now on. In addition to the obvious reason for communicating risks to the public, it is especially important in managing lead. By the nature of the problem, we will be living with lead in our home environments for a long time. Therefore, it is particularly important that we are vigilant.

The members of the first panel are here to explain efforts to repair missed communications with the public, to review lessons learned to date and to explain intended efforts or policy changes for better communication of risks in the future. Every Senator and staff member knows the challenge of communicating risks because we have been evacuated, some of us twice, from our offices when attacked with anthrax and ricin. Since those episodes, we have installed an announcement procedure by which we hear immediately of every suspicious substance found in our buildings. Even though most of these announcements are followed by an all clear message, we are prepared for the sight of a safety team wearing protective clothing as they hurry to investigate. People should have the information they need to judge risks for themselves.

In addition to the issue of communication, we also want to hear the latest developments in finding and eliminating lead. Also, I am specifically interested in how the public will be included in deliberations and decisions about this problem.

The second panel is here to describe health risks of lead, a personal experience with this issue, and professional advice about how communications can be improved.

I appreciate the witnesses from every perspective and their commitment to join us today. I strongly encourage the first panel to remain to hear what the second panel has to say. To all those who are following this issue, remember that this situation is a specific and serious example of a national issue. Depending on where you live and work, your water infrastructure is anywhere from 40 to

140 years old. That means many Americans are already experiencing either the problems of an aging system or the limits of a small system.

All systems need to work reliably everywhere and for everyone. To accomplish this will require more money than we currently have. In 2000, the Water Infrastructure Network estimated that current infrastructure needs could cost around $1 trillion over the next 15 to 20 years. That is around $20 billion per year more than current spending.

The EPA's own GAP analysis from 2002 estimates almost $300 billion in infrastructure resource shortfalls over 20 years. I raised this issue on the Senate floor and won unanimous approval to increase available spending authority for water infrastructure and I am pushing to retain this amendment in the budget resolution conference.

Today's hearing is about Washington's particular reasons for a new effort to upgrade the Nation's water systems. I encourage all cities to heed the warning and to listen to the call.

[The prepared statement of Senator Crapo follows:]

STATEMENT OF HON. MICHAEL D. CRAPO, U.S. SENATOR FROM THE STATE OF IDAHO

Today's hearing will review the detection of lead in DC drinking water; specifically on needed improvements in communication and the status of immediate actions and long-term solutions.

Mayor Anthony Williams of the District of Columbia and Councilmember Carol Schwartz were among those who requested that we hold this hearing. I appreciate their efforts and look forward to working with them and others to address the immediate risks of this situation.

Mayor Williams joined me yesterday in a meeting with city residents and Councilmember Schwartz has been very helpful and would have come with us but for a regularly scheduled legislative session.

Councilmember Schwartz has also forwarded to my attention the letter that she and Mayor Williams wrote to the Appropriations Committee last week.

Many members of this committee also advocated for this hearing.

OVERVIEW ON THE ISSUE

First, let us recognize the obvious: clean water is everyone's need and everyone's priority, even though we may take it for granted.

Second, we must appreciate that this subject is both complex and emotional. We must proceed accordingly, without coloring facts with hard feelings, and without disregarding hard feelings with factual arguments. There is a lot of work to do: some technical and some digging up of service lines. In order to do these jobs correctly we need clear heads, clear messages, and clear agreements. We need to fix this problem, and we must fix it now.

An important fact already in evidence is that lead is toxic, but historically was used for plumbing and as an ingredient in paint and automobile fuel. Because plumbing, paint chips and dust, and exhaust fumes surround most Americans, lead is very troubling. We have made progress by phasing out leaded gasoline and—more slowly—rehabilitating lead-painted homes. Lead in plumbing represents an enormous part of the Nation's need to replace and rehabilitate its water system.

Health risks of lead are generally widely accepted, and a recent study may add new concerns. Lead poisoning delays physical and mental development in children and, in adults, causes increases in blood pressure and—after long-term exposure—damages kidneys.

Another important fact is that many people were surprised in January of this year when they read in the newspaper that lead levels were high—in some cases very high—in many homes in Washington. The fact that people were surprised means that to communicate effectively from now on we must communicate differently from now on.

In addition to the obvious reason for communicating risks to the public, it is especially important in managing lead. By nature of the problem, we will be living with lead in our home environments for a long time; therefore, it requires particular vigilance.

CHARGE TO THE FIRST PANEL

The members of the first panel are here to explain efforts to repair missed communications with the public, to review lessons learned to date, and explain intended efforts or policy changes for better communicating risks in the future.

Every Senator and staff member knows the challenge of communicating risks because we have been evacuated from our offices—some of us twice—when attacked with anthrax and ricin. Since those episodes we have installed an announcement procedure by which we hear immediately of every suspicious substance found in our buildings. Even though most of these announcements are followed by an "all clear" message, we are prepared for the sight of a safety team wearing protective clothing as they hurry to investigate. People should have the information they need to judge risks for themselves.

In addition to the issue of communication, we also want to hear of the latest developments in finding and eliminating the lead. Also, I am specifically interested in how the public will be included in deliberations and decisions about this problem.

CHARGE TO SECOND PANEL

The second panel is here to describe the health risks of lead, relate personal experiences with this issue, and offer professional advice about how communications could be improved. I appreciate your commitment to join us today. I strongly urge the first panel to remain to hear what the second panel has to say.

THE NATIONAL NEED

To all who are following this issue, remember: this situation is a specific and serious example of a national issue. Depending on where you live and work, your water infrastructure is anywhere from 40 to 140 years old. That means many Americans are already experiencing either the problems of an aging system or the limits of a small system. All systems need to work reliably everywhere and for everyone.

To accomplish this will require more money than we currently have. In 2000, the Water Infrastructure Network estimated that current infrastructure needs could cost around $1 trillion over the next 15–20 years.[1] This is around $20 billion per year more than current spending. The EPA's own "Gap Analysis" from 2002 estimates almost $300 billion in infrastructure resource shortfalls over 20 years.[2]

I raised this issue on the Senate floor and won unanimous approval to increase available spending authority for water infrastructure—and I am pushing to retain this amendment in the Conference on the Budget Resolution.

Today's hearing is about Washington's particular reason for a new effort to upgrade the Nation's water systems. I encourage all cities to heed the warning and answer the call.

Senator CRAPO. At this point, we will turn to our Ranking Member, Senator Jeffords, for his opening statement.

OPENING STATEMENT OF HON. JAMES M. JEFFORDS, U.S. SENATOR FROM THE STATE OF VERMONT

Senator JEFFORDS. Thank you, Mr. Chairman, and good afternoon to everyone.

I would like to start by thanking Senator Crapo and Chairman Inhofe for granting the Minority's request to hold this hearing. The residents of Washington, DC deserve to get answers from Federal and local officials on why there is lead in the DC water and why residents were not notified that safe drinking is a right, not a privilege.

---

[1] Water Infrastructure Network. April 2000. Clean and Safe Water for the 21st Century. Link from *http://www.win-water.org/*; direct from: *http://www.amsa-cleanwater.org/advocacy/winreport/winreport2000.pdf*.

[2] EPA. 2002. The Clean Water and Drinking Water Infrastructure Gap Analysis. Link: *http://www.epa.gov/owm/* (click "Featured Information"); direct: *http://www.epa.gov/owm/gapreport.pdf*.

This committee has oversight responsibilities for the Army Corps of Engineers, the Environmental Protection Agency, as well as, the Safe Drinking Water Act. Each of us in the Senate has a special oversight responsibility for the District and its residents. I have lived in Washington for a long time and I take this responsibility seriously. At one time, I was kind of de facto Mayor for a while but that was a long time ago.

Many of us have switched to bottled water. I am disturbed because bottled water is not regulated in that manner, the tap water is. We cannot even find out what is in bottled water.

Yesterday, Senator Crapo and I met with a group of Washington parents. Their outrage and sadness at the effect on their children was unanimous. Their charges to us were, "fix this situation and don't let it happen again." I am committed to doing everything in our power to solve this problem and I know the Chairman of the subcommittee agrees with me.

My overriding question today for our witnesses is how did we get here? How did we get to the point where the future of the children living in our Nation's capital are threatened every day by the water in their faucets and bath tubs? How did we get to the point where water tests were conducted revealing startlingly high lead levels, but yet that information was never provided to residents who unnecessarily exposed themselves, their unborn children and their children to lead contaminated water? How did we get to the point where it takes congressional hearings and newspapers to expose this action? How did we get to the point where 2 years after the fact, EPA announces that WASA did not comply with the requirements of the lead and copper rule? How did we get to the point where the research from over a year ago showed that lead exposure levels below the current standard of 10 ppb have an adverse effect on children's intelligence level and yet the Federal Government has not responded?

Lead is a serious health threat to children and to pregnant women. It is particularly dangerous for children who retain about 68 percent of the lead that enters their bodies while adults retain only about 1 percent. Children exposed to lead experience low birth weight, growth retardation, mental retardation, learning disabilities and other effects. It is an also particularly harmful drug for women in pregnancy.

I have already mentioned our meeting yesterday with a group of DC parents and I want to take this chance to share a few more thoughts from other concerned parents. I ask unanimous consent that a letter and petition from PureWater DC, an Internet-based site for parents concerned about ongoing water issues in DC be placed in the record—13,077 people signed this petition expressing their concern and the expectation for District officials to take action quickly to fix the problem.

Senator CRAPO. Without objection, the petition will be made a part of the record.

[The referenced document follows:]

PUREWATERDC.COM
citizens for clean water

info@purewaterdc.com

PRESS RELEASE
FOR IMMEDIATE RELEASE
April 7, 2004

Contact: Josh Levinson
Pure Water DC
202/746-8106

1,377 Residents to Public Officials: "We Are Concerned About Lead Contamination!"

Two DC residents create website as public service, gather petition signatures

Washington, D.C. – Today the Co-Founders of PUREWATERDC.com are submitting a petition signed by 1377 individuals to public officials involved in the ongoing lead contamination crisis in the District of Columbia. The 1377 individuals who signed the petition took it upon themselves to go to www.purewaterdc.com and pledge the following:

"I am concerned by recent reports that tap water in thousands of District homes contains dangerous levels of lead contamination. I am worried about my health and the health of others. I expect District officials to take swift action to remedy this serious problem."

PUREWATERDC.com was launched by two concerned DC residents – Paul McKay (Ward 1) and Josh Levinson (Ward 6) – in response to reports that the D.C. Water and Sewer Authority had found dangerously elevated lead readings in samples of District drinking water.

The website, which Paul and Josh created and are maintaining at their own expense, provides a clearinghouse of information regarding lead contamination in general, steps residents can take to test lead levels in water and blood, preventative measures, and relevant local news, as well as a mechanism – specifically, a petition – for District residents to show public officials that they are concerned and expect swift remedial action.

The petition will be submitted to the following individuals: Sen. James M. Inhofe (R-OK), *Chairman, Senate Committee on Environment and Public Works*, Sen. James M. Jeffords (D-VT), *Ranking Member, Senate Committee on Environment and Public Works*, Sen. Mike Crapo (R-ID), *Chairman, Environment and Public Works Subcommittee on Fisheries, Wildlife, and Water*, Rep. Thomas M. Davis III (R-VA), *Chairman, House Committee on Government Reform*, Rep. Paul Gillmor (R-OH), *Chairman, House Energy and Commerce Subcommittee on Environment and Hazardous Materials*, Rep. Eleanor Holmes Norton (D-DC), Anthony A. Williams, *Mayor, District of Columbia*, Robert C. Bobb, *Deputy Mayor and City Administrator, District of Columbia*, Carol Schwartz (R) At-Large, *Council of the District of Columbia, Chair, Committee on Public Works and the Environment*, Donald Welsh, *Administrator, Region III, U.S. Environmental Protection Agency*, Jerry N. Johnson, *General Manager, District of Columbia Water and Sewer Authority*, and Thomas P. Jacobus, *General Manager, Washington Aqueduct, Baltimore District, US Army Corps of Engineers*.

PUREWATERDC.COM
citizens for clean water

info@purewaterdc.com

To: Sen. James M. Inhofe (R-OK), *Chairman, Senate Committee on Environment and Public Works*
Sen. James M. Jeffords (D-VT), *Ranking Member, Senate Committee on Environment and Public Works*
Sen. Mike Crapo (R-ID), *Chairman, Environment and Public Works Subcommittee on Fisheries, Wildlife, and Water*
Rep. Thomas M. Davis III (R-VA), *Chairman, House Committee on Government Reform*
Rep. Paul Gillmor (R-OH), *Chairman, House Energy and Commerce Subcommittee on Environment and Hazardous Materials*
Rep. Eleanor Holmes Norton (D-DC)
Anthony A. Williams, *Mayor, District of Columbia*
Robert C. Bobb, *Deputy Mayor and City Administrator, District of Columbia*
Carol Schwartz (R) At-Large, *Council of the District of Columbia, Chair, Committee on Public Works and the Environment*
Donald Welsh, *Administrator, Region III, U.S. Environmental Protection Agency*
Jerry N. Johnson, *General Manager, District of Columbia Water and Sewer Authority*
Thomas P. Jacobus, *General Manager, Washington Aqueduct, Baltimore District, US Army Corps of Engineers*

From: Paul McKay and Josh Levinson, *Co-Founders, PUREWATERDC.com*

Date: April 7, 2004

Re: Lead Contamination Petition from PUREWATERDC.com

Please find attached a petition signed by 1377 individuals who took it upon themselves to go to www.purewaterdc.com and pledge the following:

"*I am concerned by recent reports that tap water in thousands of District homes contains dangerous levels of lead contamination. I am worried about my health and the health of others. I expect District officials to take swift action to remedy this serious problem.*"

PUREWATERDC.com was launched by two concerned DC residents – Paul McKay (Ward 1) and Josh Levinson (Ward 6) – in response to reports that the D.C. Water and Sewer Authority had found dangerously elevated lead readings in samples of District drinking water. The website, which Paul and Josh created and are maintaining at their own expense, provides a clearinghouse of information regarding lead contamination in general, steps residents can take to lead levels in water and blood, preventative measures, and relevant local news, as well as a mechanism – specifically, a petition – for District residents to show public officials that they are concerned and expect swift remedial action.

PUREWATERDC.COM
citizens for clean water

info@purewaterdc.com

"I am concerned by recent reports that tap water in thousands of District homes contains dangerous levels of lead contamination. I am worried about my health and the health of others. I expect District officials to take swift action to remedy this serious problem."

No.	First Name	Last Name	City	State
1	Tim	Abrams	Washington	DC
2	Shawn	Adams	Washington	DC
3	Dan	Adams	Washington	DC
4	Dorcas	Adkins	Washington	DC
5	martha	adler	Washington	DC
6	Bill	Adler	Washington	DC
7	David	Adler	Washington	DC
8	Alicia	Aebersold	Washington	DC
9	Chris	Agharabi	Washington	DC
10	Kate	Ahmann	Washington	DC
11	Margaret	Ahmann	Washington	DC
12	Lasana	Akachi	Washington	DC
13	Nilgun	Akselioglu	Washington	DC
14	Adam	Alban	Washington	DC
15	Suzanne	Alberga	Washington	DC
16	Virginia	Alexander	Washington	dc
17	Marleen	Alexander	Temple Hills	MD
18	Joyce	Alexander Bailey	Washington	DC
19	ann	alger	Alexandria	VA
20	Edmee	Allain	Washington	DC
21	Ms. Schuyler	Allen	New York	NY
22	George	Allen	Washington	DC
23	Ewen	Allison	Washington	DC
24	Mary-Ann	Allison	Washington	DC
25	Mariah	Almond	Washington	DC
26	Katharine	Anderson	Washington	DC
27	Elizabeth	Anderson	Washington	DC
28	John	Anderson	Washington	DC
29	Susan	Anderson	Washington	DC
30	Akel	Andrea	Washington	DC
31	Mary	Antrim	Washington	DC
32	Andrea	Arnold	Washington	DC
33	Cameron	Arterton	Washington	DC

PUREWATERDC.COM
citizens for clean water

info@purewaterdc.com

34	susan	ascher	washington	DC
35	Angela	Ashe	Washington	DC
36	carrie	askew swann	washington	dc
37	Liane	Atlas	Washington	DC
38	Erin	Atwell	Washington	DC
39	Rose	Audette	Washington	DC
40	Sweena	Aulakh	Washington	DC
41	Florence	Auld	Washington	D.C.
42	Richard	Ayres	Washington	DC
43	Jarrod	Backous	Washington	DC
44	Beverley	Bailey	Washington	D.C.
45	Jan	Bailey	Washington	DC
46	Sidney	Bailin	Washington	DC
47	Erica	Baker	Washington	DC
48	Jacqueline	Baker	washington	DC
49	Dean	Baker	Washington	DC
50	Marilyn	Baker	Washington	DC
51	Carl	Ballard	Washington	DC
52	Michael	Balmoris	washington	dc
53	Mary	Baluss	Washington	DC
54	Kenneth	Bamberger	Washington	DC
55	Jonathan	Banks	Washington	DC
56	Alicia	Bannon	Washington	DC
57	Rachel	Barbanel-Fried	Washington	DC
58	Joy	Barksdale	Washington	DC
59	Amanda	Barlow	Washington	DC
60	Richard	Barnes	Washington	DC
61	Paula	Barnes	Washington	DC
62	Richard	Barnes	Washington	DC
63	Robert	Barry	Washington	DC
64	Mila	Bartos	Washington	DC
65	Peter	Bass	Washington	DC
66	Brenda	Battle	Washington	DC
67	Stephanie	Bauer	Washington	DC
68	Tony	Bayne	Washington	DC
69	David	Beckel	Washington	DC
70	Peter	Beebe	Washington	DC
71	Michael	Behm	Washington	DC
72	Bryna	Belitsky	Washington	DC
73	Jennifer	Bell	Washington	DC
74	Kim	Bell	Washington	DC
75	Elizabeth	Bell	Washington	DC

PUREWATERDC.COM
citizens for clean water

info@purewaterdc.com

76	Sharon	Belliveau	Washington	DC
77	Ann	Beman	Washington	DC
78	Jeremy	Ben-Ami	Washington	DC
79	Lillian	Benjamin	Washington	DC
80	Laura	Bennett	Washington	DC
81	William	Bennett	Washington	DC
82	Robert	Benson	Washington	DC
83	Sarah	Benton	Washington	DC
84	Marta	Beresin	Washington	DC
85	Ellen	Berg	Washington	DC
86	Kimberly	Berger	Washington	DC
87	Laura	Berger	Washington	DC
88	Richard L.	Berger	Washington	DC
89	lisa	Bergman	Washington	DC
90	Carl	Bergman	Washington	DC
91	Marta	Berhane	Washington	D.C.
92	Amy	Berman	Washington	DC
93	Ali	Bhanji	Washington	DC
94	Shannon	Bieter	Washington	DC
95	Christie	Billingsley	Washington	DC
96	Alisa	Biran	Washington	DC
97	Rob	Black	Washington	DC
98	Richard	Black	Washington	DC
99	allie	Black	Washington	DC
100	Rob	Black	Washington	DC
101	Nathalie	Black	Washington	DC
102	althea	Black	Washington	DC
103	Jessica	Blake Hawke	Washington	DC
104	Sarah	Boasberg	Washington	DC
105	Ariella	Bock	Washington	DC
106	Catherine	Bohigian	Washington	DC
107	Nicolas	Bohn	Washington	DC
108	Maggie	Boland	Washington	DC
109	Matthew	Bond	Washington	DC
110	Judith	Bonderman	Washington	DC
111	Susan	Bonfield	Washington	DC
112	Jennifer	Bonilla	Washington	DC
113	Robert	Boorstin	Washington	DC
114	Sidney	Booth	Washington	DC
115	Marc	Borbely	Washington	DC
116	Gloria	Borland	Washington	DC
117	ben	borns	washington	dc

PUREWATERDC.COM
citizens for clean water

info@purewaterdc.com

118	VASSILIS	BOROVAS	Washington	DC
119	Dirk	Bos	Washington	DC
120	Mary Ellen	Bou	Washington	DC
121	bridgette	bourge	Washington	DC
122	Evelyn	Bourne-Gould	Washington	DC
123	Heather	Boushey	Washington	DC
124	Jean	Boutcher	Washinton	DC
125	Meredith	Bove	Washington	DC
126	Paul	Bove	Washington	DC
127	Ann	Bowles	Washington	DC
128	Elizabeth Winn	Bowman	Washington	dc
129	Christian Winn	Bowman	Washington	dc
130	Varna	Boyd	Washington	DC
131	Angela	Bradbery	Washington	DC
132	Tony	Bradley	Washington	DC
133	Barbara	Bradley	Washington	DC
134	Dorothea	Brady	Washington	DC
135	Lael	Brainard	Washington	D.C.
136	Martha	Brant	Washington	DC
137	Patricia	Bray	Washington	DC
138	Shelley	Brazier	Washington	DC
139	Kathy	Brenneman	Washington	DC
140	Marc	Breslaw	Rockville	MD
141	Joseph	Breslow	washington	DC
142	Steven	Briggerman	Washington	DC
143	amable	brito	Washington	dc
144	Juanita	Britton	Washington	DC
145	shelley	brody	Washington	DC
146	Miranda	Bronsgeest	Washington	DC
147	Stefan	Brooks	Washington	DC
148	Yvonne	Brooks-Little	Silver Spring	MD
149	cecile	brou	washington	dc
150	Sheila	Brown	Washington	DC
151	stuart	brown	washington	dc
152	Shawn	Brown	Washington	DC
153	Tracy	Brown	Washington	DC
154	Ethan	Brown	Washington	DC
155	Lisa	Brown	Washington	DC
156	wendy	brown	silver spring	md
157	Rebecca	Brown	Washington	DC
158	Allison	Brown	Washington	DC
159	Sarah	Browning	Washington	DC

PUREWATERDC.COM
citizens for clean water

info@purewaterdc.com

160	Lori	bruce	Washington	DC
161	Daniel	Brumberg	Washington	DC
162	Patrick	Brundage	Campbell	CA
163	Patrick	Brundage	Campbell	CA
164	Elizabeth	Buchanan	Washington	DC
165	Wliam	Buchman	Washington	DC
166	Lucy	Buckley	Washington	DC
167	Eric	Bucsela	Washington	DC
168	Shay	Bugajinsky	Washington	DC
169	Meaghan	Burdick	Washington	DC
170	john	bursk	washington	dc
171	Brent	Bushey	Washington	DC
172	Gabrielle	Bushman	Washington	DC
173	LaVerne	Butler	Washington	DC
174	jane	byerley	washington	Dc
175	Rebecca	Cague	Washington	DC
176	elizabeth	caiafa	Washington	DC
177	Kitty	Calvert	Washington	DC
178	Elizabeth	Calvit	Washington	DC
179	Darrin	Cambridge	Washington	DC
180	Jessica	Camellakis	Washington	DC
181	Megan	Campbell	Washington	DC
182	Adrienne	Cantio	Washington	dc
183	Bryan	Cantio	Washington	dc
184	Brent	Cantley	Washington	DC
185	Albert	Caprioglio	Washington	DC
186	Nora	Carbine	Washington	DC
187	Thomas	Carmody	Washington	DC
188	Patrick	Carome	Washingtion	DC
189	Camilla	Carpenter	Washington	DC
190	Eric	Carr	Washington	DC
191	Angela	Carson	Washington	DC
192	David	Carter	Washington	DC
193	Beth	Cartland	Washington	DC
194	Sharon	Cascone	Washington	DC
195	Jill	Cashen	Washington	DC
196	peter	cashion	washington	dc
197	riccardo	cassiani-ingoni	washington	DC
198	Jill	Castleman	Washington	DC
199	Elizabeth	Cattaneo	Arlington	VA
200	Jane	Caulton	Washington	DC
201	Machaela	Cavanaugh	Washington	dc

PUREWATERDC.COM
citizens for clean water

info@purewaterdc.com

202	Karl	Celarier	Washington	DC
203	Kimberly	Cestari	Washington	DC
204	Zoe	Chafe	Washington	DC
205	Marc	Chafetz	Washington	DC
206	Anna	Chamberlin	Washington	DC
207	Daryl	Chamblee	Washington	DC
208	Linda	Chan	Washington	DC
209	Audrey	Chang	Washington	DC
210	Lorig	Charkoudian	Washington	DC
211	Pablo	Chavez	Washington	DC
212	Katherine	Chenard	Washington	DC
213	SImona	Cherif	Washington	DC
214	Ann	Chernicoff	Washington	DC
215	Matthew	Chester	Washington	DC
216	Patricia	Chittams	Washington	DC
217	Fran	Chock	Washington	DC
218	Jill	Chodorov	Washington	DC
219	Jacqueline	Chun	Washington	DC
220	Bartosz	Chwalek	Washington	DC
221	Leah	Clapman	Washington	DC
222	John	Clark	Washington	DC
223	margaret	clark	washington	dc
224	Martha	Clarke	Washington	DC
225	Casey	clines	Washington	DC
226	Margaret	Cloherty	Washington	D. C.
227	Christen	Clute	Washington	DC
228	Mary	Clute	Washington	DC
229	Ed	Coe	Washington	DC
230	Andrew	Coelho	Washington	DC
231	Alan	Cohen	Washington	DC
232	Zohara	Cohen	Washington	DC
233	Carrie	Cohen	Washington	D.C.
234	Marlene	Cohen	Washington	D.C.
235	Christopher	Cole	Washington	DC
236	Tina	Coleman	Washington	DC
237	Charlotte	Collins	Elkridge	MD
238	Eileen	Collins	Washignton	DC
239	susan	comfort	washington	dc
240	Tracey	Conaty	Washington	DC
241	Richard	Condit	Washington	DC
242	Catherine	Conger	Washington	DC
243	Paula	Conhain	Washington	DC

PUREWATERDC.COM
citizens for clean water

info@purewaterdc.com

244	Annemargaret	Connolly	Washington	DC
245	Annmarie	Contaste	Washington	DC
246	shaw	cora	washington	dc
247	Kami	Corbett	Washington	DC
248	Martha	Corbett	Washington	DC
249	Leila	Corcoran	washington	d.c.
250	Leslie	Cordes	Washington	DC
251	Nigel	Cosh	Washington	DC
252	Cathrine	Cotman	washington	DC
253	Mary	Cotton	Washington	DC
254	Aline	Coudouel	Washington	DC
255	Ayesha	Court	Washington	DC
256	Karen	Cowden	Washington	DC
257	Elizabeth	Cox	Washington	DC
258	Billy	Cox	Washington	DC
259	Kristen	Cox	Washington	DC
260	Ellen	Coyne	Washington	DC
261	Dorothy	Craft	Washington	DC
262	Sally	Craig	Washington	DC
263	Joan	Crandall	Washington	DC
264	Emma	Cravey	Washington	DC
265	celia	crawford	washington	d.c.
266	Charles	Crettier	Washington	DC
267	DeLanson	Crist	Washington	DC
268	Crystal	Crittenden	Washington	DC
269	Brian	Crowley	Washington	DC
270	Ellen	Cull	Washington	DC
271	Daphne	Cunningham	Washington	DC
272	Alisa	Cunningham	Washington	DC
273	Keith	Cunningham	Washington	DC
274	Jeffrey	Curry	Washington	DC
275	Umekki	Curry	washington	DC
276	Lisa	Cutler	Washington	dc
277	George	d'Adhemar	Washington	DC
278	Karin	Dalichow	Washington	DC
279	D. Michael	Daly	Washington	DC
280	Adrienne	Danforth	Washington	DC
281	Henry	Danforth	Washington	DC
282	Judith	Daniel	Washington	DC
283	Giovanni	Daniele	Washington	DC
284	Robert	Dardano	Washington	DC
285	Melissa	Dasgupta	Washington	DC

PUREWATERDC.COM
citizens for clean water

info@purewaterdc.com

286	Maria	Davidis	Washington	DC
287	Nestor	Davidson	Washington	DC
288	Robert R.	Davis	Washington	DC
289	Andrew	Davis	Washington	DC
290	Carol	Davis	Washington	DC
291	Karen	Davis	Washington	DC
292	Kerri	Davis	Washington	DC
293	Denise	Dawkins	Washington	DC
294	Erica	De Vos	Washington	DC
295	Carole	Dean	Washington	DC
296	Lisa	Deane	Washington	DC
297	Kim	Deane	Washington	DC
298	Paul	DeFilippes	Washington	DC
299	Mia	Dell	Washington	DC
300	seth	dematties	washington	D.C
301	Catalan	deMatties	Washington	DC
302	Jennifer	DeMayo	Washington	DC
303	Nicholas	D'Emilio	Washington	DC
304	Dorothy	Deng	Washington	DC
305	Mimi	Dennis	Washington	DC
306	Maria	DePaul	Washington	DC
307	Denise	Dermody	Washington	DC
308	Catherine	Derr	Washington	DC
309	Beth	Dessel	Washington	DC
310	Frederic	Deutsch	New York	NY
311	morris	deutsch	Washington	dc
312	Todd	Devorsetz	Washington	DC
313	Beth	Dewhurst	Washington	DC
314	Marie-Elise	Diamond	Washington	DC
315	Jason	Dick	Washington	DC
316	Diane	Dickey	Washington	DC
317	Wallace	Dickson	Washington	DC
318	Susan	Dienelt	Washington	DC
319	Jane	Dietze	Washington	DC
320	Burke	Dillon	Washington	DC
321	Crystal	DiPietro	Washington	DC
322	Delphia	Dirks	washington	DC
323	Craig	Disher	Washington	DC
324	Chuck	Dittrich	Washington	DC
325	Kristin	Diwan	Washington	DC
326	Christian	Dixon	Washington	DC
327	Virginia	Dize	Washington	DC

PUREWATERDC.COM
citizens for clean water

info@purewaterdc.com

328	Tony	Domenico	Washington	DC
329	Laura Grant	Dong	Washington	DC
330	carolyn	donnelly	washington	dc
331	Laura	Donney	Los Angeles	CA
332	Carol	Dorsett	Washington	DC
333	Ryan	Doyle	Washington	DC
334	JR	Drabick	Washington	DC
335	June	Drew	Washington	DC
336	kayla	drogosz	washington	dc
337	Theresa	DuBois	Washington	DC
338	Audrey Drew	Duff	Washington	DC
339	Tricia	Duncan	Washington	DC
340	Jennie	Dunham Smith	Washington	DC
341	C.J.	Dyess	Washington	DC
342	Lisa	Eady	Washington	DC
343	Janae	Eason	Washington	DC
344	Katherine	Ebersole	Washington	DC
345	Sharon	Eberstein	Washington	DC
346	Maria	Echaveste	washington	dc
347	Ryan	Edelstein	Washington	dc
348	Elizabeth	Edmonds	Washington	DC
349	Emily	Eigen	Takoma Park	MD
350	Tobias	Eigen	Takoma Park	MD
351	Marc	Eisenberg	Washington	DC
352	Jonathan	Eisenman	Washington	DC
353	Joshua	Eisenman	Washington	DC
354	Regina	el Arculli	Washington	DC
355	Jeremiah	Eliason	Washington	DC
356	Michael	Ellert	Washington	DC
357	MYKE	ELLIS	Washington	DC
358	William	Emnett	Washington	DC
359	Jodi	Enda	Washington	DC
360	Erin	Engle	Washington	DC
361	Elisabeth	Ensley	Washington	DC
362	Marquez	Equalibria	Washington	DC
363	Matthew	Erickson	Washington	DC
364	Kendra	Ericson	Washington	DC
365	Kymberly	Escobar	Washington	DC
366	Sherri	Etheredge	Washington	DC
367	Stephanie	Evans	Washington	DC
368	Patrick	Evans	Washington	DC
369	john	everett	Washington	DC

PUREWATERDC.COM
citizens for clean water

info@purewaterdc.com

370	Lor	Everhart	Washington	DC
371	Danielle	Ewen	Washington	DC
372	M Chris	Fabricant	New York	NY
373	David	Fathi	Washington	DC
374	Ian	Faigley	Washington	DC
375	Dionisios	Favatas	Washington	DC
376	Paul	Fekete	Washington	DC
377	Laura	Feldman	Washington	DC
378	Dan	Feldman	Washington	DC
379	Roxanne	Felpausch	Washington	DC
380	cynthia	ferranto	washington	dc
381	R Geoffrey	Ferrell	Washington	DC
382	martha	fessenden	washington	dc
383	Elizabeth	Festa	Washington	DC
384	Amy	Fettig	Washington	DC
385	Trisha	Fields	Washington	DC
386	Christina	Figueras	Washington	DC
387	Sam	Figuli	Washington	dc
388	Maj	Fiil-FLynn	Washington	DC
389	Eileen	Findlay	Washington	DC
390	LIse	Fink	Washington	DC
391	Bonnie	Finkelman	Washington	DC
392	Christina	Finkelstein	Washington	DC
393	Kathleen	Finn	Washington	DC
394	Jennifer	Fiore	Washington	DC
395	Nora	Fischer	Washington	DC
396	Michael	Fisher	Washington	DC
397	Charles	Fishman	Washington	D.C.
398	Margaret	Fishman	Washington	D.C.
399	Marcia	Fitzmaurice	Washington	DC
400	Dianne	Fiumara	Washington	DC
401	Catherine	Flanagan	Washington	DC
402	Gillian	Flory	Washington	DC
403	Eileen	Flowers	Washington	D.C.
404	Patrick	Flynn	washington	DC
405	Chris	Fortune	Washington	DC
406	Max	Fose	Washington	DC
407	Eric	Foster	Washington	DC
408	Phi	Fostvedt	Washington	DC
409	Nils	Fostvedt	Washington	DC
410	Suellen	Foth	Washington	DC
411	Jennifer	Fowler	washington	dc

PUREWATERDC.COM
citizens for clean water

info@purewaterdc.com

412	Jeremy	Fowler	Washington	DC
413	Susannah	Fox	Washington	DC
414	Thomas	Fox	Washington	DC
415	Patricia	Francis	Washington	DC
416	Karlene	Francis	washington	dc
417	Robert	Frankel	Washington	DC
418	Gloria	Frankel	Washington	DC
419	Ruth	Franklin	Washington	DC
420	John	Franzen	Washington	DC
421	Joshua	Freed	Washington	DC
422	Michal	Freedhoff	Washington	dc
423	Catherine	Freeman	Washington	DC
424	Lindsey	Freeman	Washington	DC
425	Stephanie	Freymann	Washington	DC
426	Amy	Frick	Washington	DC
427	Karen	Friedman	Washington	DC
428	Vincent	Frillici	Washington	DC
429	Brian	Fruchey	Washington	DC
430	Amanda	Fuchs	Washington	DC
431	Paul	Fugitt	Washington	DC
432	odelia	funke	washington	dc
433	Ileana	Futter	Washington	DC
434	Joanne	Gallo	Washington	DC
435	Sarah	Gammage	Washington	DC
436	patrice	gancie	Washington	DC
437	Sajit	Gandhi	Washington	DC
438	Teresa	Gardiner	Washington	DC
439	Harry	Gates	Washington	DC
440	Michael	Geglia	Washington	DC
441	Julie	Gelsinger	Washington	DC
442	Luis	Genao	Washington	DC
443	shari	gendzel	washington	dc
444	Todd	Genger	Washington	DC
445	Christa	George	Washington	DC
446	Donetta	George	washington	DC
447	Jim	Gerstein	Washington	DC
448	Rick	gersten	washington	dc
449	Tamar	Gertner	washington	dc
450	Richard	Gervase	Washington	DC
451	Aimee	Gibbons Zanovello	Washington	DC
452	Susan	Gibbs	Washingon	DC
453	Andrew	Gibby	Arlington	VA

PUREWATERDC.COM
citizens for clean water

info@purewaterdc.com

454	jennifer	gibson	washington	dc
455	Paula	Gibson	Washington	DC
456	Margaret	Gill	Washington	DC
457	Carole	Gill	Washington	DC
458	Carole	Gill	Washington	DC
459	Anne	Gilson	Washington	DC
460	Steve	Ginther	Washington	DC
461	Christopher	Girardot	Washington	DC
462	ron	given	washington	dc
463	Laura	Glaser	Washington	DC
464	Luke	Glover	Washington	DC
465	Amelia	Goerlitz	Washington	DC
466	laura	goertzel	Washington	DC
467	Mary	Goetz	Washington	DC
468	Meg	Goetz	Washington	DC
469	Monica	Goldberg	Washington	DC
470	Eleana	Gomez	Washington	DC
471	Janis	Goodman	Washington	DC
472	Pat	Goslee	Washington	DC
473	Catherine	Graham	Washington	DC
474	Barbara	Graham	Washington	DC
475	Amanda	Graham	Washington	DC
476	Charles	Graham	Washington	DC
477	Barbara	Graham	Washington	DC
478	Lorraine	Graham-Brown	Washington	DC
479	Milton	Grant	Washington	DC
480	Barbara	Green	washington	DC
481	Carol	Green	Washington	DC
482	Robert	Green	Washington	DC
483	Marcia	Greene	Washington	DC
484	charles	greene	washington	dc
485	Leslie	Greene	Washington	DC
486	Lisa	Greenman	Washington	DC
487	Mary Ann	Grena Manley	Washington	DC
488	Jill	Gross	Washington	DC
489	Jane	Gross	Washington	DC
490	John	Guagh	Washington	DC
491	Kelda	Gunderson	Washington	DC
492	Corliss	Gunn-Ray	Alexandria	VA
493	Margaret	Guroff	Washington	DC
494	Susan	Gushue	Washington	DC
495	Michael	Gushue	Washington	DC

PUREWATERDC.COM
citizens for clean water

info@purewaterdc.com

496	John	Gutierrez	Washington	DC
497	Staci	Haag	Washington	dc
498	Linda	Haan	Washington	DC
499	Satu	Haase-Webb	Washington	DC
500	Jennifer	Haefeli	Washington	DC
501	Elizabeth	Haile	Washington	DC
502	Kathryn	Haines	Washington	DC
503	Marilou	Hakuta	Washington	DC
504	Matthew	Hall	Melville	NY
505	Eric	Halperin	Washington	DC
506	Nicole	Hamam	Washington	DC
507	Ann	Hamilton	Washington	DC
508	Heather	Hamilton	Washington	DC
509	Jeff	Hamond	Washington	DC
510	Cora	Han	Washington	DC
511	John	Hance	Washington	DC
512	Joanna	Hanes-Lahr	Washington	DC
513	Teryn	Hanggi	Washington	DC
514	Stefan	Hankin	Washington	DC
515	Sean	Hanley	Washington	DC
516	K	Hannah	Washington	DC
517	Laura	Hansen	Washington	DC
518	Alexandrea	Harbold	Washington	DC
519	Michael	Harbold	Washington	DC
520	Erin	Harcourt	Washington	DC
521	Christina	Hardesty	Washington	DC
522	Michael	Hargreaves	Washington	DC
523	Nkenge	Harmon	Washington	DC
524	Kimberly	Harrington	Washington	DC
525	Tasha	Harris	Washington	DC
526	Jamie	Harris-Gershon	Washington	DC
527	Nancy	Harter	Washington	DC
528	Jahna	Hartwig	Washington	DC
529	Cynthia	Hatch	Washington	DC
530	Dina	Haynes	Washington	DC
531	Mark	Hazan	washington	dc
532	Kathy	Healy	Washington	DC
533	Michelle	Heelan	Washington	DC
534	Brian	Heilman	Washington	DC
535	Pascale	Heimonet	Washington	DC
536	Steven	Heller	Washington	DC
537	Patricia	Helmetag	annapolis	MD

PUREWATERDC.COM
citizens for clean water

info@purewaterdc.com

538	Margaret	Hemingway	Washington	DC
539	Allen	Hengst	Washington	DC
540	Linda	Henry	Washington	DC
541	Liz	Henry	Washington	DC
542	Kara	Henschel	Washington	DC
543	William	Herrmann	Washington	DC
544	James	Hershberg	Washington	DC
545	Annie	Hershberg	Washington	DC
546	arthur	hessel	washington	dc
547	Terrence	Heubert	Washington	DC
548	christina	heyniger	Washington	DC
549	Maggie	Heyward	Washington	DC
550	Peter	Heyward	Washington	DC
551	Scott	Hibbard	Washington	DC
552	A	Hicks	Washington	DC
553	Lauren	Higgins	Washington	DC
554	margel	highet	washington	dc
555	Margel	Highet	Washington	DC
556	Katherine	Hill	washington	DC
557	Koji	Hirota	Washington	DC
558	Phyllis	Hirschkop	Washington	D.C.
559	Amy	Hoang	Washington	DC
560	Cathy	Hoban	Washington	DC
561	Lisa	Hodge	Washington	DC
562	Marie	Hoffman	Washington	DC
563	Ann	Hoffman	Washington	DC
564	Brendan	Hoffman	Washington	DC
565	Symantha	Holben	Washington	DC
566	Jennifer	Holloway	Washington	DC
567	Mei-Ling	Hopgood Reel	Washington	DC
568	Lise	Howard	Washington	DC
569	Marc	Howard	Washington	DC
570	Chamene	Howard	Washington	DC
571	K. Marie	Howe	Washington	DC
572	Julia	Howell	Washington	DC
573	Dan	Howells-Schafroth	Washington	DC
574	Cecil	Howze Jr.	Washington	DC
575	Amy	Hubbard	Washington	DC
576	Amelia	Hube	Washington	DC
577	Stacey	Hudson	Washington	DC
578	Erynn	Huff	Washington	DC
579	Gary	Hughes	Washington	DC

PUREWATERDC.COM
citizens for clean water

info@purewaterdc.com

580	shannon	hughs	Washington	dc
581	Nathan	Hultman	Washington	DC
582	Gary	Humfelt	Washington	DC
583	Roger	Hunwicks	Washington	DC
584	Lia	Hutton	Washington	DC
585	Elizabeth	Iadarola	Washington	DC
586	Ajowa	Ifateyo	Washington	DC
587	Sarah	Ingersoll	Washington	DC
588	Geoffrey	Irwin	Washington	DC
589	Lisa	Jackson	Washington	DC
590	Ginger	Jacobs	Washington	dc
591	Murrey	Jacobson	Washington	DC
592	Ur	Jaddou	Washington	DC
593	Sue	Jagger	Washington	DC
594	Balchander	Jayaraman	Washington	DC
595	Diana	Jensen	Washington	DC
596	sus	jensen	washington	dc
597	Kenneth	Jeruchim	Washington	DC
598	Krissi	Jimroglou	Washington	DC
599	Allan	Jirikowic	Washington	dc
600	Kristen	Johnson	Washington	DC
601	Linda	Johnson	Washington	DC
602	Meghan	Johnson	Washington	DC
603	Sara	Johnson	Washington	DC
604	Jennifer	Jones	Washington	DC
605	Emily	Jones	Washington	DC
606	Keith	Jones	Washington	DC
607	Jackie	Jones	Washington	DC
608	Kim	Jordan	Washington	DC
609	Samuel	Jordan	Washington	DC
610	Kim	Jordan	Washington	DC
611	William	Jordan	Washington	DC
612	Marianne	Josem	Washington	DC
613	Julian	Josephs	Washington	DC
614	Celia	Josephs	Washington	DC
615	Peter	Kadzik	Washington	DC
616	Barbara	Kagan	Washington	DC
617	Rob	Kaler	Washington	DC
618	Laura Sheehan	Kallen	Washington	DC
619	Joel	Kalvesmaki	Washington	DC
620	Chelsea	Kammerer	Washington	DC
621	Karen	Kampa	Washington	DC

PUREWATERDC.COM
citizens for clean water

info@purewaterdc.com

622	Christopher	Kandik	Washington	DC
623	Preeti	Kanodia	Washington	DC
624	Katharine	Kaplan	Washington	DC
625	Sheila	Kaplan	Washington	dc
626	Betsy	Karel	Washington	DC
627	Frank	Karel III	Washington	DC
628	Jill	Kasper	Washington	DC
629	David	Kass	Washington	DC
630	Linda	Katz	Washington	DC
631	Judith	Katz-Leavy	Washington	D.C.
632	Gregory	Kaufman	Washington	DC
633	Steve	Kaufman	Washington	DC
634	Andrew	Kay	Washington	DC
635	Ann	Kay	Washington	DC
636	Michael	Keblin	Washington	DC
637	Sherri	Keene	Washington	DC
638	Marilyn	Keefe	Washington	DC
639	Katherine	Kelly	Washington	DC
640	Michael	Kende	Washington	DC
641	Lori	Kenepp	Washigton	DC
642	Julian	Keniry	Washington	DC
643	Ann	Kenkel	Washington	DC
644	Christina	Kennedy	Washington	DC
645	Elizabeth	Kennedy	Washington	DC
646	Joseph	Kenny	Washington	DC
647	helene	kenny	Washington	DC
648	Kellee	Khan	Washington	DC
649	Sakib	Khan	Washington	DC
650	Michael	Kharfen	Washington	DC
651	Pamela	Khinda	Washington	DC
652	Kevin	Kiger	Washington	DC
653	Christine	Kim	Washington	DC
654	Jennifer	Kimball	Washington	DC
655	Amber	King	Washington	DC
656	Charlene	King	Washington	DC
657	Kendall	King	Washington	D.C.
658	Rod	King	Washington	DC
659	Leonard	King	Washington	DC
660	Audrey	Kissel	Washington	DC
661	John	Kleiderer	Washington	DC
662	Phyllis	Klein	Washington	DC
663	Allison	Kletter	Washington	DC

PUREWATERDC.COM
citizens for clean water

info@purewaterdc.com

664	Deborah	Kline	Washington	DC
665	Stephanie	Klodzen	Washington	DC
666	Jane	Kochersperger	Washington	D.C
667	Kathy	Koerner	Washington	DC
668	Kay	Kohl	Washington	DC
669	Allison	Koles	washington	dc
670	mike	kopetski	washington	D.C
671	Michael	Koppenheffer	Washington	DC
672	Leonid	Kopylev	Washington	DC
673	Karen	Kornbluh	Washington	DC
674	Stacey	Kornegay	Washington	DC
675	Theresa	Kossler	Washington	DC
676	John	Kostyack	Washington	DC
677	Maria	Kozloski	Washington	DC
678	Maria	Kozloski	Washington	DC
679	Kevin	Kraham	Washington	DC
680	Rebecca	Kramer	Washington	DC
681	joshua	Kranzberg	Washington	DC
682	Catherine	Krebs	Washington	DC
683	Anthony	Kreindler	Bethesda	MD
684	Donald	Kreuzer	Washington	DC
685	Jody	Kris	bethesda	MD
686	Brent	Kruse	Washington	DC
687	Cecilia	Kullman	Washington	DC
688	Amy	Kurtz	Washington	DC
689	Jane	Kuuskraa	Washington	DC
690	Wendy	Lader	Washington	DC
691	Meredith	Lahey	Washington	DC
692	Yanna	Lambrinidou	Washington	DC
693	Julia	Lamm	Washington	DC
694	Lesley	Lamothe	Washington	DC
695	Melissa	Lane	Washington	DC
696	Tracey	Lane	Washington	DC
697	Heather	Langdon	Washington	DC
698	Gilah	Langner	Washington	DC
699	Terri	Langston	Washington	DC
700	edward	lanpher	washington	dc
701	Miriam	LaPointe	Washington	DC
702	Paul	Laporte	Washington	DC
703	Anne	Large	Washington	DC
704	Suzanne	Larkins	Washington	DC
705	Mike	Larson	Washington	DC

PUREWATERDC.COM
citizens for clean water

info@purewaterdc.com

706	Linda	Latham	Washington	DC
707	Lari	Lavigne	Washington	DC
708	Steve	Lawrence	Washington	DC
709	Jason	Lawrence	Washington	DC
710	Kim	Lazor	Washington	DC
711	James	Lazor	Washington	DC
712	Allison	Leader	Washington	DC
713	Michael	Leaf	Washington	DC
714	Stacey	Leavandosky	Washington	D.C.
715	Jody	Lee	Washington	DC
716	Emily	Leffler	Washington	DC
717	Paul	Legere	Washington	DC
718	Christine	Lehman	Washington	DC
719	dale	leibach	Washington	DC
720	Abby	Leibman	Los Angeles	CA
721	Sharon	Lemberger	Washington	DC
722	Margaret	Lenzner	Washington	DC
723	Alena	Leon	Washington	DC
724	Michael	Leonard	Washington	DC
725	Devon	Leppink	Washington	DC
726	Rene	LeRouzes	Washington	DC
727	Jeffrey	Levi	Washington	DC
728	Josh	Levinson	Washington	DC
729	Elila	Levinson	Washington	DC
730	Molly	Levinson	Washington	DC
731	Joseph	Levinson	Cincinnati	Ohic
732	Leslie	Levy	Washington	DC
733	Janine	Lewis	Bethesda	MD
734	fay	lewis	washington	dc
735	Deborah	Lewis	Washington	DC
736	frances	lewis	washington	dc
737	Laura	Lewis	Washington	DC
738	Eleanor	Lewis	Washington	DC
739	Ryan	Lewkowski	Washington	DC
740	Marjorie	Lidoff	Washington	DC
741	Anne	Lindenfeld	Washington	DC
742	Sara	Lindstrom	Washington	DC
743	kathryn	linehan	Washington	DC
744	Ronald	Lipton	Washington	DC
745	Audrey	Little	Washington	DC
746	Rita	Litvack	Plainview	NY
747	Lin	Liu	Washington	DC

PUREWATERDC.COM
citizens for clean water

info@purewaterdc.com

748	Christopher	Lively	Washington	DC
749	Megan	Livingston	Washington	DC
750	Lia	Lockert	washington	dc
751	Amy	Longsworth	Washington	DC
752	Patricia	Lovera	Washington	DC
753	Tana	Lorah	Washington	DC
754	fran	lowenstein	washington	dc
755	Wendie	Lubic	Washington	DC
756	Angie	Lundberg	Washington	DC
757	Sarah	Lynch	Washington	DC
758	Laurie	Lyons	Washington	DC
759	Jane	Lyons	Washington	D.C.
760	Clark	Madigan	Washington	DC
761	Lisa	Magnino	Washington	DC
762	Auke	Mahar-Piersma	Washington	DC
763	Amy	Mall	Washington	DC
764	Brian	Mandrier	Washington	DC
765	Janet	Mann	Washington	DC
766	Anne	Marchand	Washington	DC
767	cindy	margolies	washington	dc
768	Richard	Margolies, Ph.D.	Washington	DC
769	Kristie	Mark	Washington	DC
770	Susan	Markham	Washington	DC
771	Laura	Marks	Washington	DC
772	Sandra	Marks	Washington	DC
773	Mariuccia	Marolo	Washington	dc
774	Karen	Marsh	Washington	DC
775	Lisa	Marshall	Washington	DC
776	Eric	Marshall	Washington	DC
777	ralph	martin	Washington	DC
778	Jennifer	Martin	Washington	DC
779	Sara	Martin	Washington	DC
780	Marina	Martin	Washington	DC
781	Irene	Martinez	Washington	DC
782	Juan	Martinez-Sanchez	Arlington	VA
783	Meredith	Matthws	Washington	dc
784	Kristin	May	Washington	DC
785	Katherine	Maybank	Washington	DC
786	Sarah	McCarthy	Washington	DC
787	Stephanie	McCarty	Westerville	OH
788	Elena	McCollim	Washington	DC
789	Mary	McComb	Washington	DC

PUREWATERDC.COM
citizens for clean water

info@purewaterdc.com

790	Candace	McCombs	Rockville	MD
791	M.P.	McCoy	Washington	DC
792	Charles	McCrary	Washington	DC
793	Kevin	McCurdy	Washington	DC
794	James	McDonald	Washington	DC
795	Jeffrey J	McElhaney	Washington	DC
796	Liam	McElhone	Washington	DC
797	Margaret	McGowen	Washington	DC
798	Michelle	McGregor	Washington	dc
799	Patricia	McInnis	Washington	DC
800	Mary	McIntosh	Washington	DC
801	Paul	McKay	Washington	DC
802	Heather	McKay	Washington	DC
803	David	McKay	Flagstaff	AZ
804	Gibson	McKay	Phoenix	AZ
805	Cassandra	McKee	Washington	DC
806	Paul	McKenzie	Washington	DC
807	Ian	McKinley	Washington	DC
808	Katherine	McLean	Washington	DC
809	Kathleen	McLynn	Washington	DC
810	Sherry	McMahon	Washington	DC
811	Jim	McNeill	Washington	DC
812	Alpha	McPherson	Washington	DC
813	Elizabeth	McReady	Washington	DC
814	David	McWhorter	Washington	DC
815	Holly	Mead	Washington	DC
816	Wanda	Means-Harris	Washington	DC
817	Lara	Meisner	Washington	DC
818	Les'	Melnyk	Washington	DC
819	Mary	Melnyk	Washington	DC
820	Jon	Melzer	Washington	DC
821	Tina	Mendelson	Washington	DC
822	Lea	Mesner	Washington	DC
823	carol	messito	washington	DC
824	Craig	Metrick	Washington	DC
825	Jeanne	Metzger	Washington	DC
826	Ajia	Meux	Washington	DC
827	Jane	Meyers	Washington	D.C.
828	mari	michaud	Washington	dc
829	Alison	Michel	Washington	DC
830	Mary Fran	Miklitsch	Washington	DC
831	Douglas	Miller	Washington	DC

PUREWATERDC.COM
citizens for clean water

info@purewaterdc.com

832	Rachel	Miller	Washington	DC
833	betty	miller	washington	dc
834	Jennifer	Miller	Washington	DC
835	Jon	Minkoff	Washington	DC
836	Jill	Minneman	washington	dc
837	Theresa	Minni	Washington	DC
838	Joshua	Mintz	Washington	DC
839	Ranit	Mishori	Washington	DC
840	Justin	Mixon	Washington	DC
841	jeannine	mjoseth	Washington	dc
842	Meg	Molloy	Washington	DC
843	Maggie	Molloy	Washington	DC
844	James	Molloy III	Washington	DC
845	Jimmy	Molloy IV	Washington	DC
846	John	Monahan	Washington	DC
847	Daniel	Montoya	Washington	DC
848	Demian	Moore	Washington	DC
849	SARAH	MOORE	Washington	DC
850	Chauncey	Moore	Washington	DC
851	Genevieve	Moreland	Washington	DC
852	Lisa	Morelli	Washington	DC
853	kimberly	Morgan	Washington	DC
854	Michael	Morgan-Gaide	Washington	DC
855	Jill	Morrison	Washington	dc
856	Sonya	Morshauser	Washington	DC
857	Douglas	Morton	Washington	DC
858	Lindsay	Morton	Washington	DC
859	Holly	Moskerintz	Washington	DC
860	Bonnie	Moss	Washington	DC
861	Ginger	Moss	Washington	DC
862	Cristina	Mossi	Washignton	DC
863	Nicola	Mousset-Jones	Washington	DC
864	Anthony	Muh	Washington	DC
865	Megan	Mullins	Washington	DC
866	Pat	Munoz	Washington	DC
867	kita	murdock	washington	dc
868	Mary	Murphy	Washington	DC
869	Jill	Murphy	Washington	DC
870	Royce	Murray	Washington	DC
871	ellen	myer	Washington	DC
872	sumner	myers	washington	dc
873	alix	myerson	Washington	dc

PUREWATERDC.COM
citizens for clean water

info@purewaterdc.com

874	Michael	Nalepa	Washington	DC
875	Tina	Namian	Washington	DC
876	Jeffrey	Napolitano	Washington	DC
877	Andrea	Nash	Washington	DC
878	Andrianna	Natsoulas	Wahington	DC
879	Joseph	Neale	Washington	DC
880	Joe	Neel	Washington	DC
881	Kevin	Neil	Washington	DC
882	sylvia	ness	washington	dc
883	Zoe	Neuberger	Washington	DC
884	Gayle	Neufeld	Washington	DC
885	Johanna	Neuhaus	Washington	DC
886	Patricia	Neuman	Washington	DC
887	Dale	Newfield	Washington	DC
888	Arthur	Newmyer	Washington	DC
889	Carol	Newmyer	Washington	DC
890	Bev	Nissen	Washintgon	DC
891	Doris	Noble	Washington	DC
892	Jeffrey	Norman	washington	DC
893	Christina	Novinskey	Washington	DC
894	MATTHEW	NOWAKOWSKI	Washington	DC
895	Jacqueline	Nowell	Washington	DC
896	Frederick	Nunley	Washington	DC
897	Maya	O'Connor	Washington	DC
898	Margaret	O'Connor	Washington	DC
899	Karen	O'Connor-Floman	Washington	DC
900	Andrew	O'Dell	washington	DC
901	Maureen	O'Leary	Washington	DC
902	John Peter	Olinger	Washington	DC
903	Aimee	Olivo	Washington	DC
904	ERIK	OLSON	Washington	DC
905	Lisa	Olson	Washington	DC
906	George	Olson	Washington	DC
907	Win	Oo	Washington	DC
908	Laura	Oradei-Bayz	Washington	DC
909	David	Ordoobadi	Washington	DC
910	Laurie	O'Reilly	Washington	DC
911	Catherine	O'Riordan	Washington	DC
912	susan	orlins	Washington	dc
913	Peter	Orlowski	Bethesda	MD
914	Anne	O'Rourke	Washington	DC
915	Berit	Oskey	Washington	DC

PUREWATERDC.COM
citizens for clean water

info@purewaterdc.com

916	Karen	Ostlie	Washington	DC
917	William	O'Sullivan	Washington	DC
918	Susan	Ousley	Washington	DC
919	Idil	Oyman	Washington	DC
920	Tulin	Ozdeger	Washington	DC
921	Leslie	Pace	Washington	DC
922	Agnieszka	Paczynska	Washington	DC
923	Lee	Paige	Washington	DC
924	Julie	Paller	Washington	dc
925	Michelle	Palma	Washington	DC
926	Pho	Palmer	Washington	DC
927	Danielle	Pappas	Washington	D.C.
928	Katie	Parker	Washington	DC
929	kirstin	parsons	washington	dc
930	Sarah	Partridge	Washington	dc
931	barry	passett	washington	dc
932	Sandra	Pastorino	Washington	DC
933	Manny	Pastreich	Washington	DC
934	Kathy	Patterson	Washington	DC
935	Andrea	Patterson	Washington	DC
936	barbara	patterson	washington	dc
937	James	Paugh	Washington	DC
938	Jennifer	Pauk	Washington	DC
939	Frederick	Paxton III	washington	dc
940	Dina	Paxenos	Washington	DC
941	Bill	Pearce	Washington	DC
942	Yvette	Pearson	Washington	DC
943	Benjamin	Peasant	Washington	DC
944	MICHAEL	PEASANT	Washington	DC
945	Kristin	Pechinsky	Washington	DC
946	Elizabeth	Pelcyger	Washington	DC
947	Nancy	penn	Washington	dc
948	margery	perko	washington	dc
949	julie	perlmutter	Washington	DC
950	Wade	Perry	miami	FL
951	Lisa	Perry	Washington	DC
952	Shonda	Pertilla	Washington	DC
953	Veronika	Pesinova	Washington	DC
954	Ruth	Peterson	Washington	D.C.
955	Noel	Petrie	Washington	DC
956	John	Pfeiffer	Washington	DC
957	Christine	Philp	Washington	DC

PUREWATERDC.COM
citizens for clean water

info@purewaterdc.com

958	Lia	Phipps	Washington	DC
959	Heather	Phipps	Washington	DC
960	Theodore	Piccone	Washington	DC
961	Carol	Pier	Washington	DC
962	Jim	Pierce	Washington	DC
963	Cornelia	Pillard	Washington	DC
964	Lidia	Pines	washington	dc
965	Michael	PIPITONE	Washington	DC
966	Kristi	Plahn-Gjersvold	Washington	DC
967	David	Plant	Washington	DC
968	Mary Carroll	Platt	Washington	dc
969	Robert	Pohl	Washington	DC
970	Gina	Polidoro	Washington	DC
971	Ruth	Pollak	Washington	DC
972	Virginia	Pommer	Washington	DC
973	Colleen	Popson	Washington	DC
974	Carla	Post	Washington	DC
975	Elizabeth	Post	Washington	DC
976	Mary Susan	Powell	Washington	DC
977	Susan	Prahinski	Washington	DC
978	Katherine	Pratt	Washington	DC
979	Chris	Pressley	Washington	DC
980	Angela	Preston	Washington	DC
981	Cecelia	Prewett	Alexandria	VA
982	Brenda K.	Price	Washington	dc
983	Gaylea	Prichard-Silvers	Washington	DC
984	Maya	Primorac	Washington	DC
985	Elizabeth	Pringle	Washington	DC
986	Andrea	Pringle	Washingotn	DC
987	Elizabeth	Pringle	Washington	DC
988	Michael	Proulx	Washington	DC
989	Jonathan	Pruett	Washington	DC
990	Jacek	Pruski	Iowa City	Iowa
991	Alyssa	Przystawik	Washington	DC
992	Rimas	Puskorius	Washington	DC
993	Jill	Quinn	Washington	DC
994	Kris	Raab	Washington	DC
995	Stacey	Rabbino	Washington	DC
996	Steve	Radanovic	Washington	DC
997	lawrence	radanovic	washington	d¸c.
998	Joan	Rajah	Washington	DC
999	Corey	Ramdsen	Washington	DC

PUREWATERDC.COM
citizens for clean water

info@purewaterdc.com

1000	Firoze	Rao	Washington	DC
1001	David	Rappaport	Washington	DC
1002	David	Rappaport	Washington	DC
1003	Carol	Rasco	Washington	DC
1004	heidi	rauch	washington	dc
1005	Molly	Rauch	Washington	DC
1006	Shay	Raugh	Washington	DC
1007	Susan	Rausch	Washington	DC
1008	Lawrence	Rausch	Washington	DC
1009	Dustin	Rawlins	Washington	DC
1010	Julie	Ray	Washington	DC
1011	Sumanta	Ray	Washington	DC
1012	Carr	Rebecca	Washington	DC
1013	Nicole	Reece	Washington	D.C.
1014	Deno	Reed	Washington	DC
1015	Angela	Reid	Washington	DC
1016	Kasey	Reisman	Washington	DC
1017	Ron	Renchard	Washington	DC
1018	TJ	Resler	Washington	DC
1019	Sheila	Resnick	Washington	DC
1020	Louisa	Reynolds	Washington	dc
1021	Bridgette	Rhodes	Washington	DC
1022	Sidney White	Rhyne	Washington	D.C.
1023	Tiffany	Ricci	Washington	DC
1024	Susan	Rice	Washington	DC
1025	Carol	Rice	Washington	DC
1026	Lawrence	Rich	Washington	DC
1027	Rebecca	Richards	Washington	DC
1028	Christie	Richart	Washington	DC
1029	Pamela J.	Richart	Chicago	IL
1030	Eleanor & James	Rider	Washington	DC
1031	Constance	Ridgway	Washington	DC
1032	Bernard	Ries	Washington	D.C.
1033	Amy	Rifkind	Washington	DC
1034	john	riley	arlington	VA
1035	Deborah	Ringel	Washington	DC
1036	Rae	Ringel	Washington	DC
1037	Jocarol	Robb	Washington	DC
1038	Eilene	Robb	Washington	DC
1039	Amaya	Roberson	Washington	DC
1040	Joel	Roberson	Washington	DC
1041	Michael	Roberts	Washington	dc

PUREWATERDC.COM
citizens for clean water

info@purewaterdc.com

1042	jerome	roberts	washington	dc
1043	Mary Esselman	Roberts	Charlottesville	VA
1044	Scott	Roberts	Washington	DC
1045	Michael	Robertson	Washington	DC
1046	Windy	Robertson	Washington	DC
1047	Peggy	Robin	Washington	DC
1048	Gloria	Robinson	Washington	DC
1049	Eric	Robinson	Washington	DC
1050	Earl	Robinson-EL	Washington	D C.
1051	Daniel	Robison	washington	dc
1052	Katherine	Rodi	Washington	DC
1053	Felice	Roggen	Washington	DC
1054	Karen Wilson	Roman	Washington	DC
1055	Edgar	Roman	Washington	DC
1056	Vivian	Ronay	Washington	DC
1057	Joiwind	Ronen	Washington	DC
1058	Amit	Ronen	Washington	DC
1059	David	Roodman	washington	dc
1060	Ingrid	Roper	Washington	DC
1061	Nicole	Rosen	Washington	DC
1062	Deborah	Rosen	Washington	DC
1063	Simon	Rosenberg	Washington	DC
1064	Brian	Rosenbloom	Washington	DC
1065	Mark	Rosenman	Washington	DC
1066	Gregory	Ross	Washington	DC
1067	Judith	Ross	Washington	DC
1068	Juliet	Ross	Washington	DC
1069	Karen	Rossler	Washington	DC
1070	Susan	Roth	Washington	DC
1071	Collins	Roth	Washington	DC
1072	Jessica	Roth	Washington	DC
1073	Brigette	Rouson	Washington	DC
1074	Mary E.	Rowse	Washington	DC
1075	Louisa	Rubinfien	Washington	DC
1076	Lenore	Rubino	Washington	DC
1077	Matthew	Ruest	Washington	DC
1078	Elizabeth	Ruppert Bulmer	Washington	DC
1079	Thomas	Russell	Washington	DC
1080	Mary Ann	Ryan	Washington	DC
1081	David	Sachdev	Washington	DC
1082	Rebecca	Sachs	Washington	DC
1083	Billy	Sahm	Washington	DC

PUREWATERDC.COM
citizens for clean water

info@purewaterdc.com

1084	Nancy	Sanders	Washington	DC
1085	Alexis	Santi	Washington	DC
1086	Suzanne	Saunders	Washington	dc
1087	Sherry	Saunders	Washington	DC
1088	Timothy	Savage	Washington	DC
1089	Loretta	Schaeffer Guarda	Washington	DC
1090	Valerie	Schaeublin	Washington	D.C.
1091	Kristina	Schafer	Washington	DC
1092	alison	schafer	washington	dc
1093	Willie	Schatz	Washington	DC
1094	Kimberly	Scheckner	Washington	DC
1095	katey	schein	washington	dc
1096	William	Scher	Washington	DC
1097	Angie	Schiavoni	Washington	DC
1098	Melissa	Schiffman	Washington	DC
1099	Anne	Schleicher	Washington	dc
1100	Megan	Schmidt	Washington	DC
1101	Christy	Schmidt	washington	dc
1102	Lauren	Schmierer	Washington	DC
1103	Beth	Schmierer	Washington	DC
1104	John	Schmitt	Washington	DC
1105	Robin	Schneider	Washington	DC
1106	charlotte	schoeneman	washington	dc
1107	Sara	Schreiber	Washington	DC
1108	Rachel	Schumacher	Washington	DC
1109	Michael	Schur	Washington	dc
1110	margot	schwadron	washington	dc
1111	Norma	Scogin	Washington	DC
1112	Josephine	Scott	Washington	DC
1113	Marianne	Scott	Washington	DC
1114	Zandria	Scott	Washington	DC
1115	Julie	Scott	Washington	DC
1116	Stephanie	Scott-Melnyk	Washington	DC
1117	Deborah	Seabron Dickens	Washington	DC
1118	Desa	Sealy Ruffin	Washington	DC
1119	Danny	Sebright	Washington	DC
1120	Patty	See	Washington	DC
1121	Sandra	Seestedt	Washington	DC
1122	Debbie	Segor	Washington	DC
1123	Lenore	Sek	Washington	D.C.
1124	Thomas	Selden	Washington	DC
1125	Jennifer	Seltzer Stitt	Washington	DC

PUREWATERDC.COM
citizens for clean water

info@purewaterdc.com

1126	Ethel	Sennhauser	Washington	DC
1127	Steve	Seuser	Washington	DC
1128	Skip	Seward	Washington	DC
1129	Robert	Seward	Washington	DC
1130	Mary	Shaffner	Washington	DC
1131	Keivan	Shahrokhi	Washinton	DC
1132	Elizabeth	Shapiro	Washington	D.C.
1133	William	Sharp	Washington	DC
1134	Kat	Sharp	Washington	DC
1135	Barbara Lee	Shaw	Washington	DC
1136	Shirley	Shearman	Steamboat Springs	CO
1137	Erin	Sheehy	Washington	DC
1138	Melanie	Shepherdson	washington	dc
1139	Kevin	Shertz	Washington	DC
1140	Melissa	Shriver	Washington	DC
1141	Kirsten	Shumway	WAshington	DC
1142	Geraldine	Shuster	Washington	DC
1143	Valerie	Silensky-Lowe	Washington	DC
1144	Diana	Simon	Washington	DC
1145	Philip	Simon	Washington	DC
1146	Thomas	Sinclair	Washington	DC
1147	Elizabeth	Singer	Washington	DC
1148	Linda	Sinoway	Washington	DC
1149	Linda	Sirica	Washington	DC
1150	Brian	Sisolak	Washington	dc
1151	T.F.	Skelton	Washington	D.C.
1152	Francesca	Skelton	Washington	DC
1153	Earl	Skelton	Washington	DC
1154	Benjamin	Slade	Washington	DC
1155	Alexander	Slater	Washington	DC
1156	Laura	Slattery	Washington	DC
1157	Melanie	Sloan	Washington	DC
1158	Estelle	Slon	Washington	DC
1159	cressida	slote	Washington	DC
1160	Kent	Slowinski	Washington	DC
1161	Edna	Small	Washington	D.C.
1162	Evita	Smedley	Washington	DC
1163	Melissa	Smith	Washington	DC
1164	Nancy	Smith	Washington	DC
1165	Carole	Smith	Washington	DC
1166	Lee	Smith	Washington	DC
1167	Sara	Smith	Washington	DC

PUREWATERDC.COM
citizens for clean water

info@purewaterdc.com

1168	Eugene	Smith	Washington	DC
1169	Peter	Smith	Washington	dc
1170	stevan	smith	Washington	DC
1171	Carole	Smith	Washington	DC
1172	David	Smole	Washington	DC
1173	Robin	Snyder	Washington	DC
1174	Sabrina	Sojourner	Washington	DC
1175	Tom	Soles	Washington	DC
1176	Julie	Song	Washington	DC
1177	Jesus	Soriano	Washington	DC
1178	Richard	Sperling	Washington	DC
1179	Michael	Spevak	Washington	DC
1180	William	Spinnell	Washington	DC
1181	Laurie	Sprung	Washington	DC
1182	Margaret	Squires	Washington	DC
1183	Greg	Squires	Washington	DC
1184	Joan	Stallard	Washington	DC
1185	Megan	Stanley	Washington	DC
1186	Juliette	Steadman	Washington	DC
1187	Sue	Stefurak	Washington	DC
1188	Scott	Steinke	Washington	DC
1189	Anita	Stephen	Wasington	D
1190	David	Stern	Washington	DC
1191	Renee	Stetzer	Washington	DC
1192	Kiersten	Stewart	Washington	DC
1193	Andrew	Stitt	Washington	DC
1194	Steven	Stone	Washington	DC
1195	Roland	Stottlemyre	Washington	DC
1196	Wanetta	Stottlemyre	Washington	DC
1197	Barbara	Straub Williams	Washington	DC
1198	Tim	Strecker	Washington	DC
1199	Marina	Streznewski	Washington	DC
1200	James	Stromseth	Washington	DC
1201	Zubaidah	Sudah	Washington	dc
1202	Carl	Sukenik	Washington	DC
1203	kathi	sullivan	Washington	DC
1204	Andy	Sullivan	Washington	DC
1205	Mark	Sullivan	Washington	DC
1206	barry	summer	Washington	d.c.
1207	Mark	Sundberg	Washington	DC
1208	Alexandra	Sundman	WAshington	DC
1209	Mona	Sur	Washington	DC

PUREWATERDC.COM
citizens for clean water

info@purewaterdc.com

1210	Hadar	Susskind	Takoma Park	MD
1211	Tim	Svoboda	Washington	DC
1212	Rajesh	Swaminathan	Washington	D.C.
1213	mandy	swann	Washington	DC
1214	LIsa	Swanson	Washington	DC
1215	Stephan	Sylvan	Washington	DC
1216	Michael	Tacelosky	Washington	DC
1217	Yoram	Tanay	Washington	DC
1218	C. B.	Tannenwald	Washington	DC
1219	Barrie Lynn	Tapia	Washington	DC
1220	Mark	Tarallo	Washington	DC
1221	Mike	Tavilla	Washington	DC
1222	james	taylor	washington	dc
1223	Delia	Taylor	Washington	DC
1224	P.	Taylor	Wqshington	DC
1225	Pat	Taylor	Washington	DC
1226	cheryl	teare	washington	dc
1227	MARYANN	TERRANA	Washington	DC
1228	Allyson	Terry	Washington	DC
1229	Katherine	Tew	Washington	D.C.
1230	Neal	Tew	Washington	DC
1231	Leah	Thayer	Washington	DC
1232	Dana	Thomas	Washington	DC
1233	Nancy C,	Thomas	Washington	DC
1234	Turkessa	Thompson	Washington	DC
1235	abb	tibbs	Washington	DC
1236	David	Tillotson	Washington	DC
1237	Michele & Rick	Tingling-Clemmons	Washington	DC
1238	Kelly	Tobin	Washington	DC
1239	Tolu	Tolu	Washington	DC
1240	Tasha	Toms	Washington	DC
1241	Jason	Townsend	Washigton	DC
1242	Paul	Tracey	Pacific Palisades	CA
1243	Sarah	Tracey	Pacific Palisades	CA
1244	Jessica	Trevelyan	Washington	DC
1245	Hewan	Treworgy	Washington	DC
1246	John	Tsapogas	Washington	DC
1247	Tina	Tummonds	Washington	DC
1248	Diana	Turk	Washington	DC
1249	Melinda	Turner	Washington	DC
1250	Lydia	Turnipseed	Washington	DC
1251	Thomas	Turqman	Washington	DC

PUREWATERDC.COM
citizens for clean water

info@purewaterdc.com

1252	Christine	Turqman	Washington	DC
1253	Mario	Urdaneta	Hyattsville	MD
1254	Desiree	Urquhart	Washington	DC
1255	Ximena	Vaca	Washington	dc
1256	Amy	Vaccari	Washington	DC
1257	Christopher	Vaden	Washington	DC
1258	Margarita	Valencia	Washington	DC
1259	Lona	Valmoro	Washington	DC
1260	Anna	van der Heijden	Washington	DC
1261	Michelle	Van Lare	Washington	DC
1262	Kevin	Varney	Washington	D.C.
1263	Barbara	Varsa	Washington	DC
1264	Cleotilde	Vatter	Washington	DC
1265	Sarah W	Veerhoff	Washington	DC
1266	Anne	Veysey	Washington	DC
1267	Rick	Virgin	Washington	DC
1268	Danielle	Virgin	washington	DC
1269	Gaylin	Vogel	Washington	DC
1270	Kate	Vogel	Washington	DC
1271	Jon	Vogel	Washington	DC
1272	Lynne and Eric	Vogelbacher	Washington	DC
1273	Steven	Volkers	Washington	DC
1274	Jim	Volle	Washington	DC
1275	Alexandra	vonBallmoos	washington	dc
1276	Dana	Voorhees	Washington	DC
1277	Chari	Voss	Washington	DC
1278	Julie	Waddell	Washington	DC
1279	Beth	Wadsworth	Washington	DC
1280	Barbara	Wagner	Washington	DC
1281	John	Walke	Washington	DC
1282	Chris	Walker	Washington	DC
1283	Christine	Walker	Vienna	VA
1284	Tom	Walker	Washington	DC
1285	Naomi	Walker	Washington	DC
1286	Roxanne	Walker	Washington	DC
1287	Margy	Waller	Washington	DC
1288	katherine	wallestad	washington	dc
1289	Susan	Wallin	Washington	DC
1290	karin	walser	arlington	va
1291	Maura	Walsh Rogan	Washington	Dc
1292	Darryl	Walter	Washington	DC
1293	Sonja	Walti	Washington	DC

PUREWATERDC.COM
citizens for clean water

info@purewaterdc.com

1294	Chris	Walz	Washington	DC
1295	Mitch	Wander	Washington	DC
1296	Ellen	Ward	Washington	DC
1297	Jessica	ward	WDC	DC
1298	Jennifer	Ward	Washington	Dc
1299	Carolyn	Ward	Washington	DC
1300	Kathleen	Warner	Washington	DC
1301	Georgiana	Warner Kaempfer	Washington	DC
1302	Jessica	Wasserman	Washington	DC
1303	Hunter	Waters	washington	DC
1304	Hunter	Waters	washington	dc
1305	Patricia	Watkins	Washington	DC
1306	Timothy	Waxenfelter	Washington	DC
1307	Scott	Wayne	Washington	DC
1308	Liz	Weatherly	Washington	DC
1309	James	Weatherly	Washington	DC
1310	carolyn	weaver	washington	dc
1311	Todd	Webster	Alexandria	VA
1312	Will	Weems	Washington	DC
1313	Ronald	Weich	Washington	DC
1314	Marilyn	Weigand	McLean	VA
1315	Andrew	Weishaar	Washington	DC
1316	Chris	Weiss	Washington	DC
1317	Gayle	weiswasser	Washington	DC
1318	Stacey	Wells	Washington	DC
1319	Donna	Welton	Washington	DC
1320	Lisa	Wendel	Washington	DC
1321	Veronica	Wetherill	Washington	DC
1322	Mike	Wevrick	Washington	DC
1323	Daryl	White	Washington	DC
1324	Jeffrey	White	Washington	DC
1325	David	Whitman	Washington	D.C
1326	James	Whittemore	Washington	DC
1327	Robin	Wiener	Washington	DC
1328	Suzanne	Wiggins Buck	Washington	DC
1329	Elizabeth	Wilcox	Washington	DC
1330	Melissa	Wiley	Washington	DC
1331	Morna	Willens	Washington	DC
1332	Roslyn	Williams	Washington	dc
1333	Ann	Williams	Washington	DC
1334	Kenneth	Williams	Washington	DC
1335	Carole	Williams	Washington	DC

PUREWATERDC.COM
citizens for clean water

info@purewaterdc.com

1336	manley	williams	alexandria	va
1337	Kristin	Willsey	Washington	dc
1338	Marla	Wilson	Washington	DC
1339	Todd	Wincup	Washington	DC
1340	Amy	Wind	Washington	DC
1341	HC	Winslow	Washington	DC
1342	Diana	Winterson-Russell	Washington	DC
1343	Philip	Wirtz	Washington	DC
1344	Scott	Wiskoski	Washington	DC
1345	Alicia	Wittink	Washington	DC
1346	Carrie	Wofford	Washington	DC
1347	Katherine	Wolf	Washington	DC
1348	Erin	Wolf	washington	dc
1349	Rebecca	Wolsk	Washington	DC
1350	Caroline	Wolverton	Washington	DC
1351	Kevin	Wood	Washington	DC
1352	Tom	Wood	Washington	DC
1353	Robert	Wozniak	Washington	DC
1354	Susan Longo	Wright	Washington	DC
1355	Mary	Wright	Washington	DC
1356	Alta	Wright-Booth	Washington	DC
1357	Elizabeth	Wulkan	washiington	dc
1358	Jenny	Yang	Washington	DC
1359	Patricia	Yates	Washington	DC
1360	Patricia	Yates	Washington	DC
1361	Alida	Yath-Cruces	Washington	DC
1362	Nicole	Yohalem	Washington	DC
1363	Anna Marie	Yombo	Washington	DC
1364	Karen	Yourish	Washington	DC
1365	Brian	Yourish	Washington	DC
1366	Sylvia	Yu	Washington	DC
1367	Janine	Zacharia	Washington	DC
1368	Esther	Zaff	Washington	DC
1369	Jonathan	Zaff	Washington	DC
1370	Sandra	Zahn-Oreck	Washington	DC
1371	Seth	Zimmerman	Washington	DC
1372	Rebecca	Zimmerman	Washington	DC
1373	Lori	Zirkle	Washington	DC
1374	Todd	Zirkle	Washington	DC
1375	Joan	Zorza	Washington	DC
1376	Monica	Zubler	Washington	DC
1377	Leslie	Zucker	Washington	DC

Senator JEFFORDS. I believe it is imperative that during each moment of today's hearing, we all remember that real parents, children and babies are being affected by this situation as we speak.

Today's hearing is just the first step in what I hope will be a long list of actions that we can take to help solve DC's lead problem and prevent this from occurring elsewhere in this Nation. Today, I requested from my colleague, Senator Graham of Florida, and Representatives Dingell and Solarz that the General Accounting Office conduct an investigation into the Environmental Protection Agency's enforcement of the Safe Drinking Water Act lead provisions, using Washington, DC as a case study.

During the questioning for EPA today, I will urge the Agency to immediately initiate nationwide testing to ensure that we do not have an undetected national lead problem. In the coming days, I will be introducing legislation that will take action to overhaul the current regulatory regime for lead in drinking water. My bill will modify the Safe Drinking Water Act to improve public communications, to require immediate notification of all homes with elevated lead level results, to require public water systems to provide in-home filters where lead is a problem, to prohibit lead in plumbing fixtures, to require immediate nationwide testing of public water systems, to eliminate lead service lines and lead pipes and to increase water infrastructure funding.

I was struck by the question posed by one resident. Can you actually help fix this problem? I hope the answer to that will be a resounding yes. Today is step No. 1 in that direction.

I look forward to hearing from our witnesses today.

Thank you for this opportunity.

Senator CRAPO. Thank you.

[The prepared statement of Senator Jeffords follows:]

STATEMENT OF HON. JAMES M. JEFFORDS, U.S. SENATOR FROM THE STATE OF VERMONT

I would like to start by thanking Senator Crapo and Chairman Inhofe for granting the minority's request to hold this hearing. The residents of Washington, DC deserve to get answers from Federal and local officials on why there is lead in DC water and why residents were not notified. Safe drinking water is a right, not a privilege.

This Committee has oversight responsibilities for the Army Corps of Engineers, the Environmental Protection Agency, as well as the Safe Drinking Water Act.

Each of us in the Senate has a special oversight responsibility for the District and its residents. I've lived in Washington for a long time, and I take this responsibility seriously.

Many of us live in Washington. We certainly all work in Washington. Our family, friends, children and grandchildren drink the tap water here daily.

Many of us have switched to bottled water. I am disturbed that because bottled water is not regulated in the same manner that tap water is, we cannot even find out if our bottled water is safe.

Yesterday Senator Crapo and I met with a group of Washington parents. Their outrage and sadness at the effect on their children was unanimous. Their charge to us was: Fix this situation and don't let it happen again. I am committed to doing everything in our power to solve this problem.

My overriding question today for our witnesses is—How did we get here? How did we get to the point where the futures of children living in our Nation's capital are threatened every day by the water in their faucets and bathtubs? How did we get to the point where water tests revealed startlingly high lead levels, but yet that information was never provided to residents who unnecessarily exposed themselves, their unborn children, and their children to lead-contaminated water?

How did we get to the point where it takes Congressional hearings and newspaper exposés to get action? How did we get to the point where 2 years after the fact, the

EPA announces that WASA did not comply with the requirements of the Lead and Copper Rule?

How did we get to the point where research from over a year ago showing that lead exposure at levels below the current standard of 10 parts-per-billion have an adverse effect on children's intelligence levels, and yet the Federal Government has not responded?

Lead is a serious health threat to children and pregnant women. It is particularly dangerous for children, who retain about 68 percent of the lead that enters their bodies, while adults retain about 1 percent. Children exposed to lead experience low birth weight, growth retardation, mental retardation, learning disabilities, and other effects. It is also particularly harmful during pregnancy.

I have already mentioned our meeting yesterday with a group of DC parents, and I want to take this chance to share a few more thoughts from some other concerned parents. I ask unanimous consent that a letter and petition from PureWater DC, an internet-based site for parents concerned about the ongoing water issues in DC. Thirteen-hundred and seventy-seven people signed this petition expressing their concern and the expectation for District officials to act quickly to fix the problem.

I ask unanimous consent that the many letters and e-mails I have received from DC residents be included in the record, and I ask that the record remain open for 2 weeks to allow more people to provide their views.

I believe it is imperative that during each moment of today's hearing, we all remember that real parents, children, and babies are being affected by this situation as we speak.

Today's hearing is just the first step in what I hope is a long list of actions that we can take to help solve DC's lead problem and prevent this from occurring elsewhere in the Nation.

Today, I requested with my colleagues Senator Graham of Florida and Representatives Dingell and Solis that the General Accounting Office conduct an investigation into the Environmental Protection Agency's enforcement of the Safe Drinking Water Act's lead provisions, using Washington, DC as a case study.

During the questions for EPA today, I will urge the Agency to immediately initiate nationwide testing to ensure that we do not have an undetected national lead problem.

In the coming days, I will be introducing legislation that will take action to overhaul the current regulatory regime for lead in drinking water.

My bill will modify the Safe Drinking Water Act to improve public communication, to require immediate notification of all homes with elevated lead test results, to require public water systems to provide in-home filters where lead is a problem, to prohibit lead in plumbing fixtures, to require immediate nationwide testing of public water systems, to eliminate lead service lines and lead pipes, and to increase water infrastructure funding.

I have requested a hearing on the childhood lead poisoning in the Health, Education, Labor, and Pensions Committee on which I sit to ensure that the Centers for Disease Control is aggressively addressing childhood lead poisoning.

I was struck by the question posed by one resident—can you actually help fix this problem? I hope to answer that question with a resounding "yes."

Today is step No. 1. I look forward to hearing from our witnesses.

Thank you, Mr. Chairman.

United States Senate
COMMITTEE ON ENVIRONMENT AND PUBLIC WORKS
WASHINGTON, DC 20510-6176

February 26, 2004

The Honorable James Inhofe
Chairman
Senate Committee on Environment and Public Works
410 Dirksen Building
Washington, DC 20510

Dear Chairman Inhofe:

We are writing to you to request a hearing before the Environment and Public Works Committee regarding lead contamination in portions of the Washington, D.C. drinking water system.

In the last several years, testing has shown that significant numbers of homes in the District of Columbia have lead levels that exceed the 15 parts per billion limit established by the Environmental Protection Agency (EPA), in some cases by excessive amounts. Washington, D.C. water is managed by the District of Columbia Water and Sewer Authority (WASA) and supplied by the Washington Aqueduct, operated by the Army Corps of Engineers. There are significant questions regarding the cause of the lead contamination, the actions taken by the District of Columbia to reduce the contamination and to inform the public, the role of the EPA, and the ongoing public health impacts. As the Committee with oversight responsibilities for the EPA and the Army Corps of Engineers as well as the Safe Drinking Water Act, we believe we must take this contamination incident seriously and evaluate the role of each responsible party to determine if any Congressional action is warranted.

Lead exposure has significant health effects, particularly for children who retain about sixty-eight percent of the lead that enters their body. Children exposed to lead experience effects such as low birth weight, growth retardation, mental retardation, learning disabilities, muscle cramps, stomach cramps, anemia, and kidney and brain damage. Lead is also particularly harmful during pregnancy, impacting the unborn child as well as causing miscarriages and stillbirths. It is of great concern to us that potentially hundreds of people were exposed to lead-contaminated drinking water for what may have been an extended period of time. We ask you to hold a full Committee hearing on this matter as quickly as possible.

Sincerely,

Tom Carper

Max Baucus

Hillary Rodham Clinton

Harry Reid

(Lieberman)

Barbara Boxer

United States Senate

COMMITTEE ON ENVIRONMENT AND PUBLIC WORKS
WASHINGTON, DC 20510–6175

February 27, 2004

The Honorable James Inhofe
Chairman
Senate Environment and Public Works Committee
410 Dirksen Building
Washington, DC 20510

Dear Chairman Inhofe:

I am writing to you to join my colleagues in requesting a full committee hearing on the Washington, DC lead contamination issue. It is imperative that this Committee, with jurisdiction over the Environmental Protection Agency, the Safe Drinking Water Act, and the Army Corps, which provides drinking water to Washington, DC, take this issue seriously and evaluate what has occurred to determine if any Congressional actions are required.

Thank you for your consideration of this request.

Sincerely,

Ron Wyden
Ron Wyden
U.S. Senator

Senator CRAPO. We will now turn to the Chairman of our committee, Senator Inhofe, who has been very strong in encouraging us to hold these hearings.

Senator INHOFE. Thank you very much, Mr. Chairman, for holding this hearing.

In deference to our witnesses, I will submit my statement for the record and move on.

[The prepared statement of Senator Inhofe follows:]

STATEMENT OF HON. JAMES M. INHOFE, U.S. SENATOR FROM THE STATE OF OKLAHOMA

I like to thank Chairman Crapo for holding this hearing, like most of my colleagues, when I'm not back in my home State, I stay here in the District of Columbia. And while the lead levels in the drinking water probably won't shorten my life expectancy, I do have grand kids who come to visit and don't want to put them at any added risk.

First I think that it's important to put the risk from lead exposure into perspective. While high blood lead levels are a cause for concern, the blood lead levels of the children in the District of Columbia are far lower than those we experienced nationwide just a generation ago.

The Centers for Disease Control (CDC) has established guidelines for lead exposure, their level of concern for blood lead is 10 micrograms per deciliter. Twenty years ago, the vast majority of children in America would have exceeded that level. A nationwide study conducted by CDC in the early 1980s, 88 percent of the children in the survey exceeded the current CDC level of concern.

From the mid-1920s until the mid-1980s motor gasoline contained an additive, tetraethyl-lead, that improved fuel performance by preventing pre-ignition in the cylinders of the engine. This lead was released as a gas and in the form of a very fine dust. Even today, areas around busy roads and highways may contain elevated levels of lead.

Because of concern over lead in the drinking water the CDC recently conducted a study of residents in the District whose tap water had the highest lead levels, above 300 parts per billion. Not one person had lead levels in their blood above CDC's level of concern.

The current tempest over DC's drinking water raises several pressing questions. First and foremost is, What caused the sudden jump in lead levels seen in the water samples? Hopefully our witnesses will have some answers.

But equally important, is the longer-term question of whether or not our system properly responded to the incident. It has been almost 2 years since the first water samples tested high in lead.

What we know for certain is that somewhere between the source and the spigot, something went wrong. What we need to know is why is it taking 2 years to solve this problem.

Senator CRAPO. Thank you very much.

Senator Clinton.

OPENING STATEMENT OF HON. HILLARY RODHAM CLINTON, U.S. SENATOR FROM THE STATE OF NEW YORK

Senator CLINTON. I want to thank you and Chairman Inhofe for this hearing. I particularly want to thank Senator Jeffords for his strong interest and leadership on this issue.

With your consent, I will submit the entire statement for the record but I want to make a few additional points.

There is no safe level of lead and recent studies, one concluded last April published in the *New England Journal of Medicine*, followed 172 children in Rochester, NY and measured blood lead levels at 6, 12, 18, 24, 36, 48 and 60 months and tested their IQs at 36 and 60 months. The study found that most of the reduction in IQ attributable to lead occurred at blood levels below 10 mpd which is the level that the Centers for Disease Control considers to be the threshold level for health effects.

The researchers found that IQ scores of children who had blood lead levels of 10 mpd were about seven points lower than for children with levels of 1 mpd. An increase in blood levels from 10 to 30 mpd were associated with a small additional decline in IQ. While this is only one study, there are a number of other research findings that suggest what we currently consider to be a safe level for lead is in fact too high. That underscores the seriousness of the issue we are here to talk about today.

Lead exposure comes from a variety of sources and lead in drinking water accounts for only about 20 percent of lead exposures, but if no level of lead is safe then any source of lead needs to be taken very seriously. To the DC residents who are here, and that includes many of us who serve in this body who live part-time inside the District boundaries, I think you have a right to be disturbed and have an absolute right to have your questions answered.

I don't think it is productive in this hearing to try to assess blame but at some point, we need to get specific questions answered. Why did WASA not notify residents about elevated lead levels as soon as it had the test results? Why did the DC Health Department engage in this issue when notified of the problems? Why did EPA take so long to get involved? How is it possible that WASA or no agency has accurate records about who has lead service lines and in the absence of such records, why is WASA refusing to provide water filters to homes for which it does not have information one way or the other?

These are some of the questions I have. I look forward to this hearing but I have to conclude by saying one of the great benefits of living in our country over many, many decades has been that we could count on the water we drank and the food we ate to be safe much more so than in other countries in the world. For the capital of our Nation to have this problem is deeply concerning. I am grateful we are having this hearing and I also look forward to working with Senator Jeffords on his legislation.

Senator CRAPO. Thank you very much, Senator Clinton.

Without objection, full statements of any of the Senators will be made a part of the record.

At this point, we would like to call our first panel. We thank you all for joining us.

Before we begin, I would like to address a few of the rules of the committee to all witnesses today. You will notice there is a clock in front of you. You should have been informed you should keep your oral testimony to 5 minutes. We thank you for your written testimony, those of you who provided it, and we assure you we will carefully review your written testimony. However, 5 minutes goes by quickly and if you are like most witnesses, you will find your 5 minutes goes by before you are done saying everything you have to say. We encourage you to watch the clock. If you do forget, I will lightly tap the gavel to remind you to look down at it.

The reason for that is because we want to have the opportunity to have dialog and discussion with you. You will find you will have an opportunity to make a lot of your points that you didn't get to in your first 5 minutes as we get engaged in that dialog. So please pay attention to the clock.

With that, let me introduce our first panel. We will ask you to speak in the order I introduce you. We first have Benjamin H. Grumbles, Acting Assistant Administrator, Office of Water, EPA; Donald Welsh, Director, Region III, EPA in Philadelphia; Jerry Johnson, general manager, District of Columbia Water and Sewer Authority; Dr. Daniel Lucey, interim chief health officer, District of Columbia Department of Health; and Thomas B. Jacobus, general manager, Washington Aqueduct, Baltimore District, U.S. Army Corps of Engineers.

Gentlemen, we thank you for coming and for your preparation. Mr. Grumbles, please proceed.

STATEMENT OF BENJAMIN H. GRUMBLES, ACTING ASSISTANT ADMINISTRATOR, OFFICE OF WATER, ENVIRONMENTAL PROTECTION AGENCY

Thank you. Thank you for having this hearing and for putting together such a balanced presentation of witnesses to cover all the perspectives and issues we are all facing here.

I want to say on behalf of EPA that we, like you, are asking a lot of tough questions of ourselves as well as questions of our colleagues. We want to focus on collaborating together, not finger pointing, coming up with solutions, concrete actions, restoring the quality of the water and the confidence of the public here in the District and making sure that this situation doesn't happen in other places throughout the country.

I would like to touch on a few things, then I will turn to Don Welsh, who is the Regional Administrator, and has a more local perspective in terms of what is happening in the District itself. I just wanted to touch on a few items from the national perspective.

The first thing I would like to say is that EPA places a very high priority on reducing exposure to lead. As you all pointed out very eloquently, it is a neurotoxin, a very dangerous poison and it is all of our jobs to ensure that exposure to lead is prevented or reduced.

The next point I would like to make is that the 1991 Lead and Copper Rule signaled a fundamental change in that we went from having a 50 ppb MCL at the treatment plant itself to a different approach which would try to take into account the bad things that can happen once the water leaves the treatment plant and goes through the distribution systems to the homes and buildings in communities. That resulted in an action level and the focus of the rule is on corrosion control, monitoring, public education and if necessary, lead service line replacement.

I am asked repeatedly whether this a national problem. I would say from the data we have, it is not so much a national problem as it is a localized problem and a national opportunity. This is an opportunity to look hard at the existing regulatory framework, to look particularly hard at the monitoring and public education requirements and really focus on what we can all do, not just from a local perspective with respect to the District of Columbia, but also from the national perspective.

EPA is doing several things. One is that we have initiated a national compliance review to determine whether or not there is a national problem, what success there has been since the 1991 rule, and also to determine how well the current rule is being implemented, focusing particularly on monitoring, public education and communication. We are also reviewing within our Office of Water, various aspects of current policy and also issues associated with the rule to determine whether or not the policies or the regulation itself should be revised. We are establishing several work groups with the benefit of experts to look at simultaneous compliance issues, sampling protocols and possibly also public education, how best to advance what we did in 2002 and that was to issue guidance on public education for lead and drinking water, how to communicate and keep communities informed.

We are very concerned and want to be very proactive with respect to lead and drinking water at schools and day care facilities. I have written to all the State Environmental and Public Health Commissioners asking them to share with us what they are doing on that front. Do they test regularly for lead in drinking water at schools, what protocols they are following, what results they are finding, what EPA could do recognizing our limited legal authorities when it comes to schools and day care facilities, to provide additional guidance or leadership?

The last thing I would say is that from a national perspective, if there is a silver lining in this lead problem in the District, it is that it gives us all an opportunity to focus on areas such as public education, communication and monitoring and the importance of water infrastructure.

So we look forward to working with you, your colleagues and all the stakeholders and concerned citizens on this issue.

Thank you.

Senator CRAPO. Thank you very much, Mr. Grumbles.

Mr. Welsh.

STATEMENT OF DONALD WELSH, DIRECTOR, REGION III, ENVIRONMENTAL PROTECTION AGENCY

Mr. WELSH. Good afternoon. I am Don Welsh, the Regional Administrator for USEPA, Region III.

Thank you for the opportunity to appear before you today to discuss the important issue of lead in the tap water of District of Columbia residents and the steps that EPA and other agencies are taking to resolve the problem on a short- and long-term basis.

Elevated levels of lead in the environment, whether in drinking water or lead paint, can pose significant risks to health, particularly to pregnant women and young children. Reducing exposure to all sources of lead is vital to protecting the health of our citizens.

It is unacceptable to us that many families in the District continue to live with fear and uncertainty over the quality of the water they drink. At EPA, we will not be satisfied until all aspects of this problem are resolved. There is no higher priority for my office than to work with the city to protect the residents.

To that end, EPA and the District of Columbia have directed, and are closely monitoring, a series of interim measures with firm deadlines to ensure that residents have safe drinking water and a proper precautionary guidance. At the same time, a multi-agency

Technical Expert Working Group is acting as quickly as possible to identify and correct the cause of the elevated lead levels.

The city and EPA have had regular meetings and conversations to monitor progress and to ensure necessary actions are being taken.

By way of background, EPA's Lead and Copper Rule requires systems to optimize corrosion control to prevent lead and copper from leaching into drinking water. To assure corrosion control is effective, the rule establishes an action level of 15 parts per billion for lead. If lead concentrations exceed the action level in more than 10 percent of the taps sampled, the system must intensify tap water sampling and undertake a number of additional actions to control corrosion and to educate the public about steps they should take to protect their health. If the problem is not abated, the system must also begin a lead service line replacement program.

The results of DC's required tap monitoring exceeded the 15 ppb action level for 10 percent of taps monitored during 6 of 15 reporting periods since January 1992, 3 times prior to 1994 and 3 times since 2002. An Optimal Corrosion Control Treatment limit implemented by the Aqueduct appeared to be effective in minimizing lead levels until the sampling period between July 2001 and June 30, 2002. According to reports filed by WASA, the 90th percentile value had increased to 75 ppb during that period and registered at 40 ppb and 63 ppb for 2 subsequent monitoring periods.

On October 27, 2003, EPA Region III received results from a separate lead service line sampling program conducted by WASA indicating that 3,372 of 4,613 service lines tested had numbers exceeding the action level, many by a large margin. The information was reviewed by our technical staff with an eye toward determining how to address the underlying cause of the corrosion problem.

As indicated, WASA and the District of Columbia Government are undertaking a series of actions outlined by EPA to address the public health threat posed by lead in drinking water. WASA will ensure delivery by April 10 of certified water filters and consumer instructions to occupants and homes and buildings with lead service lines. Periodic replacement of the filters will also be assured.

Additional tap water sampling has begun in schools as part of a broader program to test a representative group of facilities that are not served by lead service lines to determine the full scope of the problem. WASA has committed to an accelerated schedule for physically replacing lead service lines in the District. WASA is expediting notification to customers of the results of water sampling at their residences, committing to providing results in 30 days or less.

As EPA, the District and WASA continue to expand outreach efforts to provide important information to consumers, WASA is providing an enhanced public education plan to reach all sectors of the population in an effective way. EPA is undertaking a compliance audit of WASA's lead service line and public education actions. In letters to WASA last week, EPA asserts instances in which requirements were not met, ,and as part of the enforcement process, requires WASA to provide information to EPA responding to those findings.

In a separate initiative, an internal EPA team is evaluating WASA's prior outreach efforts, a process to be completed by month's end that involves a review of materials, interviews with residents and public officials and a survey of best practices from public water systems around the country.

It is clear that WASA was ineffective in informing the public of the magnitude of the lead problem in drinking water and in conveying the steps families and individuals should take to protect themselves. Mass media tools, including direct contact with media representatives, as is recommended in EPA guidance, were not used effectively.

The Region is also taking a critical look back at how we could have done a better job in our oversight capacity to ensure the public interest is being served by WASA's actions. There will continue to be lessons learned that will benefit the agency in the future. The Technical Expert Working Group made up of representatives from the public and private sectors is making progress in identifying the cause of the elevated lead levels. By next week, EPA is scheduled to receive a proposal from the technical team for a water chemistry change to reduce corrosion and maintain other protections. Under the proposed timetable, a partial system test is currently planned for June 1 followed by full system implementation on or about September 1. EPA has formed an independent peer review group to check the team's findings.

In closing, working closely with the District of Columbia, our public service partners and concerned citizens, we will continue to aggressive act to protect residents and resolve the lead problem. We are taking action to hasten the day when the citizens of the District of Columbia can once again be confident in the safety of their drinking water.

Thank you for the opportunity to present this information this afternoon. I am pleased to answer any questions that you have.

Senator CRAPO. Thank you very much, Mr. Welsh.

Mr. Johnson.

STATEMENT OF JERRY N. JOHNSON, GENERAL MANAGER, DISTRICT OF COLUMBIA WATER AND SEWER AUTHORITY

Mr. JOHNSON. Good afternoon.

I am Jerry Johnson, general manager of the District of Columbia Water and Sewer Authority. I am pleased to be here to provide testimony about WASA's endeavors relative to lead replacement program and the issues of elevated lead levels in some homes in the District of Columbia.

It goes without saying that these issues are of the utmost importance and this is an excellent opportunity to reassure this panel and the residents of the District of Columbia that working expeditiously to find lasting solutions is absolutely WASA's highest priority.

There are several critical areas I wish to cover today and answer the questions members of the committee have. I will attempt to follow the outline as put forth in your letter of invitation. First, WASA has undertaken an aggressive effort to distribute filters to residents it believes have lead service lines in the District of Columbia. As of today, WASA has distributed filters to all of these households, some 27,000 of them, and will provide replacement filters for a 6-month supply. In addition, over 200 filters have been distributed to home day care centers with lead service lines.

Second, WASA is working tirelessly to keep customers informed regarding all facets of the lead issue. WASA recently expanded its lead service hotline, a program we initiated in January 2003, to facilitate direct communication with our customers. We have added personnel to allow us to staff operations for 12 hours a day, Monday through Friday and 9 hours a day on weekends. Since February, the hotline has fielded over 45,000 calls and 6,200 e-mails. Our website, which is continuously updated, averages about 1,700 visits per day. The March and April edition of our monthly newsletter, "What's on Tap," which is distributed to 125,000 plus residents of the District, focuses on the lead issue and provides advice to customers.

Also, since February 2004, WASA has supported 10 joint public information meetings across the city, along with the DC Department of Health and the Washington Aqueduct and another 15 to 16 ANC and civic association groups where we have also had joint appearances. It is estimated that approximately 1,500 residents have attended these meetings. In addition, in February WASA sent mailings to every address in the District of Columbia, over 360,000, residents regarding the lead issue.

The mailings, which were multilingual, contained a Department of Health fact sheet, general information about the subject of lead and water and contained precautions that residents should take regarding the use of water. WASA has also made available brochures and maps in all libraries and community centers throughout the center. We are currently in the process of contacting by mail, residents we believe are served by lead service pipes to provide additional information on flushing and encourage those residents who have not already done so, to avail themselves of tap water testing at no cost to them.

We are also contacting approximately 21,000 as opposed to 37,000 residents for which there is no record of the pipe materials our customers have in the data base. That is simply because we are maintaining historical records that date back to 1901 and they are just not all in place. In addition, a direction communication with customers, WASA has conducted media briefings and representatives for the board and management have appeared before congressional committees, hearings and briefings of the DC Council, the Mayor and attended regular mayoral press briefings.

Additionally, WASA has reached agreement with the George Washington University School of Public Health, Department of Environmental and Occupational Health to provide the Authority with assistance and advice in a number of health related areas.

Third, getting to the root problem by conducting research as to why there is an increase in some homes and which specific homes are affected will continue to be a priority for WASA. Currently we are working with EPA, the Washington Aqueduct, the DC Department of Health, and respected scientists and experts on this problem.

We have also increased the number of lead service pipes in public space that are to be replaced this year, adding about 50 percent to those with an addition of $7 million that has been provided by the Board of Directors, putting the lead replacement number from 800 to 1,300 this year with an additional 300 that we expect to be undertaken by the District of Columbia through its Road Replacement Program.

In addition, the Board has been considering, through a resolution adopted at its last meeting, a $350 million proposal to replace every one of the estimated 2,300 lead service lines in the District by 2010. Community input on this proposal will be sought over the next 2 months. Since lead service pipes are in both public space and private space, we will work with homeowners to replace those in their space and are working currently with the District in an effort to find financial assistance and looking at options for these citizens having difficulty in paying that cost.

With that, Mr. Chairman, I will end my testimony. I noticed the red light is on and rather than your dropping the gavel, I would be glad to respond to any questions you have but we want you to know it is WASA's intention to address this challenge in a manner that works for our city, for our residents, for the visitors who come to our city and to implement the solution as quickly and as reasonably as possible. We at WASA are firmly committed to doing this and welcome the collaboration of our partners, some of whom are here today.

Thank you. I am prepared to answer any questions you may have.

Senator CRAPO. Thank you very much, Mr. Johnson.

Dr. Lucey.

STATEMENT OF DANIEL R. LUCEY, INTERIM CHIEF HEALTH OFFICER, DISTRICT OF COLUMBIA DEPARTMENT OF HEALTH

Dr. LUCEY. Good afternoon.

My name is Daniel Lucey. I am the interim chief medical officer for the DC Department of Health. In the next 5 minutes, prior to responding to your questions, I would like to summarize my background and list several key points about the lead issues in Washington, DC.

I am a physician trained in adult medicine and infectious diseases with a Master's Degree in Public Health. After serving in the military as a physician, I joined the Public Health Service while working at the National Institutes of Health and the Food and Drug Administration.

During 9/11 and the subsequent anthrax attacks, I was the Chairman of the Infectious Disease Service at the nearby Washington Hospital Center in DC and subsequently in 2002, was involved with the smallpox vaccination program and in 2003 with SARS, traveling to Hong Kong, working in a hospital in Toronto with patients with SARS, and in 2004, earlier this year with avian influenza.

On February 10 of this year, I began work at the DC Department of Health with a focus on biodefense. On February 13, 3 days later, I attended a Lead Task Force meeting. Every day since then, I have worked on lead issues. Although not a lead expert, I have approached learning about the lead issues through an intensive process much like learning about other previously unfamiliar to me diseases such as anthrax, SARS and avian influenza.

On President's Day, Monday, February 16, I contacted the Director of the Centers for Disease Control and Prevention, Dr. Julie Gerberding, to request advice from lead experts at the CDC. Her response was immediate that day and since then, we have received outstanding CDC assistance. In fact, even today, there are CDC experts onsite with us at the Department of Health.

On February 26, 2004, the city administrator, Mr. Robert Bob, instructed me to direct the Department of Health response to lead issues. Later that day, I completed and signed a health advisory letter from the Department of Health to the approximately 23,000 residences in DC with lead service lines. The letter is Attachment I and has been translated into six languages. The advisory contained recommendations about drinking water and measuring blood lead levels in persons most at risk for lead poisoning.

In order to assess the health impact of increased lead concentrations in the water, to knowledge no such widespread health advisory on lead in drinking water has ever been issued in the United States. Our findings may be useful to other cities if they find increased lead concentrations in their drinking water.

In order to provide blood lead level testing by the Department of Health starting on the 28th of February at DC General Hospital, we mobilized many persons within the Department of Health. In addition, on Monday, March 1, I contacted the U.S. Surgeon General, Vice Admiral Dr. Richard Carmona to request additional personnel assistance. He responded immediately that day and via Admiral Babb and the Commissioned Corps Readiness Force, provided a team of public health service officers over the next 4 weeks who worked very long hours with us in clinics all across the District of Columbia. They also went with us to several hundred homes of persons at high risk of lead poisoning in the District. On March 30, the DC Department of Health, the Commission Corps Readiness Force and the CDC published our preliminary findings on blood lead levels in the CDC's publication called "Morbidity and Mortality Weekly Report."

I would like to summarize six key points. First, none of the 201 persons we tested for blood lead levels who live in homes with the highest measured levels of lead in the drinking water, greater than 300 ppb, have had elevated blood lead levels defined as Senator Clinton said by 10 mpd for children and 25 mpd for adults.

Second, from 2000 to 2003, the percentage of children less than 6 years of age with elevated blood lead levels continued to decline in the District of Columbia, both in homes with and homes without lead service lines. The percentage of children with blood lead levels greater than or equal to 5 mpd did not decline in homes with lead service lines although this percent did decline in homes without lead service lines.

Third, only 2 of the initial 280, less than 1 percent, of children in home child care facilities with lead service lines have had elevated blood lead levels.

Fourth, of the initial 4,106 persons who came to our clinics across the District of Columbia for free blood level testing in our laboratory, 1,277 were young children less than 6, of which 16 had elevated blood lead levels. The initial 14 children have been found to live in homes with dust and/or soil lead levels exceeding EPA and HUD guidelines. The homes of the other two children are currently being evaluated.

Fifth, according to the CDC from 1976 to 1980, nearly 9 of 10, that is 88.2 percent of children at that time age 1 to 5 years of age and therefore now adults 24 to 28, had blood levels that today are considered elevated, namely at least 10 mpd.

Sixth, the EPA "action level" for lead in drinking water, 15 ppb, is not a health-based recommendation. I would like to quote from the website that has been devoted to the Washington, DC area on the drinking water issue.

> "The action level was not designed to measure health risks from water represented by individual samples. Rather, it is a statistical trigger that if exceeded requires more treatment, public education and possibly lead service line replacement."

Thank you for your time and I would be pleased to respond to your questions.

Senator CRAPO. Thank you very much, Mr. Lucey.

Mr. Jacobus.

STATEMENT OF THOMAS P. JACOBUS, GENERAL MANAGER, WASHINGTON AQUEDUCT, BALTIMORE DISTRICT, U.S. ARMY CORPS OF ENGINEERS

Mr. JACOBUS. Good afternoon.

I am Tom Jacobus, the general manager of the Washington Aqueduct. Thank you for the opportunity to be here today.

Washington Aqueduct, which is part of the Baltimore District of the U.S. Army Corps of Engineers, is a public water utility. We are regulated by the U.S. Environmental Protection Agency, Region III in Philadelphia. Washington Aqueduct provides potable water, not just to the District of Columbia Water and Sewer Authority, but also to Arlington County, VA and the city of Fall Church's service area and Virginia as well. All funds for the operation and capital improvements for the Washington Aqueduct come from its customers. The provisions of the Safe Drinking Water Act and its associated regulations are the basis for all operations concerning the production, storage and transmission of the drinking water produced and sold by Washington Aqueduct to its wholesale customers. The primary objective of the treatment process is to produce and deliver water to the tap that is free of contaminants and pleasant to drink. To achieve that objective, we do three things simultaneously. We kill harmful bacteria, we remove organic and inorganic contaminants and we provide optimal corrosion control. Corrosion control treatment is designed to reduce lead and copper leaching into drinking water and to keep the concentrations below the action level in accordance with the lead and copper rule.

For many years we have accomplished that by the use of lime to adjust the pH of the water, but the recent sampling in the District of Columbia has resulted in unexpectedly high lead levels. Therefore, the corrosion control treatment needs to be reevaluated based on intensive analysis of current operations and the use of analytical models. Our team of engineers and scientists has recommended adding a phosphate-based corrosion inhibitor to the treatment process. We anticipate that EPA will approve this change by May 1 so that by June 1 we can apply a new chemistry to a portion of the distribution system.

The full system application will begin by September 1. Our cost estimates for the work are $925,000 for research and engineering analysis and laboratory studies; $250,000 for the partial system application and $3.1 million for interim facilities for full scale application. The additional chemical costs will be about $1 million per year. While it will take several months to measure the effects, we have confidence that this change will be effective in reducing the lead leaching.

I have one additional point I think is important to mention as we move forward. Washington Aqueduct and its wholesale customers have standing financial and technical working groups that regularly address ongoing operations and evaluate capital improvements. Based on our experience in the last 8 weeks, we see opportunities to improve data sharing among the customers pertaining to lead and copper corrosion. We intend to take the necessary steps to do that.

This concludes my introductory remarks. I will be happy to respond to your questions.

Senator CRAPO. Thank you very much, Mr. Jacobus.

I will begin with questions. We are going to do 5 minute rounds but we will do a number of rounds so that all the Senators have an opportunity to get out their questions.

My first question is for you, Mr. Grumbles. You mentioned the national review you are conducting right now on lead. Can you tell me what this review has already revealed about how the rule on lead is performing, how it is working and how it is being complied with?

Mr. GRUMBLES. So far, what we have done is reviewed the data that we have in the SDWIS. What we have found is that only 4 of 199 systems serving more than 50,000 people have exceeded the 15 ppb action level since 2000. One of those was the District of Columbia. All of the systems except the District are now back below the action level. For systems serving between 3,300 and 50,000 people, 56 of the 1,761 systems have exceeded the action level since 2000 and only 14 reported to exceed the level since 2002.

I want you to know, Mr. Chairman, that we are not comfortable with the amount of data we have received to date. One of the things that is part of our national compliance review is to write to and encourage the States to provide more information on compliance pursuant to the Safe Drinking Water Act and the Lead and Copper Rule. The point is that while so far the numbers are indicating to us through the SDWIS Program that it is not a crisis, we do need to get more information from the systems. I think only 22 percent of them have provided that information and there are several States, 23, who have not provided that information on the 90th percentile reporting.

Senator CRAPO. So this is preliminary information which is certainly not complete at this point.

Mr. GRUMBLES. That is correct.

Senator CRAPO. This question is for either you, Mr. Grumbles or Mr. Welsh. I would like to go into the action level and exactly what it means. Dr. Lucey quoted it in his testimony. That quote in the EPA's statement is that,

> "The action level was not designed to measure health risks from water represented by individual samples. Rather, it is a statistical trigger that if exceeded requires more treatment, public education and possibly lead service line replacement."

Whichever of you feels most qualified to respond, explain what does it mean when we identify that an action level has been exceeded?

Mr. GRUMBLES. I think what Dr. Lucey has said is an important point. The action level is not health based in the sense of an maximum contaminant level or MCL. When it was established, when the number 15 ppb was established, there were health factors taken into account. There were also feasibility factors in terms of what steps could be taken after that was reached. The analysis in the preamble to the 1991 rule describes how health factors were considered.

It triggers actions, specific requirements for optimizing corrosion control, for carrying out additional monitoring and for doing a very specific, detailed, public education or public notification process. It also requires at the end of that process, if you are still exceeding that 15 ppb, a specific lead service line replacement study and program replacing 7 percent of your lead service lines a year.

Senator CRAPO. If I understand correctly, please be sure you correct me if I am incorrect, the level and the action level has been set at such a point that when it is triggered, there is still time for an effective response if there is a response forthcoming to avoid a serious health risk? Is that correct?

Mr. GRUMBLES. That is correct. There are dozens of cities and towns across the country that have exceeded that 15 ppb. The good news is from the data we have, most of those cities or towns have reduced the action level and we are finding it is not exceeding it in those cities and towns but again, I want to caveat that one of the lessons we are learning from this experience is that the decisions we make and the determinations of compliance are only as good as the amount of data and sampling that we have.

Senator CRAPO. Thank you. I see my first 5 minutes are up. We will turn to Senator Jeffords now.

Senator JEFFORDS. Thank you.

I have been frequently told by parents if I had only known, I would have taken precautions. I want to ask each of the witnesses to tell me how you are responding to this question, what explanation are you giving parents of children who were unnecessarily exposed to lead in their drinking water and what steps are you taking to regain the trust of the citizens of Washington, DC?

Let me give you my own personal experience. We moved here 20 some years ago. I, being a macho man, drank this water and my wife came down with the kids and said, "no, we are going to get bottled water to make sure the kids get good water". We did that. Then the other day I said, "do you test the bottled water" and I found out no. I wonder if we are doing anything about the options to make sure when I go out and pay good money for bottled water, is that healthy or do you know?

Mr. GRUMBLES. Two points. One is in 2002, EPA recognized that the success of the Lead and Copper Rule depends on the ability to communicate effectively the timely and accurate information to the citizens of the community, so we developed a Lead Public Education Guidebook, a guidance document, because of the importance of communicating on that front.

With respect to bottled water, as you know the 1996 amendments to the Safe Drinking Water Act set up a framework where the Food and Drug Administration regulates the quality of the bottled water industry and to the extent EPA has established maximum contaminant levels under the Safe Drinking Water Act, the FDA is required to impose a standard on the bottled water industry. I believe I understand that for lead, the FDA has established a 5 ppb standard for bottled water.

I think your point about ensuring consumer confidence in the country is a key one. The 1996 amendments to the Act which establishes the consumer confidence reports, the value of which continues to be realized over and over again because it is the public citizens, the mothers, the parents, the families who are actually in the best position to monitor the success of the implementation of the Lead and Copper Rule. Obviously the regulatory agencies, EPA, is entrusted and has that responsibility but the consumer confidence reports and the public education components of the Lead and Cooper Rule are critical to avoiding situations like we find ourselves in today.

Senator JEFFORDS. Is there any requirement that the bottled water has to let you know what is in it?

Mr. GRUMBLES. I honestly don't know what the requirements are on the bottled water industry. I know the Food and Drug Administration has that statutory responsibility.

Senator JEFFORDS. I understand there aren't, but I just wanted to know.

Mr. GRUMBLES. From an EPA perspective, we have an interest just like you in understanding and in being able to provide an answer to that question. We will commit to doing that.

Senator JEFFORDS. Thank you, Mr. Chairman.

Senator CRAPO. Senator Clinton.

Senator CLINTON. Thank you, Mr. Chairman.

I would like to ask each one of you to answer two questions briefly for me. No. 1, looking backward, what mistake did you or your agency make in handling this matter? No. 2, looking forward, what is the one thing that you believe should be done in order to remedy the situation we find ourselves in? Why don't we start with Mr. Jacobus?

Mr. JACOBUS. We have a wholesale/retail relationship with our customers. Every day we have great visibility over the bacteriological content of the water throughout all three distribution systems—Arlington, Falls Church service area, and the District of Columbia. We test the water in our laboratory from samples at the plant, samples from the distribution system, either that we take ourselves or are brought to us for 35,000–40,000 tests a year. We have great visibility on the bacteriological and the chemical contaminants in the water leaving the treatment plants, disinfection byproduct rule compliance, all of that.

The mistake, to answer your question, is that we did not have the same visibility for the lead and copper samples. The samples are taken in a different way. They are taken throughout a period of the year. It is not a go/no-go on each individual sample. So we did not have all the samples collected at our organization; they were at different locations in two different jurisdictions under the State of Virginia's Health Department and EPA's regulatory responsibilities for the District of Columbia Water and Sewer Authority.

We had the ability and we will take the initiative to bring that data together so that we can help all our customers and give them the benefit of our systemwide look. Even though every day, we paid attention to the optimal corrosion control treatment and that the water leaving the plant was at the specification for what we had agreed in our scientific-based study with EPA of how to treat the water and we were getting anecdotal evidence occasionally of a high reading. It wasn't until the spring of 2003 when EPA, reacting to WASA's results in 2001 and 2002, said we need to open this and start looking, so we started down that road.

We did not have perfect knowledge of the big picture. We can get that knowledge even though it is regulatory, but because we think it is a responsible thing to do. I hope that answers your question.

Senator CLINTON. It certainly does and I appreciate that. Perhaps it would help if it were required by regulation so that at least all the players, all the stakeholders are at the table, but I appreciate that very much.

Dr. Lucey.

Dr. LUCEY. Looking back, again, I have been here since February 10 but nevertheless, I think looking back what we might have done differently was to have the type of face-to-face, verbal and phone interactions within the Department of Health, with EPA and with WASA and with Washington Aqueduct to discuss the issue about the action level has been exceeded. What does that mean? Is that a health-based risk or is it not? It is not, but could there be health implications? Yes, there could be. How are we going to answer that question? For me it is a clinical or medical approach and I think the folks that initially heard about the elevated water lead concentrations within the Department of Health were not the clinical and medical folks.

As you know, there is new leadership now in the DC Department of Health. The top two people are no longer in the Department of Health as of two Fridays ago and I think the new interim leadership is very, very strong. That is one recommendation I have made to the new leadership, to have more involvement within the Department of Health of the clinical and non-clinical individuals.

As far as looking forward, I think although there are many things that could be done to try to remedy this situation and any others developed in the future, I think a major one is to have the types of interactions we have had for the past couple months within the Department of Health and outside with EPA, with WASA, with Washington Aqueduct in terms of discussion of what is the best advice we can do and how we can best communicate that to the public?

For example, briefly, how to prioritize lead service line replacements if that is what is going to happen, as we have heard it is on an accelerated basis. In my opinion, it is very important not only for the Department of Health to be involved in that decision-making but to work with the new clinical team coming on board with WASA from George Washington University to work directly with EPA in the formulation of the prioritization of lead service line replacement, as well as with the Department of Transportation and everyone else who needs to be involved. Do that right now from the beginning and we are doing that.

Senator CLINTON. Mr. Johnson.

Mr. JOHNSON. Obviously hindsight is 20/20 and as I look back over this issue and consider mistakes that were made, I think probably the one that is most resounding was our focus on trying to comply with Federal regulations as opposed to looking at a broader picture in terms of where our customers were, what they were thinking and the need to get information to them perhaps sooner.

Early on in this process, we were only working with a small sample base of 50 homes in the District and did not have a clear fix on what that meant for the broader district. I felt, and it was my decision, I assume responsibility for it, that at that point, it was not appropriate to raise a flag and begin to alarm people in the District about a problem we were having. As you know, we went forward and did a broader base of sampling than has ever been done in the United States of America. That showed some results that were concerning to us. We didn't know what the results meant, we are not the medical experts, not the regulatory experts. We think we got that information to them in a timely fashion and we believe with all sincerity we did make all the efforts that were required and I won't go through the list of things in front of me to comply with Federal regulations.

An audit subsequently conducted by EPA suggested that there were some technical issues we may not have meant requirements and we will go back and respond to those over the next 21 days. I think that would be the mistake. I think we should have been focused more on the community in the District of Columbia than we were on the regulators but if you are regulated, you have to meet certain other requirements.

As we go forward, I think the two points made by both Dr. Lucey and Mr. Jacobus are very valid. I think there needs to be a more formalized relationship established between the District of Columbia Water and Sewer Authority, the District's Health Department in order for us to grab these issues early on in the process and be able to have a stronger collaborative relationship. We had focused on the relationship between Aqueduct and their customers in Virginia who are also partners of ours in that business relationship. We began, when this issue first came up, to start looking at the water. I think we probably could have pursued that as another mistake. We probably could have pursued that more aggressively when the first 50 samples came back to take a look at that the production side of it and the chemistry of the water.

As we go forward I think that kind of collaborative relationship is very important and I really think the District ought to look at the possibility of having primacy in this area. The USEPA has primacy in two jurisdictions for water distribution systems, the District of Columbia and I believe in Wyoming, so there are different relationships that get established with the regulatory bodies when you are working with a local entity that has a better sense of what is happening in that community and how that community needs to respond.

I am sorry for being so long-winded but that is a very important question.

Mr. WELSH. I believe when there is an exceedance of the action level, the spirit of the Lead and Copper Rule is to make sure that any of the citizens who might be exposed to a higher level of lead understand that fully, know that they might be exposed and have good information in their hands about how to reduce their exposure to lead. That goal is clearly what was not met in this instance. As Jerry mentioned, WASA took actions to get information into peoples' hands and in our review, we determined that some of the specific requirements weren't met fully, but the larger issue is that the information that was put out wasn't really getting home to the folks who needed to have that information so they fully understood it was important to them, they should pay attention and follow the directions put in there so they would know what they were exposed to and understand how to limit their exposure.

Looking backward, we weren't aggressive enough, we weren't thorough enough and didn't find those deficiencies soon enough, both in the letter of what was required under the rule but also in that larger question of even if the language is put out there and made available in documents, is it a message that is getting home to folks and are they really understanding it.

Going forward, we want to change our review procedures in the region so that we do not only a thorough job of checking the letter of compliance with the Lead and Copper Rule but that we also take the time to make the judgment about whether that message is being received, much in the way marketing folks do when their commercials and advertisements go out, they can measure in the public whether that message is being received.

So we have changed our operating procedure so that it is not just the technical person in the Drinking Water Branch who reviews the reports for whether they have complied with the rule, but that we also call in the communications expertise that we have available in the region as well as if necessary to do contracts for folks on the outside who are experts in the area of communications to make sure not only in the future that the reg is fulfilled, but that also the message is being effective and that we measure out in the public do you folks who need to know this know? That is what we would like to do going forward.

Mr. GRUMBLES. I couldn't have said it better. From the national perspective, the EPA does want to also acknowledge that it is not just following the letter of the rule, it is the spirit of the rule and that is partly our responsibility too, to be proactive and help oversee that not just the letter, but the spirit, the public education and other aspects of the Lead and Copper Rule are followed. That is certainly one of our objectives, acting proactively.

The other one is this whole issue of simultaneous compliance. Providing drinking water to the public can be a tricky balancing act, given various regulations and requirements. We plan to have a workshop in May with national experts. You have the Disinfection Byproduct Rule, you have the Lead and Copper Rule. How do you ensure it is all working together and there are not unintended consequences?

Thank you.

Senator CRAPO. Let me proceed a little further. As many of you probably know, yesterday, Senator Jeffords and I and Mayor Williams and several others joined some of the constituents here in Washington, DC and discussed with them their concerns. One of the concerns they raised was that under the testing procedures, it was difficult for them as individuals in individual homes to find out the results of the tests on their own homes. Until certain statistical levels had been reached, individual findings for individual homes didn't trigger a response.

First of all, is that true? Is that the way the system works, an individual or family can have a test in their home and not be able to find out the results of those tests or not be able to get effective action on the basis of one home getting a negative test or a test that exceeds the limits?

Mr. JOHNSON. I assume that question is for me?

Senator CRAPO. Yes.

Mr. JOHNSON. I would respond that when we started doing the testing we had what we thought was an effective approach for getting test kits out to people, getting them back and it was sort of a Cadillac service where we delivered them by FedEx, we went by and picked them up physically, took them to the lab and had them sampled.

When we started moving into tens of thousands of tests as opposed to a couple hundred, we simply did not change the system rapidly enough and there were some delays in getting tests back to people. I would concede that. I think we have a very effective approach for doing that now and have corrected that. As this problem has evolved because there is no road map or blueprint, we have had to change things and learn as we have gone along. I think there are very effective measures now for getting those tests results back and we generally guarantee them within about 30 days. Usually it is about a 3 to 3½ week turnaround in reality.

With respect to explaining what they meant, we at the Water and Sewer Authority simply did not have the capacity to explain what they meant. I was around when the Lead and Copper Rule was done working in water utility back in the late 1980s and early 1990s when that rule was put in place. I knew there was not a health-based standard for that, so we were not capable of explaining what the health impacts are, what this elevated level meant. All we could do was get something back to the person and explain to them you are over the action level which is in the 90th percentile of that 15 ppb and would have relied on health experts to provide that kind of followup and information to the customer.

Senator CRAPO. So if I understand, you have a system in place now where within 30 days an individual who has their home tested, can get their results back for that home. Is that true about not only the response testing, but also the standard routine samples of 50 homes that are done on an ongoing basis?

Mr. JOHNSON. Because we have exceeded the trigger level, we have one group of homes that are regulatory samples and we treat those separately from the ones we are doing upon demand and request. The regulatory sample set is 100 in the spring and 100 in the fall and they are being managed in a totally different process than the ones we are doing as random tests or tests upon demand.

Senator CRAPO. Are the people in the regulatory sample notified about how their homes are qualified?

Mr. JOHNSON. Yes, sir. They are.

Senator CRAPO. So everyone, whether they are in the demand or the regulatory test is getting the notice as to what the results are for their home within the 30 days?

Mr. JOHNSON. Yes, sir, that is correct.

Senator CRAPO. Thank you.

Another issue that came up yesterday, you indicated you had already distributed about 27,000 filters. Those filters are distributed on the basis of those who are identified as having a lead pipe delivery system to their home, correct?

Mr. JOHNSON. That is correct, sir.

Senator CRAPO. One concern that was raised yesterday is that there are something like 37,000 homes for which there are no records as to what the delivery pipes consist of. The people in that category are not able to get a filter. They don't know whether they have a lead pipe distribution system and they are not able to get a filter or a support from WASA in terms of dealing with what they perceive to be a lead problem.

What is in place to help those who fall in that category of the 37,000 homes for which we don't have information?

Mr. JOHNSON. We have been working for a couple of months in an effort to try to clean up some historical records where information has never been compiled before in the District of Columbia with respect to service pipes. We have a period of time where everything was recorded, we know the plumber paid a dollar to make the tap connection, what the material was and who the plumber was. We have some records that only show a date that it was done. We had to go through what existed manually and after going through those, we found about 21,000 as opposed to the 37,000, I would like to correct that number and I am not sure how 37,000 got created some time ago, it is about 21,000. We are notifying those people by a letter that has already been translated into the standard six languages and those letters will be going out to those residents either late this week or early next week, along with a postage-paid return card asking them to send that to us and we will get a water test kit out to them on a priority basis so they can test their water and determine if they have high lead levels.

There are a number of those addresses that have been compiled in those historical records that simply some of them don't exist anymore. If you have been living in Washington for a while, you know there has been a great deal of redevelopment. These records go back a good ways in time. Some have changed their use. There are a number of different circumstances. We are unable, except to go out on the street, to identify those. We think this is the most efficient way.

Anyone who shows an elevated level in the second draw of that test will automatically be mailed a filter with all of the instructions. We think having a two-part process is the most efficient and effective way of pursuing that.

Senator CRAPO. One last quick followup. Is it possible, under the system you now have in place, for any resident of the District who may feel your records are inaccurate or they fall in the category of homes where there is no information or are just really concerned about this issue, for them to request a test kit, have the test conducted and if the test shows their home has high levels, get a filter?

Mr. JOHNSON. Yes, sir.

Senator CRAPO. So any resident of the District has that right?

Mr. JOHNSON. Yes, sir. We have set aside resources to do a number of tests, something in excess of $2 million and are prepared to handle it. If we get an onslaught and a heavy volume, we can't do it but we would urge those persons with lead service lines to be the first priority. We do know many of our records are very, very accurate. Anything that was built after 1950 is not likely to have a lead service line because those lead service lines were outlawed after that time. We feel comfortable with anything that came during that period.

If there is something other than a 2 inch going into a house, we are 99 percent certain that would not be a lead service line because they just didn't make them much larger than 2 inch—we aren't aware that they main service lines larger than 2 inches. We think it is a single family residential phenomena that we are working with. In the case of apartment buildings, anything that is a fourplex or better is going to have something larger than a 2 inch line. So there are a number of things we can do to eliminate certain numbers.

If you notice, we have 130,000 customers in the District and the numbers we have talked about in terms of lead is somewhere around 23,000 that we feel fairly comfortable with and we have this other group. The others, we are fairly certainly, are not lead service lines.

Senator CRAPO. Thank you.

Senator Jeffords.

Senator JEFFORDS. Mr. Grumbles, your answer to Senator Crapo's question about the 15 ppb standard, you described the evolution of that standard. Can you describe why the EPA set the maximum contaminant level goal which is a health-based standard at zero?

Mr. GRUMBLES. Senator, I can. When we regulate contaminants such as lead under the Safe Drinking Water Act, we start with the process of a maximum contaminant level goal. For lead there is no safe level, certainly no safe level we can point to and say with a margin of safety that there would not be some adverse health effect. So we did establish zero as the MCLG for lead.

Senator JEFFORDS. As I mentioned in my opening statement, I have asked the Senate Health, Education and Labor and Pensions Committee—which I formerly chaired—to hold a hearing on childhood lead poisoning. I know we are not spending an extended time today discussing children's blood lead levels but I have one question for you on that.

The Centers for Disease Control Advisory Committee on Childhood Lead Poisoning Prevention is currently undertaking a review of its blood lead level standard to 10 mpd. Recent findings published in the April 17, 2003 edition of the New England Journal of Medicine show that blood lead levels below the standard of 10 are linked with declines in IQ. The study also shows that relative reduction of IQ is greater at lower concentrations of lead than at higher concentrations. These findings suggest that even low levels of lead can have devastating effects on children. I ask that this study be included in the record.

Senator CRAPO. Without objection.

[The referenced document follows:]

The NEW ENGLAND JOURNAL of MEDICINE

ESTABLISHED IN 1812 APRIL 17, 2003 VOL. 348 NO. 16

Intellectual Impairment in Children with Blood Lead Concentrations below 10 µg per Deciliter

Richard L. Canfield, Ph.D., Charles R. Henderson, Jr., M.A., Deborah A. Cory-Slechta, Ph.D., Christopher Cox, Ph.D., Todd A. Jusko, B.S., and Bruce P. Lanphear, M.D., M.P.H.

ABSTRACT

BACKGROUND

Despite dramatic declines in children's blood lead concentrations and a lowering of the Centers for Disease Control and Prevention's level of concern to 10 µg per deciliter (0.483 µmol per liter), little is known about children's neurobehavioral functioning at lead concentrations below this level.

METHODS

We measured blood lead concentrations in 172 children at 6, 12, 18, 24, 36, 48, and 60 months of age and administered the Stanford–Binet Intelligence Scale at the ages of 3 and 5 years. The relation between IQ and blood lead concentration was estimated with the use of linear and nonlinear mixed models, with adjustment for maternal IQ, quality of the home environment, and other potential confounders.

RESULTS

The blood lead concentration was inversely and significantly associated with IQ. In the linear model, each increase of 10 µg per deciliter in the lifetime average blood lead concentration was associated with a 4.6-point decrease in IQ ($P=0.004$), whereas for the subsample of 101 children whose maximal lead concentrations remained below 10 µg per deciliter, the change in IQ associated with a given change in lead concentration was greater. When estimated in a nonlinear model with the full sample, IQ declined by 7.4 points as lifetime average blood lead concentrations increased from 1 to 10 µg per deciliter.

CONCLUSIONS

Blood lead concentrations, even those below 10 µg per deciliter, are inversely associated with children's IQ scores at three and five years of age, and associated declines in IQ are greater at these concentrations than at higher concentrations. These findings suggest that more U.S. children may be adversely affected by environmental lead than previously estimated.

From the Division of Nutritional Sciences (R.L.C.) and the Department of Human Development (C.R.H.), College of Human Ecology, Cornell University, Ithaca, N.Y.; the Departments of Environmental Medicine (D.A.C.-S.) and Biostatistics and Computational Biology (C.C.), University of Rochester School of Medicine, Rochester, N.Y.; the Division of Epidemiology, Statistics, and Prevention, National Institute of Child Health and Human Development, National Institutes of Health, Department of Health and Human Services, Bethesda, Md. (C.C.); the Department of Epidemiology, School of Public Health and Community Medicine, University of Washington, Seattle (T.A.J.); and Cincinnati Children's Environmental Health Center, Children's Hospital Medical Center, Cincinnati (B.P.L.). Address reprint requests to Dr. Canfield at the Division of Nutritional Sciences, College of Human Ecology, Cornell University, Ithaca, NY 14853, or at rlc5@cornell.edu.

N Engl J Med 2003;348:1517-26.

LEAD IS NEUROTOXIC, AND YOUNG children are at particular risk for exposure.[1] Numerous studies indicate that blood lead concentrations above 10 μg per deciliter (0.483 μmol per liter) are associated with adverse outcomes on measures of intellectual functioning and social–behavioral conduct.[2-9] Such studies led to the identification of a blood lead concentration of 10 μg per deciliter or higher as a "level of concern" by the Centers for Disease Control and Prevention (CDC) and the World Health Organization (WHO).[1,10]

It remains unclear whether lead-associated cognitive deficits occur at concentrations below 10 μg per deciliter. The CDC and WHO recognized that no evidence of a threshold existed for lead-associated deficits but noted an absence of research on the possible effects of blood lead concentrations below 10 μg per deciliter. Although some studies in which the average blood lead concentration was below 10 μg per deciliter have reported associations between the blood lead concentration and cognitive deficits, the analyses did not focus specifically on children whose concentrations remained below 10 μg per deciliter throughout life.[6,11] Other evidence suggesting lead-related deficits at concentrations below 10 μg per deciliter relied on linear extrapolation or on data unadjusted for important potential confounders such as maternal intelligence and the quality of caregiving.[12-15] We examined associations between low-level exposure to lead and children's performance on intelligence tests at the ages of three and five years in a population that included many children whose blood lead concentrations remained below 10 μg per deciliter.

METHODS

STUDY COHORT

Participants had been enrolled at five to seven months of age for a prior study of dust-control efficacy.[16] The children had been born between July 1994 and January 1995. Families were invited to participate in the current study when the children were 24 to 30 months of age. Thirty-six of the 276 children in the original study were excluded from the current study because of premature birth (less than 37 weeks' gestation), low birth weight (less than 2500 g), Down's syndrome, speech and hearing abnormalities, or death or because their parents were short-term custodians or lacked English proficiency. Of the 240 eligible participants, 54 were not assessed at the age of three years and 65 were not assessed at the age of five years because they missed appointments, relocated, declined to participate, or died. Children were tested at three and five years of age. The institutional review board of the University of Rochester Medical Center (Rochester, N.Y.) approved the study protocol, and parents or guardians of all children provided written informed consent.

ANALYSIS AND QUALITY CONTROL OF BLOOD SAMPLES

Blood lead concentrations were determined by electrothermal atomic absorption spectrometry (Wadsworth Laboratories). Lead values were calculated as the means of six analyses of each sample (SD, 0.03 μg per deciliter [0.001 μmol per liter]). The results of repeated analyses, separated by five days, were highly consistent (SD, 0.40 μg per deciliter [0.019 μmol per liter]) for blood lead concentrations below 20 μg per deciliter (0.966 μmol per liter). The limit of detection was 1.0 μg per deciliter (0.048 μmol per liter), and values below this limit were set to 1.0 μg per deciliter.[17]

ASSESSMENT OF INTELLIGENCE

Children were assessed with the Stanford–Binet Intelligence Scale, fourth edition, which tests vocabulary, spatial pattern analysis, quantitative ability, and memory. We used the composite score (mean [±SD], 100±16) to represent IQ, because it is similar to the IQ score of other intelligence tests.[18,19] A different examiner administered an abbreviated Stanford–Binet Scale at each age. Examiners were unaware of children's lead status. Scores from the abbreviated batteries are highly correlated with the Stanford–Binet full composite score (0.94 at the age of three years and 0.99 at the age of five years).[20] Because of the limited diagnostic value of Stanford–Binet subscales at these ages, the composite score was the dependent variable.[19]

LEAD EXPOSURE VARIABLES

Venous blood samples were obtained at 6, 12, 18, 24, 36, 48, and 60 months of age. Four exposure indexes were analyzed: lifetime average, peak, concurrent, and average blood lead concentration in infancy. The lifetime average blood lead concentration was estimated at 3 and 5 years of age by computing the area under the blood lead curve (AUC) from 6 through 36 months of age and from 6 through 60 months of age, respectively. Dividing the AUC by the corresponding age span yields an average concentration expressed in micrograms per deciliter.

The peak blood lead concentration is the child's highest measured lead concentration through the age of three or five years. The concurrent blood lead concentration is that measured on the day of cognitive testing. The average blood lead concentration in infancy is the AUC for values measured between 6 and 24 months of age.

The lifetime average blood lead concentration best reflects chronic exposure and was used as the primary exposure variable. The blood lead concentration was specified as an untransformed continuous variable. To compute the AUC, conditional means regression[21] was used to impute values for 72 of the 1168 age-specific lead values (6.2 percent).

COVARIATES

All analyses used the same set of prespecified covariates, which were based on established predictors of children's intellectual outcomes and those widely used in studies of pediatric lead exposure.[2-4,8,22,23] The following variables were used: the child's sex, birth weight, and iron status (defined by the serum transferrin saturation at three and five years of age) and the mother's IQ (determined with use of the abbreviated Stanford–Binet Intelligence Scale), years of education, race (self-assigned as white or nonwhite), tobacco use during pregnancy (user or nonuser), yearly household income, and the total score for the Home Observation for Measurement of the Environment Inventory.[24]

STATISTICAL ANALYSIS

Mixed-model methods[25,26] were used to estimate and test parameters in linear, polynomial, and semiparametric models that always included the child's sex and the mother's race and prenatal smoking status as fixed classification effects, and a lead measure, the child's iron status, and the mother's income, level of education, IQ, and Home Observation for Measurement of the Environment score as covariates. The child's IQ (the composite score on the Stanford–Binet Intelligence Scale) was the dependent variable. The longitudinal study design provides repeated measures of the IQ variable at the ages of three and five years, and the models also include a fixed classification factor for age and a random factor for individual children. The mother's income and level of education, the child's iron status, and all lead measures (except the infancy average) were measured at both time points and are time-varying covariates. The error structure for each child assumes different variances at each age and a covariance between ages; these were assumed to be the same for all children, and covariances between children were assumed to be negligible. All significance tests were two-tailed.

For a given lead variable, regressions were specified separately according to age, and the homogeneity of these estimates was tested (i.e., the interaction of age with lead concentration). In the absence of a difference between the age-specific estimates, their unweighted average (based on all available data) is the best estimate of the association between the blood lead concentration and IQ and is referred to as the overall estimate.

Regression diagnostics were carried out for the mixed models.[27] Only one value had a standardized residual of more than 3.0 (a child who had a low IQ and a low lead concentration). It did not pass a discordancy test[27] and was retained in all analyses.

The linear relations of IQ scores to lifetime average, concurrent, peak, and infancy average blood lead concentrations were estimated in the full sample. A second, parallel set of analyses estimated the relation between IQ and the lead concentration for children whose peak lead concentration was below 10 μg per deciliter. Observations for children who were three years of age were included in these calculations only when their maximal blood lead concentration through that age was below 10 μg per deciliter and were included at the age of five years only when their maximal concentration was below 10 μg per deciliter during the entire five-year span.

Nonlinearity in the relation between IQ and the blood lead concentration across the full range of lead values was examined with the use of the mixed models described above in two types of analyses: quadratic, cubic, and higher-degree polynomials were estimated for each lead variable; and semiparametric models were estimated with the use of parametric adjustment for covariates and penalized spline smoothing for the nonparametric relation between IQ and the blood concentration.[28] The semiparametric models estimate the regression locally and, unlike the polynomial models, do not require the restrictive assumption that the true relation between IQ and the blood lead concentration conforms to a particular parametric function. Inference is less well developed in the mixed semiparametric model, and confidence intervals are not reported.

RESULTS

A total of 198 children completed at least one assessment. Of these, 172 (86.9 percent) had complete data for all variables included in the model (305 ob-

Table 1. Characteristics of the Children at the Age of Five Years and of Their Mothers.*

Characteristic	Children with Complete Data (N=154)	Children with Incomplete Data (N=21)	Children Who Did Not Participate (N=65)
Children			
Age at testing (mo)	60.6±1.0	60.6±0.9	—
Female sex (%)	52.6	45.5	53.9
Weeks of gestation	39.5±1.2	39.8±1.0	39.4±1.2
Birth weight (g)	3295±405	3400±496	3304±473
Transferrin saturation (%)	22.5±9.4	23.5±6.6	—
Blood lead concentration (μg/dl)†			
Lifetime average	7.4±4.3	7.3±3.6	—
Peak	11.1±7.1	12.6±8.2	—
Concurrent	5.8±4.1	6.4±7.5	—
Average in infancy	7.0±3.8	7.4±3.4	7.2±4.1
IQ‡	89.8±11.4	85.6±12.2	—
Mothers			
No. of prenatal visits	11.1±4.1	10.2±5.0	10.4±3.7
HOME total score§	27.3±7.1	28.7±6.1	27.8±6.2
Yearly income >$15,000 (%)	35.7	45.5	—
Smoked during pregnancy (%)	20.1	38.1	27.7
Age at delivery (yr)	25.0±6.7	25.8±4.6	23.8±5.6
Parity	1.4±1.4	1.6±1.3	1.3±1.4
Nonwhite race (%)	73.4	68.2	66.2
Education >12 yr (%)	31.2	22.7	—
IQ‡	81.9±12.7	80.5±13.6	83.8±10.2

* Data obtained at the age of three years were similar to the data obtained at five years of age and are not shown. Differences among the groups were not significant ($P<0.05$) for any variable at the age of either three or five years. Plus–minus values are means ±SD. To convert values for lead to micromoles per liter, multiply by 0.0483.

† The lifetime average blood lead concentration was estimated at the ages of 3 and 5 years by computing the area under the blood lead curve (AUC) from 6 through 36 months and from 6 through 60 months, respectively, and then dividing the AUC by its corresponding age span to yield an average on the microgram-per-deciliter scale. The peak blood lead concentration was the child's highest measured blood lead concentration through the age of three or five years. The concurrent blood lead concentration was the concentration measured on the day of cognitive testing, and the average blood lead concentration in infancy was the AUC from 6 through 24 months.

‡ The Stanford–Binet Intelligence Scale, fourth edition (abbreviated), was used to assess IQ.

§ The Home Observation for Measurement of Environment Inventory (HOME) is an index that reflects the quality and quantity of emotional and cognitive stimulation in the home environment. The total score is the sum of 39 items, each scored as present (1) or absent (0), in six categories (maternal responsivity, acceptance of child, organization of the home environment, provision of play materials, maternal involvement with the child, and the variety of stimulation).

servations; 151 at the age of three years and 154 at the age of five years). There were no significant differences in the background characteristics among children with complete data, those with incomplete data, and those who did not participate (Table 1).

BLOOD LEAD CONCENTRATIONS

The mean blood lead concentration was lowest at the age of six months (3.4 μg per deciliter [0.164 μmol per liter]), was maximal at two years (9.7 μg per deciliter [0.483 μmol per liter]), and then decreased to 6.0 μg per deciliter (0.290 μmol per liter) at five years (Fig. 1). The lifetime average blood lead concentration was 7.7 μg per deciliter (0.372 μmol per liter) at the age of three years and 7.4 μg per deciliter (0.368 μmol per liter) at the age of five years. At three years of age, 86 children (57.0 percent) had a peak blood lead concentration below 10 μg per deciliter, as did 86 (55.8 percent) at the age of five years (71 of these children had such a concentration at both ages, and the remaining 30 had data at either three or five years).

INTELLIGENCE TEST RESULTS

The mean IQ was approximately 90 at both three and five years of age (Table 1), a value consistent with the sample demographics.[20,29] Children's IQ scores at three and five years of age were strongly correlated ($r=0.67$, $P<0.001$), and these scores were correlated with maternal IQ ($r=0.43$, $P<0.001$, and $r=0.52$, $P<0.001$, respectively), consistent with prior reports.[22,30] In other bivariate analyses, the associations among the children's IQ, the children's blood lead concentrations, and the other covariates were in the expected direction (Table 2).

BLOOD LEAD CONCENTRATIONS AND IQ

Before adjustment for covariates, all four lead measures were inversely and significantly associated with IQ at three and five years of age (Table 3). The associations did not differ significantly according to age. From the overall estimate, an increase in the lifetime average blood lead concentration of 1 μg per deciliter was associated with a decrease of 0.87 IQ point; estimates for concurrent blood lead concentrations and average concentrations in infancy were similar, whereas that for the peak lead concentration was somewhat smaller.

After adjustment for the nine additional covariates, there were significant inverse associations with IQ for all blood lead variables, with no significant differences according to age (Table 3). The overall estimate indicated that an increase in the lifetime average blood lead concentration of 1 µg per deciliter was associated with a change of −0.46 IQ point (95 percent confidence interval, −0.76 to −0.15). Estimated effects were similar for the concurrent blood lead concentration and the average blood lead concentration in infancy and smaller, but still significant, for peak lead concentrations (Table 3). Other significant predictors of the child's IQ were the same in all models: maternal IQ and income and the child's birth weight.

IQ AT BLOOD LEAD CONCENTRATIONS BELOW 10 µg PER DECILITER

To examine the relation between IQ and blood lead concentrations consistently below 10 µg per deciliter, linear models for each lead variable were estimated for the subgroup of children whose peak lead concentration was below 10 µg per deciliter. Without exception, the estimates were larger in this subgroup. Lifetime average, peak, and concurrent blood lead concentrations, but not the average in infancy, were inversely and significantly associated with IQ, both before and after adjustment for covariates (Table 4) and at both three and five years of age. The estimated overall difference in IQ for each increase in the lifetime average lead concentration of 1 µg per deciliter was −1.37 points (95 percent confidence interval, −2.56 to −0.17).

NONLINEAR ANALYSES

Nonlinear mixed models were analyzed with the use of the full range of blood lead values. Semiparametric analysis indicated a decline in IQ of 7.4 points for a lifetime average blood lead concentration of up to 10 µg per deciliter (Fig. 2). For lifetime average blood lead concentrations ranging from more than 10 µg per deciliter to 30 µg per deciliter, a more gradual decrease in IQ was estimated (approximately 2.5 points). An analysis using polynomial models confirmed this departure from linearity. The quadratic term was significant in the model for lifetime average blood lead concentration ($P=0.05$), and as the blood lead concentration increased from 1 to 10 µg per deciliter, the total change in IQ was −8.0 points (95 percent confidence interval, −12.9 to −3.2). Significant nonlinearity was also found for the relations between IQ and the peak lead concentration ($P=0.003$ for the quadratic term) and between IQ and the concurrent lead concentration ($P=0.007$ for the cubic term). The spline estimates for these variables had shapes similar to that for the lifetime average. The same covariates that were significant in the linear models were also significant in the nonlinear models.

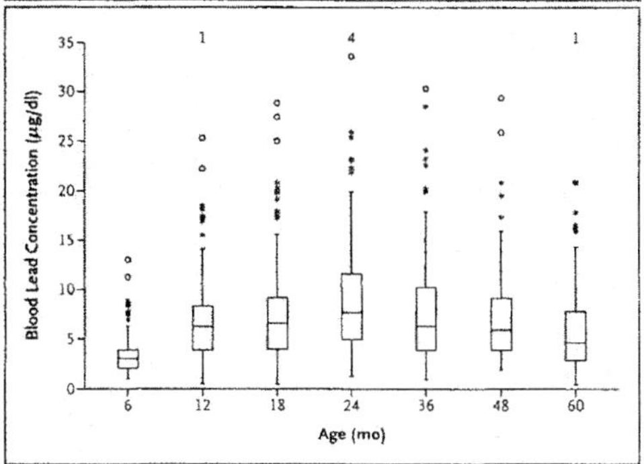

Figure 1. **Distributions of Blood Lead Concentrations at Each Assessment.**
In each box plot, the median value is indicated by the center horizontal line and the 25th and 75th percentiles are indicated by the lower and upper horizontal lines, respectively. The vertical lines represent 1.5 times the interquartile range, the asterisks represent values that are between 1.5 and 3 times the interquartile range, and circles represent values that are more than 3 times the interquartile range. The numbers at the top of the graph are the numbers of children with concurrent blood lead concentrations of more than 35 µg per deciliter. To convert values for lead to micromoles per liter, multiply by 0.0483.

DISCUSSION

Two findings from this investigation raise questions about the consequences of blood lead concentrations commonly found among U.S. children today. Of primary importance is that children's intellectual functioning at three and five years of age is inversely associated with blood lead concentrations, even when their peak concentrations remain below the CDC and WHO level of concern.[1,10] This finding was consistent for lifetime average, concurrent, and peak lead concentrations and in adjusted as well as unadjusted models. In the linear model involving the full range of lead values in this sample, the estimated IQ loss was 4.6 points for each increase in the blood lead concentration of 10 µg per deciliter, a result consistent with prior research in other cohorts.[2,11,31] In contrast, for children whose lead concentrations remained below 10 µg per deciliter, the estimated loss in IQ was considerably greater.

The second, related finding is that the relation

Table 2. Relation of Covariates to Lifetime Average Blood Lead Concentration and Mean IQ Score at Five Years of Age.*

Covariate†	No. of Children	Lifetime Average Blood Lead	IQ
		μg/dl	
Mothers			
Education level			
<12 yr	56	8.9±4.6	85.4±9.4
12 yr	50	6.4±3.5	91.2±12.4
>12 yr	48	6.6±4.1	93.4±10.8
Race‡			
Nonwhite	113	8.2±4.4	87.5±9.5
White	41	4.9±2.6	96.1±13.6
Income level			
$6,000	37	8.8±3.8	83.8±9.3
$6,001–$20,000	80	7.4±4.2	89.2±9.8
>$20,000	37	5.8±4.4	97.0±12.7
HOME total score§			
Low (<20)	24	10.1±3.2	85.8±8.1
Middle (20–30)	76	7.6±4.8	87.9±9.7
High (>30)	54	5.8±2.9	94.2±13.3
Prenatal smoking			
No	122	7.3±4.4	90.2±12.0
Yes	32	7.6±3.9	88.3±8.5
IQ¶			
Low (<75)	52	8.6±4.1	85.7±8.8
Middle (75–85)	45	7.7±5.0	86.9±8.5
High (>85)	57	5.9±3.3	95.9±12.8
Children			
Birth weight			
<3500 g	106	7.6±4.3	88.9±10.8
≥3500 g	48	6.9±4.1	91.8±12.3
Sex			
Male	73	7.6±3.9	88.3±12.5
Female	81	7.2±4.5	91.2±10.1
Transferrin saturation			
<20%	60	7.0±4.2	89.5±8.5
≥20%	94	7.6±4.3	90.0±12.9

* The lifetime average blood lead concentration was estimated at the ages of 3 and 5 years by computing the area under the blood lead curve (AUC) from 6 through 36 months and from 6 through 60 months, respectively, and then dividing the AUC by its corresponding age span to yield an average on the microgram-per-deciliter scale. Data obtained at the age of three years were similar to the data obtained at five years of age and are not shown. Plus–minus values are means ±SD. To convert values for lead to micromoles per liter, multiply by 0.0483.

† Some continuous variables were categorized for this analysis.

‡ Race was self-assigned as white or nonwhite.

§ The Home Observation for Measurement of Environment Inventory (HOME) is an index that reflects the quality and quantity of emotional and cognitive stimulation in the home environment. The total score is the sum of 39 items, each scored as present (1) or absent (0), in six categories (maternal responsivity, acceptance of child, organization of the home environment, provision of play materials, maternal involvement with the child, and the variety of stimulation).

¶ The Stanford–Binet Intelligence Scale, fourth edition (abbreviated), was used to assess IQ.

between children's IQ score and their blood lead concentration is nonlinear. The best estimate, from the semiparametric analysis, indicates a loss of 7.4 IQ points for a lifetime average blood lead concentration of up to 10 μg per deciliter. These findings suggest that the total lead-related impairment in this cohort is due largely to the initial IQ loss at blood lead concentrations of 10 μg per deciliter or less and that the linear model for children with peak concentrations of less than 10 μg per deciliter overestimates the lead-associated impairment.

Previous research is consistent with the interpretation that the effects of lead on IQ are proportionally greater at lower lead concentrations. A cross-sectional study of children with lead concentrations ranging from 3 to 34 μg per deciliter (0.145 to 1.643 μmol per liter) suggested a larger decrement in scores on ability tests over the range of 5 to 10 μg per deciliter (0.242 to 0.483 μmol per liter) than over the range from more than 10 through 20 μg per deciliter.[6] A second cross-sectional study that used data from the third National Health and Nutrition Examination Survey indicated greater possible effects on reading and math scores among children with blood lead concentrations below 5 μg per deciliter than among those with higher concentrations.[12] In addition, a prospective study[32] suggested that the effects of prenatal exposure to lead were proportionally greater at lower levels of exposure, and a meta-analysis[33] reported that studies in which average blood lead concentrations were below 15 μg per deciliter (0.725 μmol per liter) had larger slope estimates than studies in which concentrations were higher. However, we have documented this finding in children whose blood lead concentrations remained below 10 μg per deciliter, using a prospective design and adjusting for maternal intelligence and the quality of the home environment. Moreover, our findings were similar when the children were tested at three years and at five years of age.

Our results are also consistent with findings from meta-analyses that an increase in the blood lead concentration from 10 to 30 μg per deciliter is associated with a decline in IQ of 2 to 6 points.[7,33,34] Although the estimation was less precise for lead concentrations above 10 μg per deciliter in our study, the curve estimated by the semiparametric analysis suggests a loss of 2.5 IQ points as blood lead concentrations increase from more than 10 through 30 μg per deciliter. The estimates from meta-analyses reflect primarily findings from studies involving a low proportion of children with lead concentra-

Table 3. Unadjusted and Adjusted Changes in IQ for Each Increase in the Blood Lead Concentration of 1 μg per Deciliter for All Children in the Study.*

Type of Blood Lead Measurement	No. of Children	At 3 Years of Age		At 5 Years of Age		Overall	
		β±SE (95% CI)	P Value	β±SE (95% CI)	P Value	β±SE (95% CI)	P Value
Unadjusted estimate†							
Lifetime average	172	−0.74±0.18 (−1.09 to −0.39)	<0.001	−1.00±0.19 (−1.38 to −0.63)	<0.001	−0.87±0.16 (−1.19 to −0.55)	<0.001
Peak	172	−0.40±0.11 (−0.62 to −0.18)	<0.001	−0.47±0.11 (−0.70 to −0.25)	<0.001	−0.44±0.10 (−0.63 to −0.24)	<0.001
Concurrent‡	171	−0.60±0.15 (−0.89 to −0.31)	<0.001	−1.02±0.19 (−1.38 to −0.65)	<0.001	−0.81±0.14 (−1.09 to −0.53)	<0.001
Average in infancy (6–24 mo)	172	−0.73±0.21 (−1.15 to −0.31)	<0.001	−0.97±0.22 (−1.40 to −0.54)	<0.001	−0.85±0.19 (−1.23 to −0.47)	<0.001
Adjusted estimate§							
Lifetime average	172	−0.35±0.17 (−0.69 to 0.00)	0.05	−0.57±0.18 (−0.93 to −0.20)	0.003	−0.46±0.15 (−0.76 to −0.15)	0.004
Peak	172	−0.19±0.10 (−0.39 to 0.01)	0.06	−0.26±0.11 (−0.47 to −0.05)	0.02	−0.23±0.09 (−0.40 to −0.05)	0.01
Concurrent‡	171	−0.31±0.15 (−0.60 to −0.01)	0.04	−0.61±0.19 (−0.99 to −0.24)	<0.001	−0.46±0.14 (−0.74 to −0.18)	0.002
Average in infancy (6–24 mo)	172	−0.32±0.20 (−0.71 to 0.07)	0.10	−0.53±0.20 (−0.93 to −0.13)	0.01	−0.43±0.17 (−0.77 to −0.09)	0.02

* The lifetime average blood lead concentration was estimated at the ages of 3 and 5 years by computing the area under the blood lead curve (AUC) from 6 through 36 months and from 6 through 60 months, respectively, and then dividing the AUC by its corresponding age span to yield an average on the microgram-per-deciliter scale. The peak blood lead concentration was the child's highest measured blood lead concentration through the age of three or five years. The concurrent blood lead concentration was the concentration measured on the day of cognitive testing, and the average blood lead concentration in infancy was the AUC from 6 through 24 months. CI denotes confidence interval. β values are the estimated unstandardized regression coefficients.

† The unadjusted model includes only classification factors for age and for individual children.

‡ One child was lacking a concurrent blood lead measurement at the age of three years.

§ Estimates were adjusted for maternal IQ, race, level of education, use of tobacco during pregnancy, household income, and Home Observation for Measurement of Environment Inventory score, and the child's sex, birth weight, and iron status.

tions of 0 to 10 μg per deciliter. Our findings suggest that when linear estimation from such samples is extrapolated to lower blood lead concentrations, the results do not accurately reflect the greater magnitude of the lead-associated impairment at these lower concentrations.

The larger associations with IQ at lower lead concentrations may appear counterintuitive. Although we did not explore possible biologic mechanisms that could explain this finding, there is evidence that high concentrations of heavy metals may enhance cellular defense mechanisms and thereby lessen the rate at which additional damage occurs.[35]

As with any observational study, it is not possible to draw causal inferences from these findings. Instead, the plausibility of a causal interpretation must be judged by the consistency of findings from numerous epidemiologic studies and the relevant experimental studies in animals.[7,36,37] An inevitable limitation of the observational design is that it is not possible to control for all potentially confounding variables. However, the available evidence suggests that, in this area of research, a relatively small number of variables (e.g., the Home Observation for Measurement of the Environment score, socioeconomic status, and maternal IQ) are the primary confounders and that including other variables does not appreciably change the estimated

Table 4. Unadjusted and Adjusted Changes in IQ for Each Increase in the Blood Lead Concentration of 1 μg per Deciliter for Children with Peak Blood Lead Concentrations below 10 μg per Deciliter.*

Type of Blood Lead Measurement	No. of Children	At 3 Years of Age		At 5 Years of Age		Overall	
		β±SE (95% CI)	P Value	β±SE (95% CI)	P Value	β±SE (95% CI)	P Value
Unadjusted estimate†							
Lifetime average	101	−2.30±0.67 (−3.64 to −0.96)	<0.001	−2.54±0.74 (−4.01 to −1.07)	<0.001	−2.42±0.63 (−3.67 to −1.17)	<0.001
Peak	101	−2.09±0.58 (−3.25 to −0.93)	<0.001	−2.12±0.60 (−3.32 to −0.91)	<0.001	−2.10±0.53 (−3.16 to −1.04)	<0.001
Concurrent	101	−2.19±0.49 (−3.18 to −1.21)	<0.001	−2.56±0.58 (−3.71 to −1.40)	<0.001	−2.38±0.45 (−3.26 to 1.49)	<0.001
Average in infancy (6–24 mo)	105	−1.29±0.67 (−2.61 to 0.04)	0.06	−1.58±0.67 (−2.92 to −0.24)	0.02	−1.43±0.61 (−2.65 to −0.21)	0.02
Adjusted estimate‡							
Lifetime average	101	−1.22±0.66 (−2.53 to 0.09)	0.07	−1.52±0.71 (−2.94 to −0.09)	0.04	−1.37±0.60 (−2.56 to −0.17)	0.03
Peak	101	−1.36±0.55 (−2.46 to −0.27)	0.02	−1.44±0.56 (−2.55 to −0.33)	0.01	−1.40±0.48 (−2.37 to −0.44)	0.005
Concurrent	101	−1.36±0.51 (−2.37 to −0.35)	0.009	−1.79±0.60 (−3.00 to −0.60)	0.004	−1.58±0.46 (−2.50 to −0.65)	0.001
Average in infancy (6–24 mo)	105	−0.58±0.58 (−1.75 to 0.59)	0.32	−0.92±0.59 (−2.09 to 0.25)	0.12	−0.75±0.51 (−1.78 to 0.28)	0.15

* The lifetime average blood lead concentration was estimated at the ages of 3 and 5 years by computing the area under the blood lead curve (AUC) from 6 through 36 months and from 6 through 60 months, respectively, and then dividing the AUC by its corresponding age span to yield an average on the microgram-per-deciliter scale. The peak blood lead concentration was the child's highest measured blood lead concentration through the age of three or five years. The concurrent blood lead concentration was the concentration measured on the day of cognitive testing, and the average blood lead concentration in infancy was the AUC from 6 through 24 months. A total of 71 children were found to have a peak blood lead concentration below 10 μg per deciliter at both ages; an additional 15 children had a peak concentration below 10 μg per deciliter at three years of age but at five years of age had a higher concentration or were not tested, and another 15 children had a peak concentration below 10 μg per deciliter at five years but were not tested at three years. The total number of children in the analysis of the average concentration in infancy is 105 because in 4 children the peak blood lead concentration occurred after the age of 24 months. CI denotes confidence interval. β values are the estimated unstandardized regression coefficients.

† The unadjusted model includes only classification factors for age and for individual children.

‡ Estimates were adjusted for maternal IQ, race, level of education, use of tobacco during pregnancy, household income, and Home Observation for Measurement of Environment Inventory score, and the child's sex, birth weight, and iron status.

effect of lead.[11,38] For example, Tong and Lu compared the results of two empirical model-selection procedures using the Port Pirie cohort study.[38] One procedure resulted in a model with 4 covariates, and the other in a model with 14. The estimated effect of lead on IQ was nearly identical in the two models and was consistent with the linear estimates we report.

Our findings (both linear and nonlinear) for the four lead-exposure variables suggest a high degree of consistency for lifetime average, concurrent, and peak exposure. In their pattern of association with children's IQ scores, concurrent blood lead concentration was nearly identical to the lifetime average and the peak exposure. By contrast, the average blood lead concentration in infancy was less predictive of IQ, particularly for children whose lead concentrations remained below 10 μg per deciliter. We note, however, that these variables are by definition highly intercorrelated, and our results for them are not fully independent.

The results of any individual study depend, of

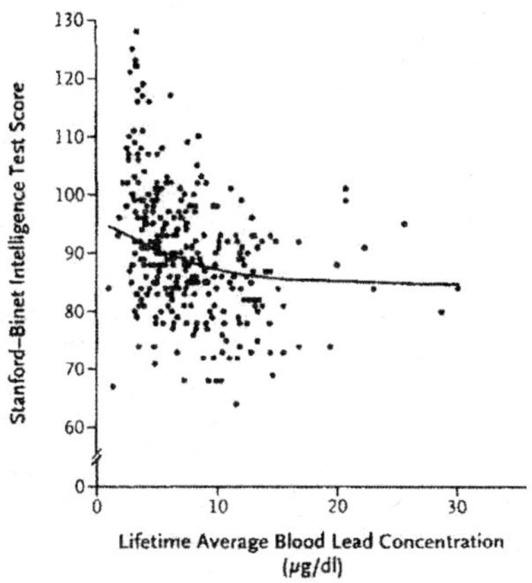

Figure 2. **IQ as a Function of Lifetime Average Blood Lead Concentration.**

IQ was assessed with use of the Stanford–Binet Intelligence Scale, fourth edition. The line represents the relation between IQ and lifetime average blood lead concentration estimated by the covariate-adjusted penalized-spline mixed model. Individual points are the unadjusted lifetime average blood lead and IQ values. To convert values for lead to micromoles per liter, multiply by 0.0483.

course, on the study population. Our study group included a cluster of children with high IQ scores and low lead concentrations, but these subjects were not unduly influential in the statistical models. Regardless, our findings should be replicated in other cohorts and with the use of other cognitive assessments.

The definition of an elevated blood lead concentration has been incrementally but consistently lowered over the past two decades. Our findings suggest that children with blood lead concentrations below 10 µg per deciliter merit more intensive investigation. These and other data suggest that there may be no threshold for the adverse consequences of lead exposure[6,7,33] and that lead-associated impairments may be both persistent and irreversible.[39-42] Furthermore, although typically investigated because of its neurotoxic properties, an elevated lead concentration is also a risk factor for other public health problems, including delinquency, cardiovascular disease, renal disease, and dental caries.[43-47]

Our findings suggest that considerably more U.S. children are adversely affected by environmental exposure to lead than previously estimated. Because there is no effective treatment for children with moderately elevated blood lead concentrations,[40] the collective evidence argues for a shift toward primary prevention of lead exposure in contrast to the current, almost exclusive emphasis on the treatment of children with elevated blood lead concentrations.[48-50]

Editor's note: Dr. Lanphear has served as an expert witness for the State of Rhode Island and the City of Milwaukee in lead-related cases, for which Children's Hospital (Cincinnati) is compensated.

Supported by a grant (R01 ES08388) from the National Institute of Environmental Health Sciences (NIEHS) and in part by grants from the NIEHS Environmental Health Sciences Center at the University of Rochester (ES01247), the Cornell University Bronfenbrenner Life Course Institute in the College of Human Ecology, the Strong Memorial Hospital Children's Research Center, and a joint research and extension program funded by the Cornell University Agricultural Experiment Station and Cornell Cooperative Extension.

Any opinions, findings, conclusions, or recommendations expressed in this publication are those of the authors and do not necessarily reflect the view of the Department of Agriculture.

Presented in part at the Pediatric Academic Societies Annual Meeting, Baltimore, April 28–May 1, 2001; at the 109th Convention of the American Psychological Association, San Francisco, August 24–28, 2001; and at the 21st Annual Meeting of the Behavioral Toxicology Society, Research Triangle Park, N.C., April 20–22, 2002.

We are indebted to Elliott G. Smith for assisting with the management and analysis of portions of these data and for valuable feedback on previous versions of the manuscript and to Keith Alexander, Kristine DiBitetto, and Karen Knauf for data collection and cohort management.

REFERENCES

1. Preventing lead poisoning in young children: a statement by the Centers for Disease Control. Atlanta: Centers for Disease Control, October, 1991.

2. Baghurst PA, McMichael AJ, Wigg NR, et al. Environmental exposure to lead and children's intelligence at the age of seven years: the Port Pirie Cohort Study. N Engl J Med 1992;327:1279-84.

3. Bellinger D, Sloman J, Leviton A, Rabinowitz M, Needleman HL, Waternaux C. Low-level lead exposure and children's cognitive function in the preschool years. Pediatrics 1991;87:219-27. [Erratum, Pediatrics 1994;93:A28.]

4. Bellinger D, Dietrich KN. Low-level lead exposure and cognitive function in children. Pediatr Ann 1994;23:600-5.

5. Dietrich KN, Berger OG, Succop PA, Hammond PB, Bornschein RL. The developmental consequences of low to moderate prenatal and postnatal lead exposure: intellectual attainment in the Cincinnati Lead Study Cohort following school entry. Neurotoxicol Teratol 1993;15:37-44.

6. Fulton M, Raab G, Thomson G, Laxen D, Hunter R, Hepburn W. Influence of blood lead on the ability and attainment of children in Edinburgh. Lancet 1987;1:1221-6.

7. Pocock SJ, Smith M, Baghurst P. Environmental lead and children's intelligence: a systematic review of the epidemiological evidence. BMJ 1994;309:1189-97.

8. Wasserman GA, Liu X, Lolacono NJ, et al. Lead exposure and intelligence in 7-year-old children: the Yugoslavia Prospective Study. Environ Health Perspect 1997;105:956-62.

9. Wasserman GA, Staghezza-Jaramillo B, Shrout P, Popovac D, Graziano J. The effect

of lead exposure on behavior problems in preschool children. Am J Public Health 1998; 88:481-6.

10. International Programme on Chemical Safety. Inorganic lead. Environmental health criteria 165. Geneva: World Health Organization, 1995.

11. Bellinger DC, Stiles KM, Needleman HL. Low-level lead exposure, intelligence and academic achievement: a long-term follow-up study. Pediatrics 1992;90:855-61.

12. Lanphear BP, Dietrich K, Auinger P, Cox C. Cognitive deficits associated with blood lead concentrations <10 microg/dl in US children and adolescents. Public Health Rep 2000;115:521-9.

13. Schwartz J. Beyond LOEL's, p values, and vote counting: methods for looking at the shapes and strengths of associations. Neurotoxicity 1993;14:237-46.

14. Winneke G, Altmann L, Krämer U, et al. Neurobehavioral and neurophysiological observations in six year old children with low lead levels in East and West Germany. Neurotoxicity 1994;15:705-13.

15. Stone BM, Reynolds CR. Can the National Health and Nutrition Examination Survey III (NHANES III) data help resolve the controversy over low blood lead levels and neuropsychological development in children? Arch Clin Neuropsychol 2003; 613:1-26.

16. Lanphear BP, Howard C, Eberly S, et al. Primary prevention of childhood lead exposure: a randomized trial of dust control. Pediatrics 1999;103:772-7.

17. Lanphear BP, Weitzman M, Winter NL, et al. Lead-contaminated house dust and urban children's blood lead levels. Am J Public Health 1996;86:1416-21.

18. McCrowell KL, Nagle RJ. Comparability of the WPPSI-R and the S-B:IV among preschool children. J Psychoedu Assess 1994; 12:126-34.

19. Assessment of children. 3rd ed. San Diego, Calif.: J.M. Sattler, 1992.

20. Thorndike RL, Hagen EP, Sattler JM. The Stanford-Binet Intelligence Scale: technical manual. 4th ed. Chicago: Riverside Publishing, 1986.

21. Buck SF. A method of estimation of missing values in multivariate data suitable for use with an electronic computer. J R Stat Soc [B] 1960;22:302-6.

22. Bouchard TJ, Segal NL. Environment and IQ. In: Wolman BB, ed. Handbook of intelligence: theories, measurements, and applications. New York: John Wiley, 1985:391-464.

23. McMichael AJ, Baghurst PA, Vimpani GV, Robertson EF, Wigg NR, Tong SL. Sociodemographic factors modifying the effect of environmental lead on neuropsychological development in early childhood. Neurotoxicol Teratol 1992;14:321-7.

24. Caldwell BM, Bradley R. Home observation for measurement of the environment. Little Rock: University of Arkansas at Little Rock, 1984.

25. Searle SR. Linear models. New York: John Wiley, 1971.

26. Henderson CR Jr. Analysis of covariance in the mixed model: higher-level, nonhomogeneous, and random regressions. Biometrics 1982;38:623-40.

27. Langford IH, Lewis T. Outliers in multilevel data. J R Stat Soc [A] 1998;161:121-60.

28. Ruppert D, Wand MP, Carroll RJ. Semiparametric regression. London: Cambridge University Press, 2003.

29. Peoples CD, Fagan JF, Drotar D. The influence of race on 3-year-old children's performance on the Stanford-Binet: fourth edition. Intelligence 1995;21:69-82.

30. Bellinger DC, Needleman HL. Lead and the relationship between maternal and child intelligence. J Pediatr 1983;102:523-7.

31. McMichael AJ, Baghurst PA, Wigg NR, Vimpani GV, Robertson EF, Roberts RJ. Port Pirie Cohort Study: environmental exposure to lead and children's abilities at the age of four years. N Engl J Med 1988;319:468-75.

32. Wasserman GA, Liu X, Popovac D, et al. The Yugoslavia Prospective Lead Study: contributions of prenatal and postnatal lead exposure to early intelligence. Neurotoxicol Teratol 2000;22:811-8.

33. Schwartz J. Low-level lead exposure and children's IQ: a meta-analysis and search for a threshold. Environ Res 1994;65:42-55.

34. Needleman HL, Gatsonis CA. Low-level lead exposure and the IQ of children: a meta-analysis of modern studies. JAMA 1990;263:673-8.

35. Bae DS, Gennings C, Carter WH Jr, Yang RS, Campain JA. Toxicological interactions among arsenic, cadmium, chromium, and lead in human keratinocytes. Toxicol Sci 2001;63:132-42.

36. Cory-Slechta DA. Relationships between lead-induced learning impairments and changes in dopaminergic, cholinergic, and glutamatergic neurotransmitter system functions. Annu Rev Pharmacol Toxicol 1995;35:391-415.

37. Rice DC. Lead-induced changes in learning: evidence for behavioral mechanisms from experimental animal studies. Neurotoxicology 1993;14:167-78.

38. Tong S, Lu Y. Identification of confounders in the assessment of the relationship between lead exposure and child development. Ann Epidemiol 2001;11:38-45.

39. Dietrich KN, Ris MD, Succop PA, Berger OG, Bornschein RL. Early exposure to lead and juvenile delinquency. Neurotoxicol Teratol 2001;23:511-8.

40. Rogan WJ, Dietrich KN, Ware JH, et al. The effect of chelation therapy with succimer on neuropsychological development in children exposed to lead. N Engl J Med 2001; 344:1421-6.

41. Tong SL, Baghurst P, McMichael A, Sawyer M, Mudge J. Lifetime exposure to environmental lead and children's intelligence at 11-13 years: the Port Pirie cohort study. BMJ 1996;312:1569-75. [Erratum, BMJ 1996; 313:198.]

42. Tong SL, Baghurst PA, Sawyer MG, Burns J, McMichael AJ. Declining blood lead levels and changes in cognitive function during childhood: the Port Pirie Cohort Study. JAMA 1998;280:1915-9.

43. Cheng Y, Schwartz J, Sparrow D, Aro A, Weiss ST, Hu H. Bone lead and blood lead levels in relation to baseline blood pressure and the prospective development of hypertension: the Normative Aging Study. Am J Epidemiol 2001;153:164-71.

44. Lin J-L, Lin-Tan D-T, Hsu K-H, Yu C-C. Environmental lead exposure and progression of chronic renal diseases in patients without diabetes. N Engl J Med 2003;348:277-86.

45. Moss ME, Lanphear BP, Auinger P. Association of dental caries and blood lead levels. JAMA 1999;281:2294-8.

46. Needleman HL, Riess JA, Tobin MJ, Biesecker GE, Greenhouse JB. Bone lead levels and delinquent behavior. JAMA 1996; 275:363-9.

47. Schwartz J. Lead, blood pressure, and cardiovascular disease in men. Arch Environ Health 1995;50:31-7.

48. Lanphear BP. The paradox of lead poisoning prevention. Science 1998;281:1617-8. [Erratum, Science 1998;282:51.]

49. Marsden PA. Increased body lead burden — cause or consequence of chronic renal insufficiency? N Engl J Med 2003;348:345-7.

50. Rosen JF, Mushak P. Primary prevention of childhood lead poisoning — the only solution. N Engl J Med 2001;344:1470-1.

The NEW ENGLAND JOURNAL of MEDICINE

Exposure to Lead in Children — How Low Is Low Enough?

Walter J. Rogan, M.D., and James H. Ware, Ph.D.

Encephalopathy in childhood due to lead poisoning was described more than 100 years ago, and cognitive sequelae after recovery were reported in 1943. During the past three decades, epidemiologic studies have demonstrated inverse associations between blood lead concentrations and children's IQs at successively lower lead concentrations. In response, the Centers for Disease Control and Prevention (CDC) repeatedly lowered its definition of an elevated blood lead concentration, which now stands at 10 μg per deciliter (0.483 μmol per liter) (see Figure). Since the removal of lead from gasoline, the median blood lead concentration in U.S. children has fallen from 15 μg per deciliter (0.724 μmol per liter) in 1978 to 2 μg per deciliter (0.097 μmol per liter) in 1999, a triumph for public health. Yet exposure to lead from deteriorating lead paint in older homes continues; of U.S. homes where children under the age of six years live, 25 percent contain hazardous lead paint. The CDC has estimated that in 2000, there were still 454,000 children in the United States with blood lead concentrations greater than 10 μg per deciliter.

In this issue of the *Journal*, Canfield et al. (pages 1517–1526) extend the unfortunately familiar relation between increased blood lead concentrations and decreased IQ to blood lead concentrations below 10 μg per deciliter. Of 172 children in whom blood lead concentrations were measured serially between the ages of six months and five years, 101 did not have a recorded blood lead concentration above 10 μg per deciliter. Strong and significant associations between blood lead concentration and IQ were observed in these children at both three and five years of age. A smoothed curve summarizing the data shows a decline in IQ of more than 7 points over the first 10 μg per deciliter of lifetime average blood lead concentration and a further decline of approximately 2 points associated with an increase from 10 to 20 μg per deciliter (0.483 to 0.966 μmol per liter).

Although most previous studies have provided little information about children with blood lead concentrations in this range, Canfield and colleagues' findings are a consistent and plausible extension of a very large epidemiologic, clinical, and experimental literature that has indicated adverse effects of lead on cognition. A previous meta-analysis suggested a 2.6-point decline in IQ for an increase in lead concentration from 10 to 20 μg per deciliter.[2] Bellinger et al.,[3] in a study of children with blood lead concentrations closest to those in the current study, estimated a decline of 5.8 points with an increase in blood lead concentration from 10 to 20 μg per deciliter. Although the remarkable steep-

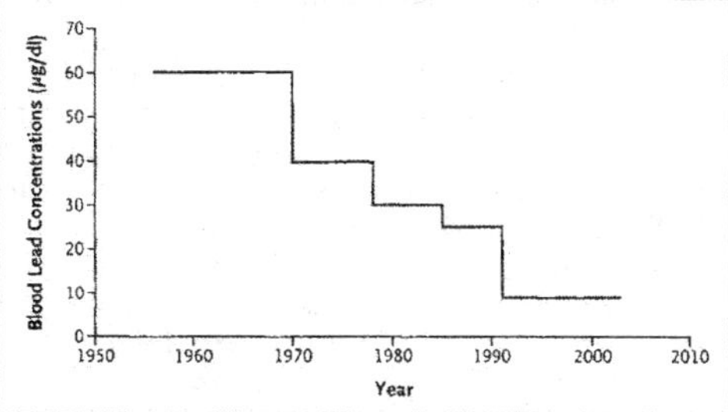

Blood Lead Concentrations Considered to Be Elevated by the Centers for Disease Control and Prevention.

To convert the values for blood lead concentrations to micromoles per liter, multiply by 0.0483. Data are from the Centers for Disease Control and Prevention, 1991.[1]

ness in the range below 10 μg per deciliter seen by Canfield et al. is influenced by results in 10 children whose blood lead concentrations were near or below 5 μg per deciliter (0.242 μmol per liter) and whose IQs were above 115, it is unlikely that the association is due only to these observations. To confirm the adverse effects of lead on IQ at these concentrations, more children whose blood lead value has never been more than 10 μg per deciliter should be studied.

Lead exposure is more common among disadvantaged children, and so confounding by maternal IQ, socioeconomic factors, and characteristics of the home environment is always a concern. In the study by Canfield and colleagues, however, the association between the blood lead concentration and IQ persisted even after the investigators accounted for several potential confounders by the best available methods. Although critics question the importance of small decrements in the IQs of individual children, these measures are blunt instruments for detecting subtle changes in brain function; any detectable effect occurring from a widespread exposure is cause for concern. Relatively small changes in the mean IQ of a large number of children will dramatically increase the proportion of children below any fixed level of concern, such as an IQ of 80, and decrease the proportion above any "gifted" level, such as 120.

In a second article in this issue, Selevan and colleagues (pages 1527–1536) report that girls from the nationally representative third National Health and Nutrition Examination Survey who had very slightly elevated blood lead concentrations at 8 to 18 years of age had evidence of delayed puberty. Unlike cognitive development, sexual maturation is a relatively new area of investigation for effects of environmental exposures in general and exposure to lead in particular. In the current report, African-American girls with blood lead concentrations of 3 μg per deciliter (0.145 μmol per liter), as compared with girls with blood lead concentrations of 1 μg per deciliter (0.048 μmol per liter), had delays of two to six months in the age at which they attained given stages of breast and pubic-hair development, and four months in the age at which they began to menstruate. Delays were also observed in white and Mexican-American girls, although those associations were not uniformly statistically significant.

Because the study by Selevan et al. was cross-sectional, it remains possible that these findings, if attributable to lead, may be the result of higher blood lead concentrations earlier in life; lead concentrations typically peak in early childhood. Nonetheless, although delays of a few months in the course of puberty are not likely to be a threat to health, these data raise the possibility of an effect on fundamental developmental processes occurring in girls from the general U.S. population at commonly encountered blood lead concentrations. Even though puberty, if delayed by lead, is not delayed by very much, these findings raise the possibility of effects on other hormonally mediated processes.

The effects of lead exposure appear to be long-lasting and irreversible. A previous report in the *Journal* indicated that chelation therapy given to lower moderately elevated blood lead levels in preschool children from environments similar to those studied by Canfield and colleagues had no beneficial effects on tests of cognition, behavior, or neuropsychological function.[4] Prevention is thus the only plausible strategy. Children should not live in housing that exposes them to hazardous amounts of lead, and children who are already exposed need to be identified and their source of exposure interrupted.

The fact that associations seen at these low lead concentrations are subtle is not reassuring; rather, it implies that there is no safety margin at existing exposures. Eliminating elevated blood lead concentrations by the end of this decade is already a federal objective.[5] The new reports underscore the importance of these goals and of the potential consequences of a delay in addressing this entirely preventable condition. They also imply that the job may not be finished even when all children have blood lead concentrations below 10 μg per deciliter.

From the National Institute of Environmental Health Sciences, Research Triangle Park, N.C. (W.J.R.), and the Harvard School of Public Health, Boston (J.H.W.).

1. Preventing lead poisoning in young children. Atlanta: Centers for Disease Control, October 1991.
2. Schwartz J. Low-level lead exposure and children's IQ: a meta-analysis and search for a threshold. Environ Res 1994;65:42-55.
3. Bellinger DC, Stiles KM, Needleman HL. Low-level lead exposure, intelligence and academic achievement: a long-term follow-up study. Pediatrics 1992;90:855-61.
4. Rogan WJ, Dietrich KN, Ware JH, et al. The effect of chelation therapy with succimer on neuropsychological development in children exposed to lead. N Engl J Med 2001;344:1421-6.
5. Department of Health and Human Services. Healthy People 2010: with understanding and improving health and objectives for improving health. 2nd ed. Washington, D.C.: Government Printing Office, November 2000. (Accessed March 18, 2003, at http://www.healthypeople.gov/document/html/objectives/08-11.htm.)

The NEW ENGLAND JOURNAL of MEDICINE

ORIGINAL ARTICLE

Blood Lead Concentration and Delayed Puberty in Girls

Sherry G. Selevan, Ph.D., Deborah C. Rice, Ph.D., Karen A. Hogan, M.S., Susan Y. Euling, Ph.D., Andrea Pfahles-Hutchens, M.S., and James Bethel, Ph.D.

ABSTRACT

BACKGROUND
Environmental lead exposure has been linked to alterations in growth and endocrine function. It is not known whether such exposure affects pubertal development.

METHODS
We analyzed the relations between blood lead concentration and pubertal development among girls (defined as females 8 to 18 years of age) who were enrolled in a cross-sectional study (the third National Health and Nutrition Examination Survey) in which race was self-reported or proxy-reported: 600 were non-Hispanic white, 805 were non-Hispanic African-American, and 781 were Mexican-American girls. Puberty was measured on the basis of the age at menarche and Tanner stage for pubic-hair and breast development.

RESULTS
Geometric mean lead concentrations were less than 3 μg per deciliter (0.144 μmol per liter) in all three groups. As compared with concentrations of 1 μg per deciliter (0.048 μmol per liter), lead concentrations of 3 μg per deciliter were associated with decreased height ($P<0.001$), after adjustment for age, race, and other factors, but not with body-mass index or weight. Blood lead concentrations of 3 μg per deciliter were associated with significant delays in breast and pubic-hair development in African-American and Mexican-American girls. The delays were most marked among African-American girls; in this group, the delays in reaching Tanner stages 2, 3, 4, and 5 associated with a lead concentration of 3 μg per deciliter as compared with 1 μg per deciliter were 3.8, 5.3, 5.8, and 2.1 months, respectively, for breast development and 4.0, 5.5, 6.0, and 2.2 months, respectively, for pubic-hair development; the associated delay in age at menarche was 3.6 months. In white girls, there were nonsignificant delays in all pubertal measures in association with a lead concentration of 3 μg per deciliter.

CONCLUSIONS
These data suggest that environmental exposure to lead may delay growth and pubertal development in girls, although confirmation is warranted in prospective studies.

From the National Center for Environme tal Assessment, Office of Research ar Development (S.G.S., D.C.R., K.A.H., S.Y.E and the Office of Pollution Preventio Pesticides, and Toxic Substances (A.P.-H Environmental Protection Agency, Was ington, D.C.; and Westat, Rockville, M (J.B.). Address reprint requests to Dr. Se van at U.S. EPA (8623D), Washington, C 20460, or at selevan.sherry@epa.gov.

N Engl J Med 2003;348:1527-36.

EXPOSURE TO ENVIRONMENTAL CONTAMINANTS may accelerate[1,2] or delay[3] pubertal development in girls. Lead is a ubiquitous environmental contaminant associated with a variety of health effects.[4] Although the average body burdens in U.S. children have decreased markedly since the removal of lead from gasoline in 1979,[5] current body burdens nonetheless remain appreciably higher than preindustrial levels.[6]

Exposure to lead could indirectly affect the timing of puberty in girls through effects on growth. Prenatal or postnatal exposure to lead is associated with growth restriction in laboratory animals[7] and humans.[8,9] In large cross-sectional studies of young children, increased lead concentrations were associated with decreased height, weight, or both.[10,11] In turn, increased height, body-mass index (the weight in kilograms divided by the square of the height in meters), and weight were directly associated with earlier pubertal development.[12,13]

In addition to effects on growth, prenatal and postnatal exposure to lead may influence pubertal development through the hypothalamic–pituitary–gonadal axis. Studies of rats exposed to lead found altered hormone concentrations and delayed puberty.[7,14,15] We assessed the relations between blood lead concentrations and puberty in girls in the third National Health and Nutrition Examination Survey (NHANES III), conducted from 1988 to 1994.[16,17]

METHODS

NHANES III was a cross-sectional, nationally representative, complex-sample survey of the U.S. noninstitutionalized population two months of age or older, designed to provide estimates of the health status of the general population.[16,17] About 40,000 civilian, noninstitutionalized persons were surveyed with use of a stratified, multistage, clustered-probability design. Race and ethnic background were self-reported or proxy-reported in NHANES III. Mexican Americans and African Americans were oversampled to provide more precise estimates for these groups.

During the household interview, 2741 girls, defined as 8 to 18 years of age at screening, were selected for the study. Of these, 2585 completed the survey, including the physical examination and biologic-sample collection. Data on pubertal development and lead concentrations were available for 2186 girls: 600 non-Hispanic white girls, 805 non-Hispanic African-American girls, and 781 Mexican-American girls. Measures of puberty included pubic-hair stage for 1964 girls, breast-development stage for 1986 girls, and age at menarche for 1796 girls (limited to those 8 to 16 years of age). Nine percent declined to be examined for pubic-hair stage, and 8 percent for breast-development stage.[17] The 113 girls belonging to other racial or ethnic groups were not analyzed owing to their small numbers. Written informed consent was obtained from all participants.

NHANES data were weighted to adjust for unequal selection probabilities and nonresponse and adjusted to match estimates of the Current Population Survey for the U.S. noninstitutionalized population.[16,17] To accommodate the complex sample design when calculating variance estimates, we used SUDAAN (release 8.0.1) and SAS (release 8.1) software. In comparing characteristics across racial and ethnic groups, we used the median test, adapted for complex survey data, for continuous data[18] and the Cochran–Mantel–Haenszel test for categorical data.[19] Since blood lead concentrations were not distributed normally, the data were log-transformed. A log (base 3) transformation was used to compare blood lead concentrations of 3 μg per deciliter (0.144 μmol per liter) with those of 1 μg per deciliter (0.048 μmol per liter), to approximate the 75th and 25th percentiles, respectively. Multivariable linear regression was used to model relations between growth and blood lead concentrations. Measures of body size included height, weight, and body-mass index.

Associations between the Tanner stage and body size and between the Tanner stage and blood lead concentrations were examined. Tanner staging for pubic-hair or breast development classifies girls into five progressively more mature stages, ranging from stage 1 (no development) to stage 5 (fully mature).[20] Tanner stages were assessed by trained clinicians without knowledge of the girls' blood lead concentrations.[16] Ordinal logistic regression (logistic regression for ordered responses) was used to assess the progression of the Tanner stage, which calculated odds ratios of the likelihood of reaching higher Tanner stages with changes in lead concentrations, after controlling for age and other factors. Odds ratios greater than 1.0 indicate accelerated pubertal development, whereas ratios of less than 1.0 indicate a delay, relative to the reference group. The mean age for each Tanner stage was estimated from the model by generating age-specific probabilities and computing the weighted mean age for that stage.

The age at menarche of the 8-to-16-year-old girls was obtained by interviewing girls who were 10 years of age or older or, for girls younger than 10 years, adults (usually the girls' mothers). To reduce the interval between the onset of menarche and a subsequent blood lead assessment, girls older than 16 years of age were not included in menarche analyses. The relation between the lead concentration and the age at menarche was analyzed with use of Cox proportional-hazards modeling after the exclusion (censoring) of data on each premenarchal girl at her age during the survey. The age at examination was recorded in months, whereas the age at menarche was recorded in years. To reduce downward bias, 0.5 year was added to age at menarche.

Pubertal onset has been reported to occur earlier in African-American girls than in whites.[21-23] In an earlier analysis of data from NHANES III, pubertal development in Mexican-American girls was intermediate between that of African-American girls and white girls.[23] Thus, analyses were performed separately for each of the three groups. Delays were estimated by comparing the modeled median age at each blood lead concentration with the median age for a blood lead concentration of 1 μg per deciliter.

In all analyses, age and potential risk factors or confounders were examined, including age at examination (in months), smoking status, dietary information (a 24-hour report of dietary calcium, iron, vitamin C, and total fat), presence or absence of anemia,[4,11,24] and measures of socioeconomic status (urban vs. rural residence and family income of less than $20,000 per year vs. $20,000 or more). Variables remained in the model if they reached borderline statistical significance ($P<0.1$) or were potential confounders (i.e., if the coefficient for the log blood lead concentration changed by more than 10 percent when that factor was included in the final model). For consistency, any factor included for one racial or ethnic group was included in analyses for the other groups.

RESULTS

The mean ages of the girls were similar in the three groups (Table 1). As compared with whites, African-American girls were significantly taller, and both African Americans and Mexican Americans had a higher average body-mass index. The average stages of breast and pubic-hair development and blood lead concentrations were highest among African Americans, a finding consistent with prior analyses of a subgroup of these girls.[23] African-American girls were significantly younger at menarche than whites, and Mexican Americans were intermediate between and not significantly different from African-American girls and white girls. There were also differences between groups in some measures of socioeconomic status (income and level of education of the head of household) and some personal factors (ever having smoked or consumed alcohol and dietary calcium levels).

Overall, blood lead concentrations declined with age (Fig. 1). The geometric mean blood lead concentrations were below 3 μg per deciliter for all groups. Few girls had blood lead concentrations of more than 5 μg per deciliter (0.24 μmol per liter): 2.7 percent of whites, 11.6 percent of African Americans, and 12.8 percent of Mexican Americans. Very few had concentrations of more than 10 μg per deciliter (0.48 μmol per liter) — the Centers for Disease Control and Prevention's level of concern for children[25] — 0.3 percent of whites, 1.6 percent of African Americans, and 2.3 percent of Mexican Americans.

Each stage of pubertal development occurred earliest among African-American girls. Among girls whose blood lead concentration was 1 μg per deciliter or below, the average ages for Tanner stages 2, 3, 4, and 5 of breast development were 10.1, 11.7, 13.2, and 15.8 years, respectively, among African Americans; 10.7, 12.5, 14.1, and 16.4 years, respectively, among Mexican Americans; and 11.2, 12.9, 14.7, and 16.4 years, respectively, among whites. African Americans were also the youngest at each stage of pubic-hair development (10.1, 11.3, 13.4, and 16.0 years at Tanner stages 2, 3, 4, and 5, respectively). The average ages for each of these stages in white girls (11.2, 12.6, 14.8, and 16.8 years, respectively) and Mexican-American girls (11.1, 12.7, 14.8, and 16.7 years, respectively) were similar. African-American girls were also the youngest at menarche, with an average age of 12.0 years, as compared with 12.4 years in whites and 12.2 years in Mexican Americans.

To explore whether pubertal development was associated with measures of growth, we assessed relations between pubertal development and body size and between lead concentration and body size. Greater height, weight, and body-mass index were each individually associated with a higher breast-development stage. When height, body-mass index, and weight were included in models simultaneously, however, only height and body-mass index

Table 1. **Characteristics of the Participants According to Racial and Ethnic Group.***

Characteristic	Non-Hispanic Whites (N=600)	African Americans (N=805)	Mexican Americans (N=781)	P Value
Age at physical examination (yr)				0.79
Mean	13.4	13.4	13.4	
95% CI	13.0–13.8	13.1–13.7	13.1–13.7	
Height — in.				<0.001
Mean	60.5	61.0	59.6	
95% CI	59.9–61.1	60.5–61.4	59.2–60.0	
Body-mass index				0.02
Mean	20.5	21.7	21.7	
95% CI	20.0–21.1	21.2–22.1	20.9–22.5	
Urban residence (%)	43.0	50.0	54.5	0.24
Family income <$20,000/yr (%)	22.0	58.6	57.7	<0.001
Head of household's level of education (yr)				<0.001
Mean	13.1	11.6	8.6	
95% CI	12.8–13.5	11.3–11.8	8.1–9.1	
Any history of smoking 100 cigarettes (%)	11.0	1.5	3.6	<0.001
No. of cigarettes smoked/day (among smokers)				0.02
Mean	11.8	5.8	7.1	
95% CI	10.6–13.0	NA	NA	
Any history of alcohol consumption (among those >12 yr of age) (%)	37.9	17.9	29.7	<0.001
Dietary calcium (mg/day)				<0.001
Mean	854.7	735.2	923.3	
95% CI	811.2–898.2	692.6–777.9	856.3–990.2	
Dietary iron (mg/day)				0.93
Mean	13.1	12.7	13.4	
95% CI	12.1–14.1	12.0–13.5	12.4–14.5	
Dietary vitamin C (mg/day)				0.18
Mean	89.6	110.0	106.2	
95% CI	80.7–98.5	100.7–119.2	97.9–114.6	

remained associated with greater breast development, owing to the high correlation between body-mass index and weight (odds ratio for each 1-in. [2.5-cm] increase in height, 1.17 [95 percent confidence interval, 1.09 to 1.25], and odds ratio for each 1-unit increase in body-mass index, 1.08 [95 percent confidence interval, 1.01 to 1.15], after adjustment for age, age squared, dietary iron, family income [less than $20,000 per year vs. $20,000 or more], and race). Increased height was associated with earlier pubic-hair development; the odds ratio for reaching a given Tanner stage associated with a 1-in. increase in height was 1.23 (95 percent confidence interval, 1.14 to 1.33), after adjustment for age, age squared, dietary vitamin C, family income, and race and ethnic group. No measure of growth was significantly associated with age at menarche. Higher blood lead concentrations (3 μg per deciliter vs. 1 μg per deciliter) were associated with decreased height (regression slope = −0.51, P<0.001) but not with changes in body-mass index or weight, after adjustment for age, age squared, race and ethnic group, family income, presence or absence of anemia at examination, and dietary vitamin C, iron, and calcium — a finding that was similar to those in girls younger than eight years of age.[11]

Higher blood lead concentrations were associated with significant delays in all pubertal measures among African Americans and in breast and pubic-hair development among Mexican Americans (Table 2 and Fig. 2 and 3), after adjustment for relevant physical, demographic, and socioeconomic characteristics. For white girls, the relations between lead concentrations and pubertal development were in

Table 1. (Continued.)

Characteristic	Non-Hispanic Whites (N=600)	African Americans (N=805)	Mexican Americans (N=781)	P Value
Anemia (%)†	4.5	23.1	7.2	<0.001
Blood lead concentration (μg/dl)				<0.001
Geometric mean	1.4	2.1	1.7	
95% CI	1.2–1.5	1.9–2.3	1.6–1.9	
Age at menarche (yr)‡				0.02
Mean	12.4	12.0	12.2	
95% CI	12.2–12.6	11.9–12.2	12.1–12.4	
Breast-development stage§				0.02
Mean	3.3	3.7	3.4	
95% CI	3.0–3.5	3.5–3.8	3.3–3.6	
Pubic-hair stage§				<0.001
Mean	3.2	3.6	3.2	
95% CI	2.9–3.4	3.5–3.8	3.1–3.3	

* P values among the three racial and ethnic groups were calculated with use of the Cochran–Mantel–Haenszel chi-square test for categorical outcomes[19] and the median test[18] for continuous variables. For outcomes significant at a P value of less than 0.05, comparisons examined differences among pairs of groups: all three groups differed from the other two with respect to the level of education of the head of household and blood lead concentration; African Americans differed from both other groups with respect to drinking status, anemia status, and dietary calcium levels; whites differed from the other two groups with respect to family income, ever having smoked 100 cigarettes, and body-mass index; Mexican Americans differed from the other two groups with respect to height; whites and Mexican Americans differed with respect to the number of cigarettes smoked per day; and whites differed from African Americans with respect to pubic-hair development and age at menarche. To convert values for height to centimeters, multiply by 2.54. To convert values for lead to micromoles per liter, multiply by 0.048. CI denotes confidence interval, and NA not available.

† The presence of anemia at examination was defined with the use of age-specific hemoglobin cutoff values.[24]

‡ The midpoint of the year of age at menarche was used for girls 8 to 16 years of age (for girls whose age was less than the midpoint of the year of age at menarche, current ages were used). Of all the girls who were 8 to 16 years of age, the following had reached menarche: 237 of 490 non-Hispanic whites (48.4 percent), 345 of 654 African Americans (52.8 percent), and 312 of 652 Mexican Americans (47.9 percent).

§ These values represent the average stage for all girls in each group. Tanner stages[20] were scored separately. For both, stage 1 represented preadolescence, with no development. Tanner stages for pubic-hair development are as follows: stage 2 is defined by sparse growth of nonadult hair in a limited area, stage 3 by darkening hair and greater coverage, stage 4 by hair of adult quality but limited to the pubic area, and stage 5 by full distribution of hair to medial aspect of thigh (fully mature). Tanner stages for breast development are as follows: stage 2 is defined by breast buds, stage 3 by further enlargement and elevation of breast and aureola, stage 4 by projection of aureola and papilla above breast, and stage 5 by fully mature breasts.

the same direction as in the other groups but were not significant.

The greatest delays in breast development associated with higher lead concentrations were in African Americans. As compared with a lead concentration of 1 μg per deciliter, a lead concentration of 3 μg per deciliter was associated with delays in reaching Tanner stages 2, 3, 4, and 5 of breast development (delays of 3.8, 5.3, 5.8, and 2.1 months, respectively). The likelihood of reaching each stage of breast development decreased as blood lead concentrations increased (Table 2). Smaller delays were observed for Mexican-American girls (for Tanner stages 2, 3, 4, and 5, the respective delays were 2.4, 2.8, 3.0, and 1.3 months). The delays were nonsignificant for white girls (respective delays of 1.7, 2.2, 2.3, and 0.7 months).

The delays in pubic-hair development associated with lead concentrations of 3 μg per deciliter as compared with 1 μg per deciliter were also largest among African Americans: 4.0, 5.5, 6.0, and 2.2 months for Tanner stages 2, 3, 4, and 5, respectively. As for breast development, the likelihood of reaching consecutive stages of pubic-hair development decreased as blood lead concentrations increased (Table 2). There were smaller, stage-specific delays in pubic-hair development for the same lead concentrations among Mexican Americans (3.5, 4.0, 3.7, and 1.5 months for Tanner stages 2, 3,4, and 5, respectively) and nonsignificant delays among whites (2.6, 3.0, 3.0, and 1.1 months, respectively).

A significant association between higher blood lead concentration and age at menarche was found only in African-American girls, with a delay of 3.6 months for blood lead concentrations of 3 μg per deciliter as compared with 1 μg per deciliter (Table

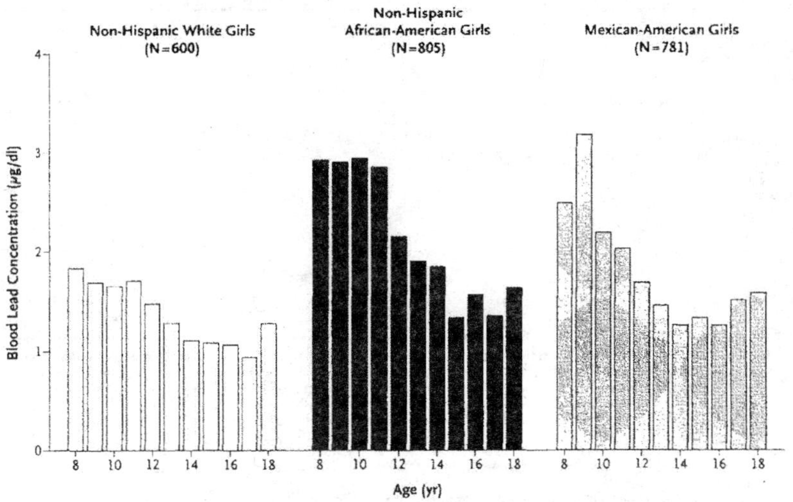

Figure 1. **Blood Lead Concentrations among Non-Hispanic White, Non-Hispanic African-American, and Mexican-American Girls.**

Age-specific geometric means were calculated with the use of sampling weights to adjust for unequal probabilities of selection. The distribution among the three groups was significantly different (P=0.006). To convert values for lead to micromoles per liter, multiply by 0.048.

2). The difference in age at menarche with a blood lead concentration of 3 µg per deciliter as compared with 1 µg per deciliter was smaller and not significant in whites (2.9-month delay) and Mexican Americans (0.4-month delay).

DISCUSSION

As compared with a blood lead concentration of 1 µg per deciliter, a blood lead concentration of 3 µg per deciliter was associated with delayed pubertal development, after adjustment for body size and other factors. Delays were statistically significant for all three measures in African-American girls and for breast and pubic-hair development in Mexican-American girls. For white girls, the relations between lead concentration and measures of puberty were in the same direction as in the other groups but were not significant.

We also found that higher blood lead concentrations were associated with decreased height, but not weight or body-mass index, after adjusting for racial and ethnic group and other factors. Our analyses of girls 8 to 18 years of age extend findings from previous studies of younger children.[10,26,27] In two studies, increased blood lead concentration was associated with reduced height but not weight or body-mass index.[10,11] In a third study, decreased height and weight were independently associated with increased blood lead concentration.[27] Prenatal and postnatal exposure to lead also increased pituitary growth hormone and suppressed plasma insulin-like growth factor 1 in rats.[15,28]

In our study, relations between measures of body size and pubertal development, adjusted for racial and ethnic group, varied with the measure of puberty. Only increased height was associated with a more advanced pubic-hair development; weight and body-mass index were not. All three body-size measures were individually associated with breast development, but in analyses including all three, only height and body-mass index were significantly associated with breast development, owing

Table 2. **Effect of Blood Lead Concentrations of 3 μg per Deciliter as Compared with 1 μg per Deciliter on Measures of Pubertal Development.***

	Non-Hispanic Whites	African Americans	Mexican Americans
8-to-18-year-old girls			
Breast development — odds ratio (95% CI)			
Adjusted for age only	0.82 (0.53–1.27)	0.60 (0.42–0.84)†	0.77 (0.61–0.97)†
Fully adjusted‡	0.82 (0.47–1.42)	0.64 (0.42–0.97)†	0.76 (0.63–0.91)†
Pubic-hair development — odds ratio (95% CI)			
Adjusted for age only	0.64 (0.36–1.13)	0.58 (0.41–0.81)†	0.60 (0.45–0.80)†
Fully adjusted§	0.75 (0.37–1.51)	0.62 (0.41–0.96)†	0.70 (0.54–0.91)†
8-to-16-year-old girls			
Age at menarche — hazard ratio (95% CI)§			
Adjusted for lead only	0.81 (0.61–1.08)	0.79 (0.64–0.99)†	0.91 (0.76–1.10)
Fully adjusted¶	0.74 (0.55–1.002)	0.78 (0.63–0.98)†	0.90 (0.73–1.11)

* Odds ratios reflect the likelihood of reaching a successive stage of pubertal development for girls with a log-transformed lead concentration of 3 μg per deciliter as compared with 1 μg per deciliter. Hazard ratios were calculated with the use of Cox proportional-hazards models. CI denotes confidence interval.

† $P<0.05$.

‡ Analyses were adjusted for age, age squared, height, body-mass index, family income (<$20,000 per year vs. $20,000 or more), ever having smoked 100 cigarettes (vs. never having done so), and dietary intake of iron, vitamin C, and calcium.

§ Analyses were adjusted for age, age squared, height, family income (<$20,000 per year vs. $20,000 or more), ever having smoked 100 cigarettes (vs. never having done so), presence or absence of anemia (defined on the basis of age-specific hemoglobin cutoff values[24]), and dietary intake of iron and vitamin C.

¶ Analyses were adjusted for height, body-mass index, family income (<$20,000 per year vs. $20,000 or more), presence or absence of anemia (defined on the basis of age-specific hemoglobin cutoff values[24]), and dietary intake of calcium.

to a high correlation between body-mass index and weight. No body-size measures were associated with the age at menarche. Body-fat measurements, including body-mass index, and height were associated with pubertal development in other studies.[12,13,29-31]

We found differences in the timing of pubertal development in girls of different racial and ethnic groups, findings consistent with the results of previous studies.[21-23] For each measure of puberty, African Americans were younger than whites at every stage of pubertal development. Mexican-American girls reached breast-development stages earlier than whites, but pubic-hair development and menarche occurred at ages similar to those of whites.

Higher blood concentrations of lead were associated with delayed puberty in African-American and Mexican-American girls, but these associations were not significant in white girls. The reasons for these differences are unclear. Nonsignificant pubertal delays in white girls with higher lead concentrations may result from a decreased power to detect an association owing to the examination of fewer white girls. Alternatively, differences among the groups may result from genetic or other environmental factors. For example, African-American girls reach skeletal maturity earlier than whites, although this may be accounted for by increased adiposity among African Americans.[13,32] White girls have longer menstrual cycles and durations of menstrual bleeding than African Americans, suggesting differences in hypothalamic–pituitary–gonadal regulation.[33,34] White and African-American girls also have differences in hypothalamic–pituitary–adrenal function, as evidenced by a greater corticotropin response,[35] which may be affected by exposure to lead.[36]

Developmental exposure to lead affects growth and sex hormones in animals. Prenatal, lactational, and prepubertal exposure to lead delayed the age at which vaginal opening and first estrus occurred in rats.[7,14,15,37] Rats exposed to lead prenatally or during lactation and rats with low postnatal lead concentrations had decreased serum concentra-

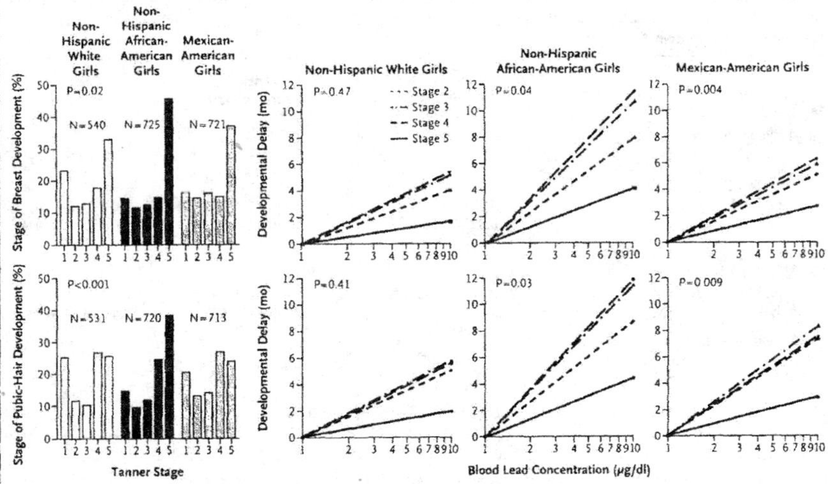

Figure 2. Association between Tanner Stage and Blood Lead Concentrations.

The two left-most panels show the Tanner stages of all girls. The other panels show the model-based estimates of the delay in each stage, calculated with use of the results of ordinal logistic regression of Tanner stages, and on solving the equation, with the use of age-specific means for other risk factors and confounders. To convert values for lead to micromoles per liter, multiply by 0.048.

tions of insulin-like growth factor 1, luteinizing hormone, and estradiol in the absence of effects on body weight.[15] Decreases in these puberty-related hormones might delay the onset or progression of puberty.

The delays in puberty associated with blood lead concentrations are striking given the low concentrations recorded during the survey. An increase in blood lead concentrations from 1 µg per deciliter to 3 µg per deciliter was associated with delays in breast and pubic-hair development ranging from two to six months among African-American girls, depending on the stage of puberty. Smaller delays were observed in these measures in the other two groups. However, we do not have data on earlier concentrations in these girls, and we cannot exclude the possibility that higher lead concentrations early in life may explain or contribute to observed pubertal delays.

In children, blood lead concentrations typically peak at about one to two years of age and gradually decrease thereafter,[5,11] data consistent with the decline in blood lead concentrations with age observed in this study. Younger children have higher blood lead concentrations than older children in the same environment because of greater exposure and lead absorption from the gastrointestinal tract.[38] In addition, environmental concentrations of lead have decreased markedly over the past two decades.[4] Some girls in this cross-sectional study would have been young children at the time of NHANES II (1976 to 1980), in which the geometric mean lead concentration among children one to five years of age was 15.0 µg per deciliter (0.72 µmol per liter), with 88.2 percent having concentrations above 10 µg per deciliter. Although our findings cannot prove a causal relation between mildly elevated lead concentrations and delayed puberty, they suggest that even a relatively low level of exposure to lead may influence growth and sexual development in girls.

In addition, other factors associated with body lead burden and pubertal development that we did

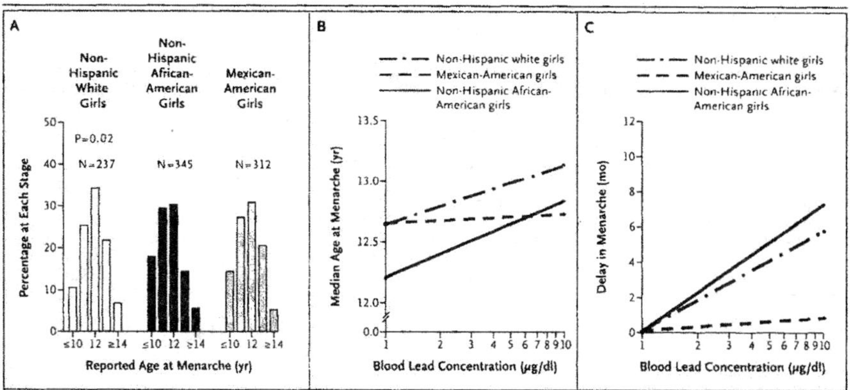

Figure 3. **Association between Age at Menarche and Blood Lead Concentrations.**

Panel A shows the reported age at menarche for all girls. Delays in menarche were estimated with the use of PROC LIFEREG software (SAS). In Panel B, the age at menarche was modeled with the use of the same potential risk factors and confounding variables in the SUDAAN PROC SURVIVAL analysis. In Panel C, model-based estimates of delays in menarche were obtained by evaluating each model at blood lead concentrations ranging from 1 to 10 μg per deciliter, with the use of mean values for each group (shown in Panel B). Then delays were calculated by subtracting the median age for a blood lead concentration of 1 μg per deciliter from the median age for lead concentrations of 2 to 10 μg per deciliter. Since the PROC LIFEREG software cannot accommodate the complex sampling weights used in the third National Health and Nutrition Examination Survey and this analysis was not available in SUDAAN software, only the ages at menarche, and not the confidence intervals or significance levels, were estimated. To convert values for lead to micromoles per liter, multiply by 0.048.

not assess may be responsible for the observed associations. As with any cross-sectional study, reporting of past events, such as age at menarche and dietary history, is subject to errors in recall. We adjusted for several potential confounders measured at the time of the study, but these factors may have differed during periods critical for pubertal development or other unmeasured confounders may have affected the results. Our study documents associations between the blood lead concentration and delayed pubertal development after adjustment for measures of body size, age, and potential confounders. These findings suggest that delays in pubertal development may be due at least in part to mechanisms independent of effects on growth, conceivably to alterations in endocrine function.

The views expressed in this report are those of the authors and do not necessarily reflect the views or policies of the Environmental Protection Agency.

We are indebted to Robin Jones of Westat for assistance with the development of the data base, to Dr. Susan Schober of the National Center for Health Statistics for assistance in working with NHANES III data, and to Dr. Deborah Winn of the National Cancer Institute and Drs. Bruce Rodan and Allan Marcus of the Environmental Protection Agency for their helpful comments during the development of the manuscript.

REFERENCES

1. Blanck HM, Marcus M, Tolbert PE, et al. Age at menarche and Tanner stage in girls exposed in utero and postnatally to polybrominated biphenyl. Epidemiology 2000;11:641-7.

2. Krstevska-Konstantinova M, Charlier C, Craen M, et al. Sexual precocity after immigration from developing countries to Belgium: evidence of previous exposure to organochlorine pesticides. Hum Reprod 2001;16:1020-6.

3. Den Hond E, Roels HA, Hoppenbrouwers K, et al. Sexual maturation in relation to polychlorinated aromatic hydrocarbons: Sharpe and Skakkebaek's hypothesis revisited. Environ Health Perspect 2002;110:771-6.

4. Toxicological profile for lead. Atlanta: Agency for Toxic Substances and Disease Registry, July 1999.

5. Prikle JL, Brody DJ, Gunter EW, et al. The decline in blood lead levels in the United States: the National Health and Nutrition Examination Survey (NHANES). JAMA 1994;272:284-91.

6. Patterson CC. British mega exposure to industrial lead. In: Rutter M, Jones RR, eds. Lead versus health. New York: John Wiley, 1983:17-32.

7. Ronis MJ, Badger TM, Shema SJ, Roberson PK, Shaikh F. Effects on pubertal growth and reproduction in rats exposed to lead perinatally or continuously throughout devel-

opment. J Toxicol Environ Health A 1998;53: 327-41.
8. Bellinger D, Leviton A, Rabinowitz M, Allred E, Needleman H, Schoenbaum S. Weight gain and maturity in fetuses exposed to low levels of lead. Environ Res 1991;54: 151-8.
9. Shukla R, Dietrich KN, Bornschein RL, Berger O, Hammond PB. Lead exposure and growth in the early preschool child: a follow-up report from the Cincinnati Lead Study. Pediatrics 1991;88:886-92.
10. Frisancho AR, Ryan AS. Decreased stature associated with moderate blood lead concentrations in Mexican-American children. Am J Clin Nutr 1991;54:516-9.
11. Ballew C, Khan LK, Kaufmann R, Mokdad A, Miller DT, Gunter EW. Blood lead concentration and children's anthropometric dimensions in the Third National Health and Nutrition Examination Survey (NHANES III), 1988-1994. J Pediatr 1999;134:623-30.
12. Rubin K, Schirduan V, Gendreau P, Sarfarazi M, Mendola R, Dalsky G. Predictors of axial and peripheral bone mineral density in healthy children and adolescents, with special attention to the role of puberty. J Pediatr 1993;123:863-70.
13. Kaplowitz PB, Slora EJ, Wasserman RC, Pedlow SE, Herman-Giddens ME. Earlier onset of puberty in girls: relation to increased body mass index and race. Pediatrics 2001; 108:347-53.
14. Kimmel CA, Grant LD, Sloan CS, Gladen BC. Chronic low-level lead toxicity in the rat. I. Maternal toxicity and perinatal effects. Toxicol Appl Pharmacol 1980;56:28-41.
15. Dearth RK, Hiney JK, Srivastava V, Burdick SB, Bratton GR, Dees WL. Effects of lead (Pb) exposure during gestation and lactation on female pubertal development in the rat. Reprod Toxicol 2002;16:343-52.
16. Plan and operation of the Third National Health and Nutrition Examination Survey, 1988–94. Vital and health statistics. Series 1, No. 32. Hyattsville, Md.: National Center for Health Statistics, 1994. (DHHS publication no. (PHS) 94-1308.)
17. National Health and Nutrition Examination Survey III, 1988-1994. NHANES III series II. No. 1A. Hyattsville, Md.: National Center for Health Statistics, 1997 (CD-ROM).
18. Mood AM. Introduction to the theory of statistics. New York: McGraw-Hill, 1950.
19. Selvin S. Statistical analysis of epidemiologic data. New York: Oxford University Press, 1991.
20. Marshall WA, Tanner JM. Variations in pattern of pubertal changes in girls. Arch Dis Child 1969;44:291-303.
21. Biro FM, McMahon RP, Striegel-Moore R, et al. Impact of timing of pubertal maturation on growth in black and white female adolescents: the National Heart, Lung, and Blood Institute Growth and Health Study. J Pediatr 2001;138:636-43.
22. Herman-Giddens ME, Slora EJ, Wasserman RC, et al. Secondary sexual characteristics and menses in young girls seen in office practice: a study from the Pediatric Research in Office Settings network. Pediatrics 1997; 99:505-12.
23. Wu T, Mendola P, Buck GM. Ethnic differences in the presence of secondary sex characteristics and menarche among US girls: the Third National Health and Nutrition Examination Survey, 1988-1994. Pediatrics 2002;110:752-7.
24. Recommendations to prevent and control iron deficiency in the United States. MMWR Morb Mortal Wkly Rep 1998;47 (RR-3):1-29.
25. Preventing lead poisoning in young children: a statement by the Centers for Disease Control. Atlanta: Centers for Disease Control, October 1991.
26. Pounds JG, Long GJ, Rosen JF. Cellular and molecular toxicity of lead in bone. Environ Health Perspect 1991;91:17-32.
27. Schwartz J, Angle C, Pitcher H. Relationship between childhood blood lead levels and stature. Pediatrics 1986;77:281-8.
28. Ronis MJ, Badger TM, Shema SJ, et al. Endocrine mechanisms underlying the growth effects of developmental lead exposure in the rat. J Toxicol Environ Health A 1998;54:101-20.
29. Vizmanos B, Marti-Henneberg C. Puberty begins with a characteristic subcutaneous body fat mass in each sex. Eur J Clin Nutr 2000;54:203-8.
30. Magarey AM, Boulton TJ, Chatterton BE, Schultz C, Nordin BE, Cockington RA. Bone growth from 11 to 17 years: relationship to growth, gender and changes with pubertal status including timing of menarche. Acta Paediatr 1999;88:139-46.
31. Wattigney WA, Srinivasan SR, Chen W, Greenlund KJ, Berenson GS. Secular trend of earlier onset of menarche with increasing obesity in black and white girls: the Bogalusa Heart Study. Ethn Dis 1999;9:181-9.
32. Russell DL, Keil MF, Bonat SH, et al. The relation between skeletal maturation and adiposity in African American and Caucasian children. J Pediatr 2001;139:844-8.
33. Harlow SD, Campbell B. Ethnic differences in the duration and amount of menstrual bleeding during the postmenarcheal period. Am J Epidemiol 1996;144:980-8.
34. Harlow SD, Campbell B, Lin X, Raz J. Ethnic differences in the length of the menstrual cycle during the postmenarcheal period. Am J Epidemiol 1997;146:572-80.
35. Yanovski JA, Yanovski SZ, Boyle AJ, et al. Hypothalamic-pituitary-adrenal axis activity during exercise in African American and Caucasian women. J Clin Endocrinol Metab 2000;85:2660-3.
36. Kim D, Lawrence DA. Immunotoxic effects of inorganic lead on host resistance of mice with different circling behavior preferences. Brain Behav Immun 2000;14:305-17.
37. Ronis MJ, Gandy J, Badger T. Endocrine mechanisms underlying reproductive toxicity in the developing rat chronically exposed to dietary lead. J Toxicol Environ Health A 1998;54:77-99.
38. Exposure factors handbook. Vol. 1. General factors. Washington, D.C.: Environmental Protection Agency, 1997:4-16. (EPA/600/P-95/002.)

IMAGES IN CLINICAL MEDICINE

The *Journal* welcomes consideration of new submissions for Images in Clinical Medicine. Instructions for authors and procedures for submissions can be found on the *Journal*'s Web site at **http://www.nejm.org**. At the discretion of the editor, images that are accepted for publication may appear in the print version of the *Journal*, the electronic version, or both.

Senator JEFFORDS. New research also shows that there may be an identifiable link between childhood lead exposure and educational achievement and social behavior. I ask unanimous consent to include several articles on this subject by Dr. Herbert Needleman.

[The referenced document follows:]

ELSEVIER

Neurotoxicology and Teratology 24 (2002) 711–717

NEUROTOXICOLOGY AND TERATOLOGY

www.elsevier.com/locate/neutera

Bone lead levels in adjudicated delinquents A case control study

Herbert L. Needleman*, Christine McFarland, Roberta B. Ness, Stephen E. Fienberg, Michael J. Tobin

Western Psychiatric Institute and Clinic, University of Pittsburgh School of Medicine, Keystone Building, Suite 310, 3520 Fifth Avenue, Pittsburgh, PA 15213, USA
Graduate School of Public Health, University of Pittsburgh, Pittsburgh, PA, USA
Department of Statistics and Center for Automated Learning and Discovery, Carnegie Mellon University, Pittsburgh, PA, USA

Received 4 February 2002; received in revised form 30 April 2002; accepted 14 May 2002

Abstract

Background: Lead exposure shares many risk factors with delinquent behavior, and bone lead levels are related to self-reports of delinquent acts. No data exist as to whether lead exposure is higher in arrested delinquents. The goal of this study is to evaluate the association between lead exposure, as reflected in bone lead levels, and adjudicated delinquency. **Methods:** This is a case-control study of 194 youths aged 12–18, arrested and adjudicated as delinquent by the Juvenile Court of Allegheny County, PA and 146 nondelinquent controls from high schools in the city of Pittsburgh. Bone lead was measured by K-line X-ray fluorescence (XRF) spectroscopy of tibia. Logistic regression was used to model the association between delinquent status and bone lead concentration. Covariates entered into the model were race, parent education and occupation, presence of two parental figures in the home, number of children in the home and neighborhood crime rate. Separate regression analyses were also conducted after stratification on race. **Results:** Cases had significantly higher mean concentrations of lead in their bones than controls (11.0 ± 32.7 vs. 1.5 ± 32.1 ppm). This was true for both Whites and African Americans. The unadjusted odds ratio for a lead level ≥ 25 vs. <25 ppm was 1.9 (95% CL: 1.1–3.2). After adjustment for covariates and interactions and removal of noninfluential covariates, adjudicated delinquents were four times more likely to have bone lead concentrations >25 ppm than controls (OR = 4.0, 95% CL: 1.4–11.1). **Conclusion:** Elevated body lead burdens, measured by bone lead concentrations, are associated with elevated risk for adjudicated delinquency.

Keywords: Lead poisoning; Delinquency; Violence; Criminality; Impulsivity; Case control

1. Introduction

Most studies of childhood lead exposure have focused on cognitive function, using IQ tests as the outcome of interest. There are a number of reasons to believe that antisocial behavior may be a more sensitive and consequential outcome. This is not a new notion. Pediatricians who treat lead poisoning have frequently been told by parents that, after recovery, their offspring became oppositional, aggressive or violent. In 1943, Randolph Byers was stimulated to design the first follow-up study of lead-poisoned children. The precipitant was his discovery that children referred to him for evaluation of violent behavior were former patients who had been treated for lead poisoning. Of the 20 subjects he studied, 19 had severe behavior problems or were learning disordered [5].

Published controlled studies of the lead-delinquency hypothesis are limited to three. Denno [9] found that the strongest predictor of arrest in Philadelphia youths enrolled in the Collaborative Perinatal Project was a history of lead poisoning. In 1996, we studied a cohort of 301 boys in the Pittsburgh School System. Bone lead levels at 12 years of age were significantly related to parents' and teachers' Child Behavior Checklist ratings of aggression, attention and delin-

* Corresponding author. Western Psychiatric Institute and Clinic, University of Pittsburgh School of Medicine, Keystone Building, Suite 310, 3520 Fifth Avenue, Pittsburgh, PA 15213, USA. Tel.: +1-412-383-3100; fax: +1-412-383-3131.

E-mail address: hlnlead@pitt.edu (H.L. Needleman).

quency. The subjects' self-reports of delinquent acts were also positively associated with bone lead concentrations [19]. Dietrich et al. [10] have recently reported that prenatal exposure to lead was associated with increased parent reports of antisocial behavior and postnatal exposure with increased reports of delinquent acts by the subjects themselves.

To more directly examine the relationship between lead exposure and criminality, we conducted a case-control study of bone lead levels in 194 male youths arrested and adjudicated as delinquent by the court.

2. Methods

2.1. Subjects

Cases were youths who resided in Allegheny County, PA, who were arrested and adjudicated by the Juvenile Court as delinquent. Controls were nondelinquent youths attending high schools in Pittsburgh. Two programs of the Allegheny County Juvenile Court serve serious delinquents: the Community Intensive Supervision Program (CISP) and the Allegheny Academy. Both programs function either as alternatives to incarceration of serious offenders or provide aftercare following incarceration. Three CISP centers located in Allegheny County participated in our study. Thirty-nine percent of the CISP enrollees were serious offenders discharged from incarceration to the program. Other enrollees were classified by the court program as drug dealers, assaulters, probation violators, firearm possessors, auto thieves or were guilty of robbery or other crimes.

Those offenders from neighborhoods in Allegheny County that did not have CISP homes were sent to the Allegheny Academy, a facility for delinquents in a neighboring township. The distribution of offenses in the Academy and CISP was similar. At the end of their public school session, students were picked up at home and taken by bus to the Academy or to the CISP center. At about 9 p.m., students were returned to their homes and, at 11 p.m., a telephone bed check made. Almost all enrollees in the CISP program were African American; in the Academy, the proportion of White youths was approximately 35%.

During the period of our study, 547 males were enrolled in the two programs. From the CISP program, we recruited 100 delinquent males and from the Academy 95 males. Of those we did not recruit, 166 were released or transferred from the programs before we could make contact; 72 candidates declined to participate and we were unable to reach the parents of 108. Two were excluded for medical conditions and four had incomplete data collection.

We initially designed our study to match controls with cases within each high school classroom. When we began the study, school administration policy had changed and we were barred from direct contact with potential controls. High school principals became the contact with potential subjects. Those high school principals who chose to cooperate sent letters describing the study to their students' families. Each letter contained a reply addressed to our laboratory. We contacted parents who expressed interest and made appointments for them and their offspring. We recruited from 6 of 11 Pittsburgh high schools. Of the five high schools that were not included, one refused, three were magnet schools offering special educational programs and one served a distinctly higher socioeconomic population. Controls were screened by telephone to eliminate those with arrest histories, seizures or taking neuroactive medications. From 283 respondents, 200 male controls were recruited; 19 refused, 25 were unable to be reached and 39 were excluded by the telephone screen.

Subjects and controls were given the Self-Report of Delinquency (SRD) [14], a 36-item inventory of antisocial acts committed over the previous 6 months, scaled from 0 to 4 depending on frequency of acts committed. Because many delinquents are not arrested and therefore are not known to the juvenile justice system, we attempted to minimize unidentified delinquents in our control group by excluding from analysis controls with either a Juvenile Court record or an SRD score ≥ 20, the 90th percentile. Of the 200 controls we recruited, 50 were excluded, 34 because of court records, 1 because of a diagnosis of autism, 13 because of high SRD scores and two because of a history of lead poisoning. To evaluate potential bias produced by excluding the latter two groups, we reintroduced them into the model and compared unadjusted odds ratios with and without them. Because a number of cases attended different high schools than controls, we evaluated the potential bias by conducting an additional logistic regression including only cases and controls who attended the same high schools. We then compared odds ratios for reduced models from the full sample with that from subjects attending the same schools.

Data for cases and controls are given in Tables 1 and 2. One case and four controls were not included in the stratified analysis because their race was indeterminate. White high school students responded to our control recruitment letter at a higher rate than African Americans, and our case-to-control ratio was different across racial groups (Whites: 36 cases/95 controls, African Americans: 158 cases/51 controls).

Subjects were studied between April 1996 and August 1998. Informed consent was obtained at the time of study, at which time they were paid US$30.00 and their parents US$20.00. This study was reviewed and approved by the University of Pittsburgh Institutional Review Board.

2.2. Bone lead measurements

Tibial bone lead concentrations were estimated with in vivo X-ray fluorescence (XRF) employing 88.03 keV photons from a ^{109}Cd source to induce characteristic lead K X-rays, measured with a backscatter counting geometry [31]. Bone lead concentrations were estimated from the lead K $\beta_{1,3}$ X-rays (84.94 and 84.45 keV). A 30-min tibia exposure

Table 1
Descriptive variables of cases and controls stratified by race

	n	African American cases[a]	n	African American controls[a]	P value*	n	White cases[a]	n	White controls[a]	P value*
Age (mean ± S.D.)	158	15.8 ± 1.4	51	15.5 ± 1.1	.0662	36	15.7 ± 1.3	95	15.8 ± 1.1	.6829
Grade (mean ± S.D.)	158	9.5 ± 1.5	51	9.8 ± 0.9	.0446	36	9.3 ± 1.2	95	10.3 ± 1.1	< .0001
Parent education (mean ± S.D.)	109	12.7 ± 2.0	51	13.0 ± 1.8	.3621	34	12.6 ± 2.2	94	12.9 ± 1.6	.4609
Parent occupation code (mean ± S.D.)	112	2.4 ± 2.6	51	2.5 ± 2.5	.7387	36	2.4 ± 2.95	94	3.4 ± 2.8	.1038
Two parental figures in home	112	–	51	–	–	36	–	94	–	–
Yes (%)	38	33.9%	28	54.9%	.0114	17	47.2%	75	79.8%	.0003
Two biological parents (% yes)	15	13.4%	11	21.6%	.1862	7	19.4%	61	64.9%	< .0001
Spouse education (mean ± S.D.)	36	12.1 ± 2.2	28	12.7 ± 2.2	.3144	17	12.4 ± 1.5	75	12.7 ± 2.1	.5704
Spouse occupation (mean ± S.D.)	38	2.3 ± 2.3	27	3.4 ± 3.1	.0869	17	2.8 ± 2.4	75	3.9 ± 2.3	.0716
No. of children living in home (mean ± S.D.)	112	2.0 ± 1.3	51	2.4 ± 1.4	.0650	36	2.1 ± 1.3	95	1.9 ± 1.3	.4294
Neighborhood crime rate (mean ± S.D.)	155	83.8 ± 41.1	50	124.5 ± 312.8	.3634	36	42.2 ± 35.3	94	53.1 ± 26.8	.1005
Self-reported delinquency score (mean ± S.D.)	158	23.9 ± 16.5	51	6.5 ± 5.0	< .0001	36	31.5 ± 21.2	95	4.5 ± 4.7	< .0001

[a] One case and four controls not included because not identified as Black/biracial or White.

* P values based on t test for means or χ^2 for frequencies.

resulted in an effective dose of 17/19 nSv, corresponding to 0.001% of the average natural environmental radiation dose (3 mSv). The XRF protocol was approved by the University of Pittsburgh Radiation Safety Committee.

Lead K $\beta_{1,3}$ X-rays and coherent scatter peak areas were obtained with a nonlinear minimization program. Spectral data were modeled as Gaussian peaks superimposed on monotonic background functions. Instrument quality control was achieved by daily monitoring lead K X-ray peak widths and locations using a lead plug check standard.

A set of bone lead phantoms (lead-doped plaster of Paris) furnished by the National Institute of Standards and Technology (NIST) was used to recalibrate our instrument and validate our spectrum analyses. A plot of lead concentration estimates obtained with our analytical protocol against NIST-reported lead values shows close agreement (Fig. 1).

Conservative criteria were used to set our minimum detectable concentration (MDC) values [1]. Instrumental MDC was estimated as 4.65θ, the standard deviation of replicate low-lead "blank" NIST phantom measurements. For 16 replicates, an MDC of 14.7μg Pb/g plaster, corresponding to an MDC of 21.5μg Pb/g bone mineral, was obtained. When the lead K X-ray signal is small, the masking effect of the Poisson-distributed background frequently produces negative bone lead concentrations. These were expected and observed. To deal with this in our analyses, we dichotomized our lead estimates at 25 ppm, the 80th percentile of the distribution, and just above the alternate MDC estimates described above. Dichotomizing at 20 and 30 ppm resulted in similar splits.

2.3. Data analysis

After data checking and verification, univariate distributions of variables were calculated (Table 1). To deal with the

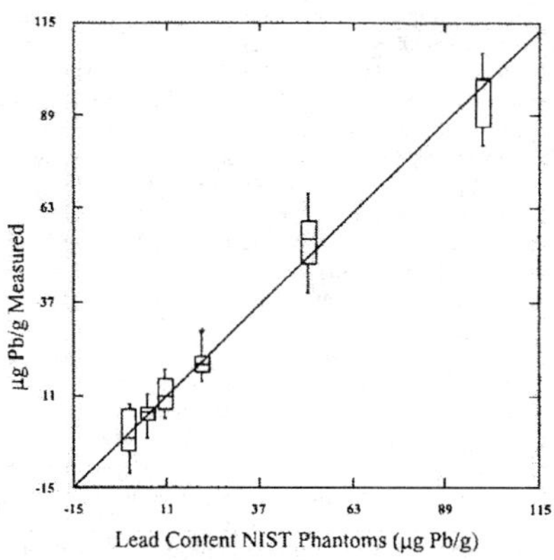

Fig. 1. XRF spectroscopy performance using NIST lead phantoms. For each of six lead-doped plaster of Paris phantoms, actual micrograms of lead per gram of plaster are presented on the abcissa. Box plots showing medians, quartiles and range of replicate ($N \geq 16$) XRF measurements are displayed on the ordinate. The diagonal line shows the Pearson correlation (r=.98).

Table 2
Bone lead concentrations for cases and controls stratified by race

	Cases		Controls		P value
	n	Mean ± S.D.	n	Mean ± S.D.	
All subjects	195	11.0 ± 32.7	150	1.5 ± 32.1	.007
African American[a]	158	9.0 ± 33.6	51	− 1.4 ± 31.9	.05
White[a]	36	20.0 ± 27.5	95	3.5 ± 32.6	.008

[a] One case and 4 controls not included because not identified as black/biracial or white.

race-related differential recruitment of cases and controls, bone lead levels were compared after stratification by race (Table 2).

Cognizant that our controls were a self-selected group within the underlying population from which our cases came, we adjusted our analysis for both individual and community potential confounders. The association of bone lead, dichotomized at 25 ppm, with delinquency was modeled by logistic regression. Covariates were selected on the basis of *a priori* information on risk factors; all but one were dichotomized. Included were race, parental education (number of years), parental occupation (Hollingshead occupational scale) [12], presence of two parental figures in the home, number of children in the home and neighborhood crime rate. Because, in many cases, the distribution of covariates was sparse or uneven, we elected to categorize most covariates using *a priori* rules creating categories that seemed logical. Two interaction terms were created on the basis of exploratory analyses employing variates in pairs. They were: lead × race and lead × single parent.

To evaluate the influence of neighborhood factors, from the 1992 Pittsburgh Police Statistical Report and the Pennsylvania Uniform Crime Report, we obtained the number of serious crimes committed in each Pittsburgh neighborhood. From census data, we calculated crimes per 1000 residents/year for each neighborhood. This neighborhood crime rate was stratified at the median (63 cases/1000) and assigned to each subject, according to neighborhood residence at time of testing.

The initial logistic model evaluated all subjects. An unadjusted odds ratio was computed and all covariates then entered into the model. Two interaction terms were then created and the model reduced, eliminating noninfluential covariates. To deal with the imbalance in proportion of controls and cases between Whites and African Americans, and to examine the effects of lead and other covariates while holding race constant, we stratified the subjects by race and analyzed two separate logistic models within racial strata. Table 3 presents the odds ratios for lead, covariates and interaction terms for all models. Table 4 presents the cross-tabulation of subjects by three covariates: race, presence of two parental figures and parental occupation.

3. Results

Table 2 shows the mean ± S.D. bone lead concentrations for all cases and controls and stratified by race. Delinquents had significantly higher bone lead levels than controls (11.0 ± 32.7 vs. 1.5 ± 32.1 ppm; $P = .007$). Limiting subjects to those attending the same high schools and comparing bone lead levels, we found almost identical bone lead levels (11.1 ± 27.6 vs. 1.5 ± 32.1 ppm). White delinquents had higher bone lead levels than African American delinquents (20 ± 27.5 vs. 9.0 ± 33.6 ppm). Both White and African American cases had achieved lower grade levels than controls at the time studied.

For all subjects, the unadjusted odds ratio was 1.9 (95% CL: 1.1–3.2) (Table 3). Adding nondelinquent controls with high SRD scores and the three lead-poisoned subjects to the model had minimal effect on the odds ratio (OR = 1.96, 95% CL: 1.1–3.4).

After covariate adjustment, the odds ratio for bone lead was 3.7 (95% CL: 1.3–10.5). The odds ratio (all subjects) in the reduced model was 4.0 (95% CL: 1.4–11.1). The odds ratio evaluating only subjects attending the same high schools was 3.2 (95% CL: 0.4–25.3).

Race and absence of two parental figures were influential covariates. After stratification, White subjects had an unadjusted odds ratio of 3.4 (95% CL: 1.4–8.1). Covariate adjustment increased the odds ratio to 3.8 (95% CL: 1.1–13.3). The odds ratio for the reduced model was 3.6 (95% CL: 1.1–12.3). African American subjects showed a similar pattern. The unadjusted odds ratio in this group of 1.5 increased to 2.2 in the full model and 2.6 in the reduced model. While African American race and single parenthood were themselves risk factors for delinquency,

Table 3
Odds ratios: full and reduced logistic models for all subjects, Whites and African Americans

Variable	All subjects		Whites		African Americans	
	Unadjusted OR = 1.9 (1.1–3.2)		Unadjusted OR = 3.4 (1.4–8.1)		Unadjusted OR = 1.5 (0.7–3.6)	
	Full model	Reduced model	Full model	Reduced model	Full model	Reduced model
Bone lead	3.7 (1.3–10.5)	4.0 (1.4–11.1)	3.8 (1.1–13.3)	3.6 (1.1–12.3)	2.2 (0.5–10.0)	2.6 (0.6–11.0)
Race	7.6 (3.6–15.8)	5.6 (3.0–10.4)	–	–	–	–
Lead × race	0.6 (0.2–2.4)	0.6 (0.2–2.0)	–	–	–	–
Single parent	3.2 (1.7–6.1)	3.4 (1.9–6.3)	5.1 (1.6–15.6)	4.5 (1.5–13.5)	3.0 (1.3–6.8)	3.0 (1.4–6.5)
Lead × single parent	0.5 (0.1–1.7)	0.5 (0.1–1.7)	0.2 (0.0–1.8)	0.4 (0.1–2.6)	0.5 (0.1–3.3)	0.5 (0.1–2.9)
Parent education	1.0 (0.8–1.1)	–	1.1 (0.8–1.4)	–	0.9 (0.7–1.1)	–
Parent occupation	1.3 (0.7–2.4)	–	2.7 (1.0–7.2)	2.4 (1.0–5.9)	0.9 (0.4–1.9)	–
No. of children	0.7 (0.4–1.4)	–	2.3 (0.8–6.1)	1.9 (0.7–5.0)	0.4 (0.2–0.9)	0.5 (0.2–1.0)
Neighborhood crime rate	0.5 (0.3–1.0)	–	0.3 (0.1–1.0)	–	0.6 (0.3–1.4)	–

Table 4
The influence of social risk factors on the number of low lead cases and odds ratios

Strata	Bone lead (ppm)	Cases	Controls	Odds Ratios
Race				
African American	≥25	35	8	1.5 (0.7–3.6)
subjects	<25	123	43	
White subjects	≥25	14	15	3.4 (1.4–8.1)
	<25	22	80	
Two parental figures				
No	≥25	25	10	1.3 (0.6–3.0)
	<25	68	35	
Yes	≥25	15	13	2.6 (1.1–5.9)
	<25	41	91	
Parent occupation				
Manual/menial/	≥25	25	11	1.7 (0.8–3.7)
unemployed	<25	67	49	
Skilled/clerical/	≥25	15	12	2.3 (1.0–5.4)
professional	<25	42	77	

White subjects and children with two parents had higher odds ratios for lead.

4. Discussion

These findings of higher concentrations of lead in delinquents and of an association of bone lead levels with delinquency after covariate adjustment support those we reported in 1996 on the relationship between bone lead and antisocial behavior [19], and are consistent with both Denno's [9] and Dietrich et al.'s findings [10]. The effect is substantial. With all subjects in the model, bone lead level was the second strongest risk factor, exceeded only by race. In the stratified models, with race eliminated, bone lead was exceeded as a risk only by single parent status.

The case for lead as a causal factor in antisocial behavior is also given credence by two recent ecological studies. Stretesky and Lynch reported positive correlations between homicide rates (National Center for Health Statistics) and air lead levels for 3111 US counties. After adjustment for 15 confounding variables, a four-fold increase in homicides in the highest lead counties compared to the lowest lead counties was found [28]. Nevin [21] reported a statistically significant association between gasoline lead sales and violent crime after adjustment for unemployment and percent of population in the high crime age group.

Selection of controls is always a critical issue in case-control studies and presents a possible challenge to validity in this study. Although we attempted to recruit controls similar in background to our delinquent sample, volunteer bias was encountered. African Americans from our potential control pool responded at a much lower rate than Whites to our recruitment letters. Within racial strata, the differences between cases and controls were few: A higher proportion of controls were White; controls also tended to be in a higher school grade (P=.04, African Americans; P<.0001, Whites), more often have two parental figures in the home and have parents with higher Hollingshead employment status. Adjustment for these factors in the logistic model increased the effect size for lead. Control of race by stratification demonstrated a lead effect within both Black and White strata, and disclosed an increased effect size for lead in White subjects. Limiting our sample to only those subjects attending the same high schools (n=43 cases, 145 controls), the odds ratio was reduced by 25% from 4 to 3.1 and the confidence limits widened, crossing 1. Although we cannot exclude the possible role of unmeasured confounders, the relative stability of the effect size after adjustment for important confounders is reassuring.

Most studies of the causes of criminal behavior have focused on social risk factors; less attention has been given to the influence of brain dysfunction [24,25]. Of seven reviews on the influence of brain lesions or other central nervous system (CNS) disorders on violent offending published between 1974 and 1989, however, six concluded that there was a positive association [17]. Some neurotoxins, notably alcohol, amphetamines and other drugs of abuse, are acknowledged as facilitators of criminal behavior. Other neurotoxins, including lead, have largely been ignored.

Abnormal CNS function is frequently found in the most serious delinquents. Those delinquents who display antisocial behavior early in life persist in their behavior, while those who start later tend to give it up in their late teens. Those offenders in the group of life-persistent delinquents have a higher rate of impaired neuropsychological function when compared to those who begin later [18]. This early onset/life persistent group, approximately 6% of the population, is responsible for 50% of the crime.

PET scans of brain function have demonstrated decreased glucose metabolism in the prefrontal lobe of murderers compared to controls [23]. Impulsivity, a critical precursor of antisocial behavior, is mediated in the prefrontal lobes of the cerebral cortex and damage in this area is associated with behavioral disturbances. Lesions in the prefrontal cortex are also associated with many characteristic behaviors of ADHD [6]. ADHD, if accompanied by conduct disorder or defiance, is a strong risk factor for antisocial behavior and later arrests [2,22,26]. In numerous studies from the United States and around the world, lead-exposed children have been reported to display impulsivity and attentional problems [4,11,26,27,30,33,34]. This observation confirms that found by Byers in his follow-up study of 20 lead-poisoned children published in 1943:

> Behavior difficulties were common throughout the series. Much of this behavior could be classified as "forced reaction to stimuli in the environment" described by Strauss and Werner as evidence of cortical damage. It was usually described as unreliable impulsive behavior, cruel impulsive behavior, short attention span and the like. [5]

The specific biological mechanisms underlying lead's effect on aggression and impulsivity are not known. Lead acts at a large number of CNS sites, some of which are involved in impulse control. Lead interferes with synaptogenesis [3], diminishes the inhibition of brain phosphokinase C [15], decreases norepinephrine-induced inhibition [29] and lowers brain levels of serotonin or 5-HIAA [13,32]. Lead exposure is associated with increased levels of D-aminolevulinic acid, which may antagonize GABA inhibition [16]. Lead also enhances both D_1 and D_2 dopamine sensitivity, and alters NMDA receptor sensitivity [7,8].

In addition to its direct action on the brain and impulse control, lead exposure can increase risk for delinquency through a separate, indirect route: impaired cognitive function and classroom performance. Early lead exposure has been shown to be associated with a seven-fold increase in the rate of high school failure and a six-fold increase in reading disabilities [20]. Students who do poorly in school, read poorly and fail out are more likely to become lawbreakers. In the sample reported here, cases had lower grade achievement than controls. This could be due to grade retention or to time spent in incarceration.

Adjustment for covariates in the logistic models in this study increased the odds ratio for bone lead. While African American race, absence of two parental figures and low status parental occupation are independent risk factors for criminality, the effect size for lead was larger in White subjects, in families with two parental figures and with parents in higher status employment. This apparent paradox occurs because nonlead social factors that raise the risk of delinquency increase the number of subjects in the low lead/delinquent case group. This becomes apparent in examination of Table 4. In each higher risk stratum, the proportion of delinquents in the low lead cell is higher and the resultant lead odds ratio lower.

Raine reported a similar finding in his PET scan study of murderers. Those subjects whose rearing histories were rated more favorably had greater impairment in prefrontal glucose metabolism than those with severe social deficits. He suggested that among violent offenders without deprived home backgrounds, the "social push" to violence is minimized and "consequently brain abnormalities provide a relatively stronger predisposition to violence" [23].

A limitation of this study is the smaller number of African American controls, resulting in wider confidence limits in this stratum. Given the sizable difference in bone lead levels, it seems unlikely that this limited control sample biased the study towards a false positive conclusion.

These data are the first reported bearing on the association between lead at asymptomatic doses and adjudicated delinquency. If other studies find a similar association between lead and delinquency, a sizable segment of this important societal problem of delinquency and violence would become accessible to primary prevention. Future epidemiological studies of the causes of criminality should include lead and other neurotoxic agents as risk factors.

Acknowledgements

This study was supported by the National Institute of Environmental Health Sciences, Grant no. ES 05015.

We thank Mr. Joseph Daugerdas, former Director of the Allegheny Juvenile Court, the staffs of the Community Intensive Supervision Program, the Allegheny Academy and the Pittsburgh Public School System for their assistance and support in providing access to subjects. Heather Schaner, Carolina Diaz-Peroza, Jennifer Skwarlo and Jennifer Bowser carefully evaluated the subjects. Loretta Kemp conducted home visits to families. Daniel Schaner was responsible for analyzing most of the XRF spectra. Sung-Jun Myung of the University Center for Social and Urban Research collected and tabulated the community crime data. We thank Drs. James Ware, David Bellinger, Alan Leviton and Raymond Neutra for critical review of the manuscript and helpful suggestions.

Data from this study were presented at the Pediatric Academic Societies meeting in Boston, MA on May 15, 2000.

References

[1] American National Standards Institute, Performance criteria for radiobioassay, Health Physics Society, 1996, N13.30.

[2] A.F. Arnsten, J.C. Steere, R.D. Hunt, The contribution of alpha$_2$-noradrenergic mechanisms to prefrontal cortical cognitive function: Potential significance for attention-deficit hyperactivity disorder, Arch. Gen. Psychiatry 53 (1996) 448–455.

[3] D. Averill, H.L. Needleman, Neonatal lead exposure retards cortical synaptogenesis in the rat, in: H.L. Needleman (Ed.), Low Level Lead Exposure: The Clinical Implications of Current Research, Raven Press, New York, 1980, pp. 201–210.

[4] D. Bellinger, A. Leviton, E. Allred, M. Rabinowitz, Pre- and postnatal lead exposure and behavior problems in school age children, Environ. Res. 66 (1994) 12–30.

[5] R.K. Byers, E.E. Lord, Late effects of lead poisoning on mental development, Am. J. Dis. Child. 66 (1943) 471–483.

[6] F.X. Castellanos, J. Giedd, W.L. Marsh, S.D. Hamburger, A.C. Vaituzis, D.P. Dickstein, et al., Quantitative brain magnetic resonance imaging in attention-deficit hyperactivity disorder, Arch. Gen. Psychiatry 53 (1996) 607–616.

[7] D.A. Cory-Slechta, M. Garcia-Osuna, J.T. Greennamyre, Lead-induced changes in NMDA receptor complex binding correlations with learning accuracy and with sensitivity to learning impairments caused by MK-801 and NMDA administration, Behav. Brain Res. 85 (1997) 61–74.

[8] D.A. Cory-Slechta, J. Pokara, D.V. Wodzowski, Postnatal exposure induces a supersensitivity to the stimulus properties of a D_2–D_3 agonist, Brain Res. 598 (1992) 162–172.

[9] D.W. Denno, Biology and Violence, Cambridge Univ. Press, New York, NY, 1990.

[10] K.N. Dietrich, M.D. Ris, P.A. Succop, O.G. Berger, R.L. Bornschein, Early exposure to lead and juvenile delinquency, Neurotoxicol. Teratol. 23 (2001) 511–518.

[11] D.M. Fergusson, J.E. Fergusson, L.J. Horwood, N.G. Kinzett, A longitudinal study of dentine lead levels, intelligence, school performance and behavior: Part III. Dentine lead levels and attention/activity, J. Child Psychol. Psychiatry 29 (1988) 811–824.

[12] A.B. Hollingshead, Four factor index of social status, Working Paper, New Haven, CT, June 1975.

[13] S.M. Lasley, R.D. Greenland, D.J. Minnema, I.A. Michaelson, Influence of chronic inorganic lead exposure on regional dopamine and 5-hydroxytryotamine turnover in rat brain, Neurochem. Res. 90 (1984) 1675–1688.

[14] R. Loeber, M. Stouthhammer-Loeber, W. VonKammen, D. Farrington, Development of a new measure of self-reported antisocial behavior for young children: Prevalence and reliability, in: M.W. Klein (Ed.), Cross National Research in Self-Reported Crime and Delinquency, Kluwer Academic Publishers, Dordrecht, 1989, pp. 203–225.

[15] J. Markovac, G.W. Goldstein, Picomolar concentrations of lead stimulate brain protein kinase C, Nature 334 (1988) 71–73.

[16] P.A. Meredith, M.R. Moore, B.C. Campbell, G.G. Thompson, A. Goldberg, Delta-aminolevulinic acid metabolism in normal and lead-exposed humans, Toxicology 9 (1978) 1–9.

[17] A. Mirsky, Biology of violence, in: A.J. Reiss, K.A. Miczek, J.A. Roth (Eds.), Understanding and Preventing Violence, vol. 2, National Academy Press, Washington, DC, 1994.

[18] T.E. Moffit, Adolescence-limited and life-course-persistent antisocial behavior: A developmental taxonomy, Psychol. Rev. 100 (1993) 674–701.

[19] H.L. Needleman, J.A. Riess, M.J. Tobin, G.E. Biesecker, J.B. Greenhouse, Bone lead levels and delinquent behavior, J. Am. Med. Assoc. 275 (1996) 363–369.

[20] H.L. Needleman, A. Schell, D. Bellinger, A. Leviton, E. Allred, The long-term consequences of exposure to low doses of lead in childhood: An 11-year follow-up report, N. Engl. J. Med. 322 (1990) 83–88.

[21] R. Nevin, How lead exposure relates to temporal changes in IQ, violent crime, and unwed pregnancy, Environ. Res. 83 (2000) 1–22.

[22] D.R. Offord, K. Sullivan, N. Allen, Delinquency and hyperactivity, J. Nerv. Ment. Dis. 167 (1979) 734–741.

[23] A. Raine, J. Stoddard, S. Bihrle, M. Buchsbaum, Prefrontal glucose deficits in murderers lacking social deprivation, Neuropsychiatry Neuropsychol. Behav. Neurol. 11 (1998) 1–7.

[24] A.J. Reiss, K.A. Miczek, J.A. Roth (Eds.), Understanding and Preventing Violence: Biobehavioral Influences, 2nd ed, National Academy Press, Washington, DC, 1994.

[25] J. Satterfield, J. Swanson, A. Schell, F. Lee, Prediction of antisocial behavior in attention-deficit hyperactivity boys from aggression/defiance scores, J. Am. Acad. Child Adolesc. Psychiatry 3 (1994) 185–190.

[26] W. Sciarillo, Lead exposure and child behavior, Am. J. Public Health 82 (1992) 1356–1360.

[27] P.A. Silva, P. Hughes, S. Williams, J.M. Faid, Blood lead, intelligence, reading attainment, and behaviour in eleven year old children in Dunedin, New Zealand, J. Child Psychol. Psychiatry 29 (1988) 43–52.

[28] P. Stretesky, M.J. Lynch, The relationship between lead exposure and homicide, Arch. Pediatr. Adolesc. Med. 155 (2001) 579–582.

[29] D. Taylor, J. Nathanson, B. Hoffer, L. Olson, A. Seiger, Lead blockade of norepinephrine-induced inhibition of cerebellar purkinje neurons, J. Pharmacol. Exp. Ther. 206 (1978) 371–381.

[30] G.O.B. Thomson, G.M. Raab, W.S. Hepburn, R. Hunter, M. Fulton, D.P.H. Laxen, Blood lead levels and children's behaviour—results from the Edinburgh lead study, J. Child Psychol. Psychiatry 30 (1989) 515–528.

[31] A.C. Todd, F.E. McNeill, J.E. Palethorpe, D.E. Peach, D.R. Chettle, M.J. Tobin, et al., In vivo X-ray fluorescence of lead in bone using K X-ray excitation with 109Cd sources: Radiation dosimetry studies, Environ. Res. 57 (1992) 117–132.

[32] H.R. Widmer, E.E. Butikofer, M. Schlumf, W. Lichtenseiger, Pre- and postnatal lead exposure affects the serotonergic system in the immature rat brain, Experientia 47 (1991) 463–466.

[33] W. Yule, R. Lansdown, M. Urbanowicz, I. Miller, The relationship between blood lead concentrations. Intelligence and attainment in a school population: A pilot study, Dev. Med. Child Neurol. 23 (1981) 567–576.

[34] W. Yule, R. Lansdown, Blood lead concentrations and school performance, Br. J. Med. 283 (1981) 13–36.

Senator JEFFORDS. I ask unanimous consent that a report by the CDC Advisory Committee be included in the record.

Senator CRAPO. Without objection.

[The referenced document follows:]

Turning Lead Into Gold: How the Bush Administration is Poisoning the Lead Advisory Committee at the CDC

A report by the staff of Rep. Edward J. Markey (D-MA) on the Removal or Rejection of Academic Experts in Lead Poisoning from and Nomination of Industry-Friendly Representatives to the Centers for Disease Control & Prevention (CDC) Advisory Committee on Childhood Lead Poisoning Prevention
October 8, 2002

Introduction:

According to recent Centers for Disease Control and Prevention (CDC) estimates, 890,000 U.S. children age 1-5 have elevated blood lead levels, and more than one-fifth of African-American children living in housing built before 1946 have elevated blood lead levels. The major sources of lead exposure are deteriorated paint in older housing, and dust and soil that are contaminated with lead from old paint and from past emissions of leaded gasoline. The CDC web site[1] states that:

- "Lead poisoning affects virtually every system in the body, and often occurs with no distinctive symptoms.
- Lead can damage a child's central nervous system, kidneys, and reproductive system and, at higher levels, can cause coma, convulsions, and death.
- Even low levels of lead are harmful and are associated with decreased intelligence, impaired neurobehavioral development, decreased stature and growth, and impaired hearing acuity."

The Federal and many State governments have monitoring programs to test blood lead levels in children, programs to increase public awareness about risks and prevention of lead poisoning, and programs to reduce the amount of lead in gasoline and to remediate lead-contaminated sites. These have greatly reduced the numbers of children with elevated blood lead levels from 3-4 million in 1978, to 890,000 in the 1990s[2].

The CDC Advisory Committee on Childhood Lead Poisoning Prevention (hereafter to be referred to as the "Advisory Committee") is charged with assessing the scientific data and recommending changes to CDC policy to prevent lead poisoning, including assessing whether the blood lead level limits are adequate. These blood lead levels are then used to determine which children are at risk for adverse health effects, and how much remediation must be done to ensure that a lead-contaminated site is safe. The Committee has guided major changes in lead poisoning policy for more than a decade. For example, in 1991, the acceptable blood lead level limits were revised from 25 µg/dL (micrograms per deciliter, the unit used to measure blood lead levels) down to 10 µg/dL in a report released by CDC and developed in part by the Advisory Committee[3]. In March 2002, the Advisory Committee issued Recommendations entitled "Managing Elevated Blood Lead Levels Among Young Children" which provides health care case managers guidance on how to assess and treat children with elevated blood lead levels[4].

This report reveals recent changes to the membership of the Advisory Committee that indicate that the nominations of renowned scientists with a long record in determining the health effects associated with childhood lead poisoning are being rejected, and that instead the vacancies are being filled by individuals who have direct ties to the lead industry, which has a financial interest in the policies adopted by the Advisory Committee; If the acceptable blood lead level limits are revised upwards, or if

---

[1] http://www.cdc.gov/nceh/lead/factsheets/leadfcts.htm
[2] http://www.epa.gov/opptintr/lead/
[3] http://aepo-xdv-www.epo.cdc.gov/wonder/prevguid/p0000029/p0000029.asp#head004000000000000
[4] http://www.cdc.gov/nceh/lead/CaseManagement/caseManage_main.htm

new scientific evidence indicating they should be revised further downwards is ignored, the health of many children in this country will be imperiled, and corporate polluters will be allowed to trade the long-term health of children for short-term commercial gain.

Documents obtained by Rep. Markey indicate that the Advisory Committee membership has been proposed for alteration in the following manner:

Reappointment rejected:

- Dr. Michael Weitzman, Department of Pediatrics, University of Rochester, and Pediatrician in Chief, Rochester General Hospital, Advisory Committee member since 1997, author of numerous peer-reviewed publications on lead poisoning.

Nominations rejected:

- Dr. Bruce Lanphear, Associate Professor, Department of Pediatrics, University of Cincinnati, Cincinnati, Ohio, currently the Sloan Professor of Children's Environmental Health, author of numerous peer-reviewed publications on lead poisoning.
- Dr. Susan Klitzman, Associate Professor of Urban Public Health at the Hunter College School of Health Sciences, author of numerous peer-reviewed publications on lead poisoning.

Nominated to the panel:

- Dr. William Banner, expert witness for the lead industry who believes that lead is harmful only at levels that are 7-10 times as high as the current CDC blood lead levels[5].
- Dr. Joyce Tsuji, principal scientist for Exponent, a company whose corporate clients include ASARCO (which is currently disputing EPA's assumptions that ASARCO is the source of elevated arsenic and lead in residential soils in El Paso and fighting Superfund designation) and King and Spalding, a DC law firm representing several large lead firms, and who has testified that the health risks of toxic plumes were not imminent.
- Dr. Kimberly Thompson, Assistant Professor of Risk Analysis and Decision Science, Harvard School of Public Health, affiliated with the heavily industry-funded Harvard Center for Risk Analysis. HCRA has 22 corporate funders with a financial interest in the deliberations of the CDC Advisory Committee on Childhood Lead Poisoning Prevention and less stringent regulation of lead.[6] Three of these funders have Superfund sites with lead contamination - Ciba-Geigy Corporation, FMC Corporation, and Monsanto.[7]
- Dr. Sergio Piomelli, Professor, Columbia Presbyterian Medical Center, researcher who reportedly disagrees with the current blood lead standard set by the CDC Advisory Committee in 1991[8].
- Tracey V. Lynn, affiliation and lead poisoning expertise not able to be determined.

---

[5] [5] Deposition of William Banner Jr, MD, Rhode Island v. Lead Industries Assoc, Ind (Sup Ct. R.I.)(No.99-5226)

[6] http://www.hcra.harvard.edu/funding.html

[7] U.S. EPA National Priorities List (2002).

[8] http://www.rainerlaw.com/pages/articles/wsj.html and http://www.leadinfo.com/MEDIA/newsleaderjune2000.htm

Changes to the Advisory Committee's Membership

Proposed changes to the Advisory Committee's membership were revealed in a September 6, 2002 email from CDC to Dr. Jennifer Sass at the Natural Resources Defense Council and obtained by Rep. Markey:

From: Shepard, Evelyn [ess8@cdc.gov]
Sent: Friday, September 06, 2002 11:18 AM
To: Sass, Jennifer
Subject: FW: Advisory Committee Members

Please let me know if you receive this email.

Thanks,
Evelyn Shepard
Branch Secretary
National Center for Environmental Health
Environmental Hazards and Health Effects
Lead Poisoning Prevention Branch
1600 Clifton Road, N.E., MS E-25, RM 1064.02
Atlanta, Georgia 30333
Phone: (404) 498-1429
Fax: (404) 498-1444
Email: ess8@cdc.gov

> -----Original Message-----
> From: Shepard, Evelyn
> Sent: Friday, September 06, 2002 11:12 AM
> To: 'Tfboyd@nrdc.org'; 'jscass@nrdc.org'
> Subject: Advisory Committee Members
>
> Per your request, below are Advisory Committee Members.
>
> List of the Advisory Members:
> 1. Ms. Carla Campbell - Acting Chair
> 2. Ms. Linda Anderson - Acting Branch Chief,
>
> Members:
> 3. Mr. Cushing Dolbeare
> 4. Ms. Anne Wengrovitz
> 5. Mr. Birt Harvey
> 6. Mr. Thomas Matte
> 7. Ms. Amy Murphy
> 8. Mr. Steve Hays
> 9. Mr. David Jacobs
> 10. Ms. Patricia McLaine
> 11. Ms. Rebecca Parkin
> 12. Ms. Karen Pearson
> 13. Mr. Routt Reigart
> 14. Mr. George Rodgers
> 15. Mr. Robert Roscoe
> 16. Mr. Joel Schwartz
> 17. Mr. Michael Weitzman[9]
> 18. Mr. Walter Rogan
> 19. Mr. Byron Bailey
> 20. Mr. Jerry Zelinger
> **Nominees are subject to change:**
> **21. Mr. Sergio Piomelli**
> **22. Ms. Kimberly Thompson**
> **23. Ms. Joyce Tsuji**
> **24. Ms. Tracey Lynn**
> **25. Mr. William Banner**
> Please let me know if I can be further assistance.

---

[9] Please note that Dr. Weitzman has been informed that he will not be reappointed to the panel

Names and Backgrounds of Individuals Removed from or Rejected by the Centers for Disease Control (CDC) Advisory Committee on Childhood Lead Poisoning Prevention

The following is a more expanded description of the backgrounds of the individuals who were not reappointed as expected to the Advisory Committee, or whose nominations were rejected.

1) Michael Weitzman, M.D. (716) 275-1544

- **Resume:** http://www.urmc.rochester.edu/gchas/fellowships/weitzman.htm
- 1990 – present Department of Pediatrics, University of Rochester, and Pediatrician in Chief, Rochester General Hospital
- 1997-2002 Member, CDC Advisory Committee on Childhood Lead Poisoning Prevention – **reappointment rejected**
- 1990-91 Member, CDC Advisory Committee on the Prevention of Childhood Disability
- 2000-01 Member, Environmental Protection Agency (EPA) Children's Health Protection Advisory Committee
- 2002 - Chair, EPA Children's Health Protection Advisory Committee on Childhood Lead Poisoning Prevention Workgroup to Review Evidence of Health Effects of Blood Lead Levels <10 micrograms per deciliter.
- Author of numerous peer-reviewed publications on lead poisoning.

2) Bruce P. Lanphear, M.D., M.P.H. (513)-636–3778

- **Resume:** http://www.cincinnatichildrens.org/Services/Faculty_And_Staff/L/Bruce+P.+Lanphear.htm
- Conducted numerous epidemiologic studies of lead-contaminated house dust and residential soil, and principal investigator on the primary study to be used by the EPA to establish federal standards for lead in residential dwellings.
- 1997-2002: Associate Professor, Department of Pediatrics, University of Cincinnati, Cincinnati, Ohio, currently the Sloan Professor of Children's Environmental Health
- 1998–2002: Member, Science and Research Work Group, EPA Office of Children's Health Protection Advisory Committee
- 1996–1998: Chair, U.S. HUD Committee on Lead-Contaminated Dust
- 2002: Nominee, CDC Advisory Committee on Childhood Lead Poisoning Prevention – **nomination rejected**

3) **Susan Klitzman,** M.P.H., Ph.D.

- Associate Professor of Urban Public Health, Hunter College School of Health Sciences
- Author of numerous peer-reviewed publications on lead poisoning.
- 2002: Nominee, CDC Advisory Committee on Childhood Lead Poisoning Prevention – **nomination rejected**

Names and Backgrounds of Individuals Nominated by the Bush Administration to the Centers for Disease Control (CDC) Advisory Committee on Childhood Lead Poisoning Prevention

The following is a more expanded description of the backgrounds of the individuals who have recently been nominated to the Advisory Committee.

1) William Banner, Jr., MD, PhD – Expert Witness for the Lead Industry

Dr. Banner, who is currently an attending physician in the pediatric intensive care unit, Children's Hospital at St. Francis Clinical Professor of Pediatrics, University of Oklahoma College of Medicine, is also currently retained by the Lead Industries Association as an expert witness, in an ongoing legal case wherein the State of Rhode Island is holding the lead paint industry responsible for childhood lead poisoning in Rhode Island.

The Rhode Island complaint cites a 1904 Sherwin-Williams article that says that "lead is poisonous in a large degree, both for the workmen and for the inhabitants of a house painted with white lead colors," but notes that the company continued to sell lead paint until 1978 when it was banned (see Appendix A). In June 2002, Banner was deposed by the plaintiff's lawyers, representing the people of Rhode Island[10]. The June 2002 deposition reveals the following, regarding Banner's qualifications and perspective on childhood lead poisoning:

A) Banner believes that blood levels of lead below 70 μg/dL do not pose a threat to children's health. The current CDC position[11] is that blood lead levels of 10 μg/dL or greater are high enough to be a health concern.

Q: (Neil Leifer, attorney for the State of Rhode Island)[12] What is the dose and time that you think is required in order for the lead to achieve its target of toxicity in the brain?
A: (William Banner, for the Lead Industries Association) Well, it's probably over, you know, as you said over 70.
Q I haven't said anything.
A Oh, well. Some of your witnesses, I'm sorry. Over 70 and closer to 100, probably. And the time factor nobody's really looked very carefully into. It appears to take a fair amount of time because everybody knows that short exposures is even well over several hundred can be tolerated with minimal effect.
Q[13] So the absence of encephalopathy, which you have indicated is possible above lead levels of 70 but more likely of lead levels above 100, you don't believe -- is it your opinion that there are no central nervous system deficits or injuries that are associated with exposure and ingestion of lead?
A Well, you're using the word "associated".
Q Okay. That's right, I am.
A And, no, I don't believe that there have been – no.

---

[10] Deposition of William Banner Jr, MD, Rhode Island v. Lead Industries Assoc, Ind (Sup Ct. R.I.)(No.99-5226)
[11] http://aepo-xdv-www.epo.cdc.gov/wonder/prevguid/p0000029/p0000029.asp#head004000000000000
[12] Deposition of William Banner, Jr., M.D. June 13, 2002. Page 133
[13] Deposition of William Banner, Jr., M.D. June 13, 2002 Page 135

B) Banner doesn't believe that lead can cause childhood cognitive disorders

Q[14] So are you familiar with literature that has found that there are cognitive, IQ, and other kinds of learning deficits that have been associated with ingestion of lead levels above 10 micrograms?
A That people have published that and claim that, yes.
Q You've read some of those articles?
A Yes, I've read many of them.
Q Do you reject those conclusions?
A Yes.
Q[15] I'm asking you for your expert opinion in this case. Do you believe that any of the epidemiological studies have established any relationship between lead ingestion and adverse cognitive, behavioral, or emotional status?
A No

C) Banner has does not appear to have done research on children and lead poisoning

Q[16] Is it your position that you have conducted and published research on the management and treatment of children with lead toxicity?
A Yeah.
Q Okay. Which ones are those, sir?
A Well, we've already been through them.
Q So the rat study is one of them?
A Well, it's a human epidemiology studies as we talked about.
Q Answer my question: Is the rat study one of them?
A These were clinical studies that were designed emulate principles of -- you know, involve in the management of children.
Q My question is: Was the rat study one of the studies that you had in mind when you wrote or approved the statement that you have conducted research and published in the area of the management and treatment of lead toxicity in children?
A That would be one of them, yes.
Q And what were the other ones?
A You know, you got the whole thing there.
Q Doctor, as far as I can tell from my notes you did a Salt Lake City survey.
A Right.
Q You did two -- you did the DMSA rat study and you did the experimental chelation study also on rats with some other compound that I can't pronounce.
A You can't pronounce.
Q And my question is: Is it your representation that those studies constitute research and publication in the area of management and treatment of lead toxicity in children? Is that your position?
A That in part. And what do you do with all the studies on extracorporal removal of toxins with chelators?
Q Did they involve lead in children?
A They involved chelators and how they--the mechanisms of action and, yes, they impact--if I admitted a kid today under certain circumstances we would use this general approach.
Q What is the basis for your position that you have conducted research and published in the area of treatment of children with elevated blood lead levels, same articles?
A Yeah. I mean, I've done -- I've published, as you pointed out, we went over the discussion of the treatment of children for the Academy of Pediatrics. Do you not consider that to be a publication on treatment of children with elevated blood lead levels?
Q Was that the product of research that you conducted?
A It's a published article.

---

[14] ibid
[15] Deposition of William Banner, Jr., M.D. June 13, 2002. Page 157
[16] Deposition of William Banner, Jr., M.D. June 13, 2002. Page 124-126

2) Joyce Tsuji, Ph.D., DABT, Principal Scientist, Exponent

- Resume: **http://www.exponent.com//leaders/bios/pdf/principal/tsuji.pdf**
- Exponent's clients[17] include corporations such as ASARCO (which is currently disputing EPA's assumptions that ASARCO is the source of elevated arsenic and lead in residential soils in El Paso and fighting Superfund designation[18]), Dow Chemical, and Dupont (named as a defendant in the Rhode Island lead lawsuit), large insurance companies such as Allstate and USAA, trade associations such as the American Chemistry Council, the National Mining Association and the American Petroleum Institute, and law firms such as Winston and Strawn and King and Spalding (which represents several large lead companies[19]).
- 31% of Exponent's 51 corporate clients have a financial interest in the deliberations of the CDC Advisory Committee on Childhood Lead Poisoning Prevention.[20] One of these companies, FMC Corporation, has a lead-contaminated Superfund site. The following 15 Exponent clients reported releases of lead or lead compounds to air, land or surface waters in EPA's 2000 Toxic Release Inventory: 3M, ASARCO, Becton Dickinson & Company, Chevron Corp., DaimlerChrysler Corporation, Exxon, FMC Corp., Ford Motor Co., General Electric, Honeywell International, Motorola, Phelps Dodge Corporation, Raychem, Whirlpool Corporation, and Zurn Industries.
- Dr. Tsuji provided testimony in a class action lawsuit regarding the "alleged" need for medical monitoring for all residents in the vicinity of a smelter living on soil with arsenic and lead levels above background levels. "Key issues included the lack of sensitivity of tests at these low exposure levels and the negligible risk of adverse effects."

---

[17] http://www.exponent.com/about/clients.html#corporations
[18] http://www.asarco.com/elpaso/asarcoresponse.htm
[19] http://www.kslaw.com/practice_areas/prac_environmental_matters.asp
[20] http://www.exponent.com/about/clients.html#corporations

3) **Kimberly Thompson Sc.D., Assistant Professor of Risk Analysis and Decision Science, Harvard School of Public Health**

- Resume: http://www.hsph.harvard.edu/insight/kmtCV.PDF
- Affiliated with the Harvard Center for Risk Analysis (HCRA). HCRA has 22 corporate funders with a financial interest in the deliberations of the CDC Advisory Committee on Childhood Lead Poisoning Prevention and less stringent regulation of lead.[21] Three of these funders have superfund sites with lead contamination - Ciba-Geigy Corporation, FMC Corporation, and Monsanto.[22] The other 19 funders reported lead releases in EPA's 2000 Toxic Release Inventory (including releases of lead or lead compounds to air, land or surface waters): 3M, Alcoa, ASARCO Inc., BASF, Bethlehem Steel Corporation, BP America Inc., ChevronTexaco, Delphi Automotive Systems, Dow Chemical Company, Eastman Chemical Company, Exxon, Ford Motor Co., General Electric Fund, Lyondell Chemical Company, Mobil Foundation, Inc., Olin Corporation, PPG, Shell Oil Company, and Volvo.
- Two HCRA funders – Atlantic Richfield Corporation and E.I. DuPont de Nemours & Company – are named as defendants in an ongoing legal case wherein the State of Rhode Island is holding the lead paint industry responsible for childhood lead poisoning.

4) **Sergio Piomelli, MD, Professor, Columbia Presbyterian Medical Center**

- **Resume:** http://www.columbiapresbyterian.com/profile.asp?ID=1377
- Referred to the 1991 Advisory Committee as being dominated by "well-meaning fanatics."[23] The 1991 CDC Advisory Committee recommended a change in the blood lead levels from 25 µg/dL down to 10 µg/dL. Dr Piomelli reportedly disputed this conclusion and has said "there is no epidemic of lead poisoning in the United States today, but some people are trying to create an epidemic by decree.[24]"

5) **Tracey V. Lynn –affiliation and lead poisoning expertise not able to be determined.**

---

[21] http://www.hcra.harvard.edu/funding.html
[22] U.S. EPA National Priorities List (2002).
[23] http://www.rainerlaw.com/pages/articles/wsj.html
[24] http://www.leadinfo.com/MEDIA/newsleaderjune2000.htm

Appendix A – Press Release on the Rhode Island Lead Lawsuit

RHODE ISLAND DEPARTMENT OF ATTORNEY GENERAL

FOR IMMEDIATE RELEASE
CONTACT:
Jim Martin
October 13, 1999
Tel: (401) 274-4400 ext. 2336
Fax: (401) 222-1302

PRESS RELEASE

A.G. Whitehouse Sues Lead Paint Industry
Seeks Treatment, Education and Abatement
Documents Nearly a Century-Long Record of Industry Culpability

PAWTUCKET — Calling on the lead paint industry "to take responsibility and clean up its mess," Rhode Island Attorney General Sheldon Whitehouse has made his state the first in the nation to file suit against eight manufacturers of lead paint and the industry's trade association.

The 10-count lawsuit seeks damages to "get the lead out of Rhode Island's children, homes and buildings, by paying to treat children with lead poisoning, meet the education needs of affected children, and abate lead paint from buildings and homes.

"Every Rhode Island child knows, if you make a mess you are supposed to clean it up," said Whitehouse at a news conference today. "However, there is one difference," he added. "Children often make a mess by accident. Lead dust is in our homes by design. This silent, invisible poison is the result of intentional, deliberate, informed decisions by the lead paint industry."

"We are doing this for the health of Rhode Island's children," Whitehouse said, citing reports that this year, one in five kindergartners in the state — and nearly one in three children in Providence, Pawtucket, Woonsocket, Newport and Central Falls combined — have elevated lead levels in their blood.

"We know now that this industry knew lead paint was toxic dating back as early as 1904, yet promoted its use and profited by that use," Whitehouse charged. "It willfully made the mess that has endangered the health of many children and imposed great burdens on Rhode Island families and the State."

The lawsuit, filed yesterday in Rhode Island Superior Court, alleges that the defendants marketed and sold lead-based paint with the full knowledge that it was toxic.

For example, the complaint cites a 1904 Sherwin-Williams article warning that "white lead is poisonous in a large degree, both for the workmen and for the inhabitants of a house painted with white lead colors," but notes that the company subsequently made and sold lead paint until 1978 when lead was banned.

The lawsuit explains that, "By at least 1912, National Lead excluded women and children from working in its white lead processes because of risk of lead poisoning." Yet in 1923, National Lead's publication, "Dutch Boy Painter," claimed, "Lead Helps to Guard Your Health."

In 1930, the Lead Industries Association (LIA) acknowledged the dangers of lead paint on children's toys and furniture, and three years later "the LIA internally suggested that its members consider discontinuing the use of Lead on children's toys and furniture," according to the complaint. Yet in 1936, LIA member Glidden was recommending lead-containing "Jap-A-Lac" paint for children's furniture.

The 10 counts in *Rhode Island v. Lead Industries Association, et als* allege Public Nuisance, Violation of Rhode Island Unfair Trade Practice and Consumer Protection Act, Strict Liability, Negligence, Negligent Misrepresentations and Omissions, Fraudulent Misrepresentations and Omissions, Civil Conspiracy, Unjust Enrichment, Indemnity, and Equitable Relief to Protect Children.

The relief demanded by the State of Rhode Island includes:

- Ø Damages to compensate the State for lead-poisoning related health, education and abatement costs;
- Ø Punitive damages;
- Ø Funding of a lead-poisoning public education campaign and "lead-poisoning detection and preventative screening programs in the State";
- "Judgment ordering the Defendants to detect and abate Lead in all residences, schools, hospitals, and public and private buildings within the State accessible to children"; and
- Ø "An order awarding the State such other extraordinary, declaratory and/or injunctive relief . . . to assure that the State has an effective remedy" to the problem of lead poisoning.

The defendants in this case are: the Lead Industries Association (a lead industry trade group), and eight manufacturers: American Cyanamid Company, Atlantic Richfield Company, E.I. duPont De Nemours and Company, The O'Brien Corporation, The Glidden Company, NL Industries, Inc., SCM Chemicals, and The Sherwin-Williams Company.

The civil law suit will be prosecuted on behalf of the State for ½ the standard contingency fee by Leonard Decof, Esq. of Decoff & Grimm (Providence), Jack McConnell of Ness, Motley, Providence, and by Special Assistant Attorney General Linn Freedman.

Senator JEFFORDS. The Heavy People 2010 Initiative of CDC calls for the elimination of childhood lead poisoning by 2010, yet the CDC website provides three reasons why blood levels should not be reduced. It states that it is difficult to measure blood lead level concentrations below 10. I ask unanimous consent to include in the record a paper by Herbert Needleman which refutes this claim and describes the technology advances that have lowered the least observable effect level until it approaches zero.

Senator CRAPO. Without objection.

[The referenced document follows:]

Annu. Rev. Med. 2004. 55:X-X
doi: 10.1146/annurev.med.55.091902.103653

0066-4219/04/0218-0000$14.00
<DOI>10.1146/annurev.med.55.091902.103653</DOI> (rev 7/30/03)

Lead Poisoning

Herbert Needleman

Professor of Psychiatry and Pediatrics, University of Pittsburgh School of Medicine, Pittsburgh, Pennsylvania 15213; email: hlnlead @pitt.edu

■ **Abstract** Understanding of lead toxicity has advanced substantially over the past three decades, and focus has shifted from high-dose effects in clinically symptomatic individuals to the consequences of exposure at lower doses that cause no symptoms, particularly in children and fetuses. The availability of more sensitive analytic methods has made it possible to measure lead at much lower concentrations. This advance, along with more refined epidemiological techniques and better outcome measures, has lowered the least observable effect level until it approaches zero. As a consequence, the segment of the population who are diagnosed with exposure to toxic levels has expanded. At the same time, environmental efforts, most importantly the removal of lead from gasoline, have dramatically reduced the amount of lead in the biosphere. The remaining major source of lead is older housing stock. Although the cost of lead paint abatement is measured in billions of dollars, the monetized benefits of such a Herculean task have been shown to far outweigh the costs.

INTRODUCTION

In recent years, the focus in lead poisoning has shifted away from adults exposed to high doses in industrial settings to the larger population of asymptomatic children with lesser exposures. This chapter surveys the past three decades of lead research and reviews the evolving knowledge of the distribution, toxicology, and remediation of lead toxicity.

EARLY HISTORY

Warnings of lead's poisonous properties extend at least as far back as the second century B.C., when Nikander, a Greek physician, described the colic and paralysis that followed lead ingestion. The early victims of lead toxicity were mainly lead workers and wine drinkers. Lead's sweet flavor made it useful in winemaking, to counteract the astringent flavor of tannic acid in grapes. Lead-sweetened wine, containing as much as 20 mg of lead per liter, was an important part of the diet of upper-class Romans. The synchronous decrease in fertility and increase in psychosis among the Roman aristocracy has raised speculation implicating lead poisoning in the fall of Rome (1).

Widespread outbreaks of colic due to lead-adulterated wine continued in Europe until as late as the sixteenth century, when Duke Ludwig of Würrtemberg,upon learning that there was an epidemic of lead colic in his duchy, banned its use in winemaking, imposing the death penalty for violators

Workers in the metals trades remain an important risk group; lead exposure remains one of the leading causes of workplace illness. In the United States, more than 320,000 American workers are occupationally exposed to lead .

DISCOVERY OF CHILDHOOD LEAD POISONING

It was only a century ago that childhood lead poisoning was recognized. The rapid growth of scientific understanding can be divided into four stages. The first reports of lead-poisoned children in Brisbane, Australia, in 1892 were greeted with widespread disbelief that lead toxicity could afflict children (2). Although the disease had reached epidemic proportions, there was considerable doubt that lead was the cause. Many of the homes in Brisbane were raised on piles, with large wooden-enclosed verandas that served as play areas for children. The rails were painted with white lead, which chalked and powdered in the hot Brisbane sun. The cause of the epidemic, lead-containing paint, was established in 1904, and lead paint was banned for household use in Brisbane in 1920.

Childhood lead poisoning was first described in the United States in 1914 (3). The prevailing belief in the second stage of knowledge was that acute poisoning had only two outcomes: death or complete recovery without any residua. This misconception was discarded in 1943 with the first follow-up of children who had recovered from acute toxicity. Nineteen of 20 surviviors had significant deficits: behavioral disorders, learning difficulties, and school failure (4). In this third stage, it was generally accepted that lead toxicity caused long-term deficits, but these deficits were thought to occur only in those children who had displayed clinical signs of encephalopathy during the acute episode. The fourth stage began in the 1970s, when studies of children with no clinical signs showed lead-related deficits.

TOXICOLOGY OF LEAD

Lead is a divalent cation, and it binds strongly to sulfhydryl groups on proteins. Of the many organs affected by lead, the most important is the central nervous system (CNS). Much of lead's toxicity can be attributed to distortion of enzymes and structural proteins, but this versatile toxicant has many other targets. Lead interferes with the development of the endogenous opiate system (5). It efficiently cleaves the ribophosphate backbone of tRNA catalytically at specific sites, with no evidence of a threshold (6). Many of lead's toxic properties are due to its ability to mimic or compete with calcium. At picomolar concentrations, lead competes successfully with calcium for binding sites on cerebellar phosphokinase C and thereby affects neuronal signaling (7). It inhibits calcium entry into cells . (8)Lead is picked up by mitochondria and produces swelling and distortion of mitochondrial cristae. Uncoupled energy metabolism, inhibited cellular respiration, and altered calcium kinetics follow .(9)Lead has a binary impact on neurotransmitter release: Spontaneous neurotransmitter release is enhanced, whereas stimulated release is inhibited (10)

Attention has also focused on the heme synthetic pathway, where many sites for lead activity are found. Delta aminolevulinic acid dehydratase is extremely sensitive to lead. Inhibition of this enzyme results in increased circulating aminolevulinic acid (ALA). ALA is a weak gamma-aminobutyric acid (GABA) agonist that decreases GABA release by presynaptic inhibition. Increased circulating ALA may account for some of the behavioral disorders seen in patients with porphyria and perhaps in lead toxicity.

Lead has diverse impacts on the CNS. Immature astrocytes are sensitive to lead, and lead interferes with myelin formation and the integrity of the blood-brain barrier (11).. Lead interferes with the synthesis of collagen and affects vascular permeability. At high enough doses, this results in brain edema and hemorrhage.(12) At lower doses, lead given to lactating rats interferes with synaptogenesis in their pups (13). Lead's interference with brain development has been demonstrated using the rodent barrel field cortex as a model (14).

Behavioral alterations secondary to lead exposure in rodents and primates are analogous to changes in humans. In one study, monkeys were dosed from birth to 200 days of age to achieve blood lead levels ranging from 3μg/dl to 25μg/dl. At 7 to 8 years, they were given a delayed alternation test, in which the critical positive stimulus was alternated Treated monkeys showed impaired ability to learn, particularly at longer intervals of delay (15). Lead-exposed primates also demonstrate impaired social function (16). Rodents given lead show deficits in learning mediated by dopaminergic and glutamatergic systems (17). In one interesting report, untreated rats found 15% solutions of alcohol aversive in a free-choice situation, but when their blood lead levels were raised to 61μg/dl, they increased their alcohol intake in both free- and forced-choice paradigms (18). The author speculated that lead increased the irritability of the rats and that they sought alcohol as a tranquilizer.

CLINICAL ASPECTS OF TOXICITY

Although adult lead poisoning is mainly of occupational origin, cases of acute lead poisoning from leaded dishware, bootlegged moonshine liquor, certain cosmetics, and folk remedies continue to be reported. Lead is still mined and smelted, although this has declined with the removal of lead from gasoline.

Lead poisoning in adults can affect the peripheral and central nervous systems, the kidneys, and blood pressure. Classical descriptions of occupational toxicity depict peripheral neuropathy with wrist or foot drop. At lesser exposures, slowed peripheral nerve conduction has been reported. (19)Patients with high blood lead levels may present with severe, intractable colic, motor clumsiness, clouded consciousness, weakness, and paralysis. Lead has adverse effects on both male and female reproduction. The fetotoxic properties of lead were known to British factory inspectors at the end of the nineteenth century; they found a high incidence of stillbirths (60%), neonatal deaths, and a decreased fertility rate in women employed in the ceramic industry (20). Males may manifest decreased sperm counts and teratospermia (21).

Hypertension has been associated with acute lead poisoning, along with renal failure. At lesser exposures, both experimental and epidemiological evidence of interference with renal function and elevations in blood pressure have been reported. Using data from the third National Health and Nutrition Examination Survey (NHANES III), a recent reevaluation of blood pressure in relation to contemporary blood lead levels found that black men and women had higher blood lead levels (5.4 μg/dl, 3.4μg/dl) than their white counterparts (4.4 μg/dl, 3.0 μg/dl). Black subjects, both men and women, had a statistically significant association of blood lead with blood pressure after covariate adjustment. The association was not seen in whites (22).

In its alkyl form, lead is a powerful neurotoxin. When tetraethyl lead (TEL) was first produced for use as a motor fuel additive in 1925, workers at all three operating plants began to die. After a brief moratorium imposed by the Surgeon General, production resumed and continued until the 1980s. TEL is fat-soluble; absorption through the skin and uptake by the brain is rapid. Because of growing evidence of neurotoxic effects at low doses, TEL was removed from gasoline in stepwise fashion beginning in 1978.

Lead has been classified as an animal carcinogen, but the data on human carcinogenesis are considered inadequate. Some recent studies of cancer rates in lead trades, e.g., smelters, painters, body and fender repairment have shown an increase in standard mortality rates, but others have not (23).

CHILDHOOD LEAD TOXICITY

Children are more sensitive to lead than adults for several reasons: Their exposure is increased by their universal hand-to-mouth activity; their gut absorbs lead more readily than an adult's; and the developing CNS is more vulnerable to toxicants than the mature CNS.

At high doses, generally blood lead levels >60 μg/dl, clinical symptoms become visible in children. Abdominal pain and arthralgia are common early complaints. Clumsiness and staggering may be seen, followed by headache and behavioral changes, which are signs of early encephalopathy. This may progress to alterations of consciousness, stupor, and convulsions.

Encephalopathy, fortunately, has become rare in the United States. A high percentage of those children who recover from clinical encephalopathy have severe cognitive, attentional, and behavioral impairments.

ASYMPTOMATIC LEAD TOXICITY

In the 1960s, the accepted toxic threshold for lead in children was 60μg/dl. Screening studies in eastern U.S. cities found that 10%--20% of inner-city children had blood lead levels over 40μg/dl. This finding raised the conjecture, first made by Byers & Lord in 1943, that a proportion of school failure and behavior disorder resulted from unrecognized lead toxicity. Five studies of lead levels and behavior in children without signs or symptoms of classical lead toxicity were undertakenin the early 1970s. Three reported an association between lead and IQ (24- 26): two did not (27,28). These early studies were constrained by a number of design flaws: The number of subjects in each study was small, and each relied on blood lead levels to rank exposure. Blood lead, a short-term marker, may misclassify earlier exposure. Some studies used relatively insensitive outcome measures, such as group or screening tests. Control of confounders and statistical procedures was limited. Some investigators studied clinic samples and their data may thus have suffered selection bias.

In 1979, we conducted a study that attempted to address these design issues. Lead concentration in deciduous teeth was selected as the marker of exposure. The sample comprised asymptomatic primary school students from the public schools of Chelsea and Somerville, Massachusetts. Subjects were classified on a large number of covariates, and these were controlled in the analysis by Analysis of Covarance Excluded were children with a history of lead exposure or toxicity. Children with elevated lead levels in the 90th %ile for lead concentration in their teeth were found, after covariate control, to be significantly impaired on the Wechsler IntelligenceScale forChildren-Revised IQ test, on language processing, and on reaction time under varying conditions of delay, a measure of attention. When teachers' ratings of 2146 children on an 11-item forced-choice scale were classified by dentine lead level, we found a dose-dependent increase in bad classroom behavior in direct relation to tooth lead level (**Figure 1, see color insert**) (29).

Figure 1 Distribution of negative teachers' ratings in relation to dentine lead concentrations.

In the 1980s, following the removal of lead from gasoline, the blood lead levels in the referent group dropped. This enabled well-designed studies, employing larger samples, better measures of outcome and lead burden, and more sophisticated statistical analyses, to discover effects of lower blood lead levels (30-336--). Three meta-analyses confirmed that low-level lead exposure was associated with IQ deficits (34--36). In response to the new data, in 1991, the Centers for Disease Control revised its limit of acceptable blood lead leveldownward in steps, from 60 μg/dl in the 1970s to its current status of 10 μg/dl.

In 2002 and 2003, two new studies found effects at levels below 10 μg/dl, further lowering the observed threshold for effect. Lanphear et al. examined psychological performance and blood lead levels in 4853 children who were NHANES III subjects. The mean blood lead in this sample was 1.9 μg/dl; 2.1% of the sample had blood lead levels in excess of 10 μg/dl. After adjustment for covariates, significant inverse relationships between blood lead and math and reading subtests and the Block Design and Digit Span of the WISC III were reported down to 2.5 μg/dl (37). Canfield et al. (38) studied 166 36-month-old children whose mean blood lead level was 7.9 μg/dl. Seventy-three percent of the subjects had blood lead levels under 10 μg/dl. Significant inverse relationships were found between IQ scores and lead after covariate adjustment. The slope of the effect was greater at the lower blood levels of lead (38). The import of the recent studies is that a threshold for lead and neurobehavioral function has not yet been demonstrated.

FOLLOW-UP STUDIES OF CHILDREN

The late effects of early-childhood lead exposure have been examined by several investigators, who found persistence of deficits over time. Our subjects, first examined in 1979, were seen again 12 years later at mean age of 18.7 years. Subjects whose dentine lead levels were in the high lead group(90th %ile) had more school failure, reading disabilities, lower class standing in their final year of high school, and disturbances in fine motor function (49) (**Figure 2, see color insert**).

Figure 2 Rate of failure to graduate from high school in relation to dentine lead levels.

Fergusson et al. followed a sample of New Zealand children into their eighteenth year and reported that elevated lead levels were associated with poorer reading scores, more failure to graduate from high school, and poorer examination scores (40)

FORWARD STUDIES OF PRENATAL AND INFANT LEAD EXPOSURE

Bellinger et al. found significant associations between umbilical blood lead levels and neurodevelopmental scores at 2 years of age (41). At later ages, the association between umbilical cord blood lead and outcome was attenuated. The 2-year blood lead concentrations, however, were significantly related to scores at 10 years of age. Exposure at 2 years had no observable threshold, demonstrating neurotoxic effects at blood lead levels below 10 μg/dl.

Dietrich et al. followed a group of 253 children from birth until 6 years of age. Postnatal blood lead levels were associated with deficits in performance scores of the WISC, after covariate adjustment, and in motor function (42).

LEAD AND BEHAVIOR

Cognitive function, measured by psychometric IQ tests, has been the major focus of most studies of lead exposure in childhood. There are persuasive reasons to believe that cognitive dysfunction may not be the most important effect of lead, and that we may be entering a fifth stage of understanding of lead's effects, in which lead is recognized to adversely affect social behavior.

This is not an entirely new notion. Parents have frequently reported that after recovery from an episode of acute lead poisoning, their child's behavior changed dramatically, and they became restless, inattentive, and aggressive. In 1943, Byers & Lord reported attentional dysfunction and aggression in a sample of lead-poisoned children on follow-up (4).

We studied 301 primary-school students and found that children with elevated bone lead levels scored higher on the attention deficit, aggression, and delinquency clusters of the Child Behavior Checklist after adjustment of covariates (43). Dietrich et al. found that prenatal lead exposure was associated with parents' reports of delinquency and aggression, and postnatal lead exposure was associated with self reports of delinquent acts (44). A recent case-control study of 195 arrested and convicted delinquent youths found an increased risk of delinquency associated with bone lead concentrations measured by X-ray fluorescence. The covariate-adjusted odds ratio was 4 (95% CL 1.4--11.1). The population-attributable risk for delinquency due to lead exposure ranged from 11% to 38% in this sample (45).

A number of recent ecological investigations correlating leaded gasoline sales or ambient lead levels with crime rates support an association between lead exposure and crime. Stretesky & Lynch compared homicide rates in 3311 counties in the United States (46) After adjustment for

15 covariates, they reported a fourfold increase in homicide rates in those counties with the highest air lead levels compared to controls. Nevin correlated sales of leaded gasoline with violent crime rates and, adjusting for unemployment and percent of population in the high-crime age group, found a statistically significant association (47). It has been speculated that one of the reasons for the recent decline in crime rates is decreased exposure to lead.

LEAD EFFECTS IN OLDER SUBJECTS

The greatest storage pool for lead is in bone, and the question of lead's fate in older subjects when bone demineralizes has attracted considerable speculation. It is estimated that 50% of trabecular bone in women is lost over a lifetime. Lead is mobilized when bone resorption begins; significantly higher blood lead levels have been measured in postmenopausal women than in premenopausal women (48). Elevated lead appears to adversely affect cognitive function in elderly subjects as well. Older women (mean age 70.5 years) with blood lead levels >8µg/dl had poorer performance on cognitive measures and slower reaction times than women with blood lead levels <3µg/dl after covariate adjustment (49).

Results of a large-cohort study of former TEL workers support a causal association between lead and dementia (50). The subjects ($n = 535$) were studied a mean of 16 years after their workplace exposure. The investigators report elevated bone lead levels and dose-related deficits in verbal and visual memory, executive ability, and manual dexterity. TEL workers exhibited a greater decline in function measured at yearly intervals than controls (). The same investigators examined the interaction between APOE genotypes with bone lead levels. An interaction effect for bone lead × APOE genotype was found for 19 of 20 regression models, indicating that the toxic effect of lead is greater in subjects with at least one APOE allele 51

Lead has induced apoptosis in a number of experimental systems, including rat midbrain (52), rat testis (53), rat fibroblasts (54), lung (55), and retinal rod cells (56, 57). These findings, and similarities in the distribution of lead exposure and the rates of Alzheimer's make the topic of lead –related dementia worthy of further study

Lead exposure may also decrease lifespan. This is borne out by a recent study of subjects from the second NHANES study (1976--1980), who were followed up in 1992. The mortality of 4292 subjects with blood lead levels of 20--29 µg/dl was compared to those with levels <10µg/dl.

Subjects with higher lead levels had a 46% increased all-cause mortality, 39% increased cardiovascular mortality, and 68% increased cancer mortality (58).

DIAGNOSIS AND MANAGEMENT OF CLINICAL LEAD POISONING

In adults, lead toxicity should be considered in the differential diagnosis of abdominal pain, arthralgia, hypertension, severe headache, increased intracranial pressure, CNS dysfunction, anemia, and renal dysfunction. An occupational history and an inventory of possible sources of exposure are useful. A blood lead level >10 µg/dl should be considered elevated, even though clinical symptoms are rarely seen below 60 µg/dl.

Any child with growth failure, abdominal pain, behavior change, hyperactivity, language delay, or anemia should have a blood lead test to rule out lead toxicity.

The cornerstone of lead toxicity management is the termination of exposure. For children, this means inspection of the home, and if this does not reveal lead, a survey of other possible sources. For lead levels >40 µg/dl, chelation therapy is effective in lowering the blood lead level. Calcium disodium edathamil (EDTA) was the preferred method until recently, when dimercaptosuccinic acid (succimer), an oral agent, was found to have equal efficacy. Both agents will reduce an elevated blood lead level to 40%--50% of its baseline. After treatment is concluded (5 days for EDTA, 19 days for succimer), body pools tend to equilibrate, and blood lead levels begin to rise, often requiring repeated courses.

EDTA has drastically reduced the mortality rate from encephalopathy, but its efficacy at lower exposures has never been systematically studied. As a result, whether it conveys any benefit to children without encephalopathy remains unknown.

After succimer had been in use for a few years, a multicenter study evaluated its efficacy in children with moderate elevations of lead (25--44 µg/dl). Blood lead levels in the treatment group were reduced to significantly lower levels than controls at the completion of treatment, but at 2 years of age, there were no differences between the two groups. At the conclusion of the study, no significant differences were found between treatment subjects and controls in cognitive, behavioral, or neuropsychological function (59). The only remedy at this time for low-level lead exposure is therefore primary prevention.

PRIMARY PREVENTION

In the early 1970s, the question of silent lead toxicity became the focus of intense controversy because of its regulatory implications. In 1973, when the Environmental Protection Agency began examining the health effects of TEL, industrial representatives claimed that the associations between lead and IQ were spurious, and that removing lead from gasoline would have no impact on body lead burdens. In 1977, after review of the health effects, the Environmental Protection Agency established an air lead standard of 1.5 $\mu g/M^3$. The stepwise removal of lead from gasoline, based on the new air standard, began in the late 1970s. **Figure 3** shows the effect of removing lead from gasoline on blood lead levels in the United States between 1975 and 1980.

Figure 3 Parallel decreases in blood lead values observed in the NHANES II study and amounts of lead used in gasoline during 1976--1980.

With the removal of lead from gasoline, a single major source remains for American children: leaded paint. Although it has been banned in household paint since 1971, 80% of the houses built before 1950, or 23,000,000 units, contain leaded paint. A cost-benefit analysis by PublicHealth Service estimated the cost of abatement in these houses over a 30-year period at $33.7 billion in 1991. The estimated benefit from avoided health care costs and increased income due to raised IQ was $61.7 billion. This cost analysis may be conservative; it does not include avoided delinquency and cardiovascular disease, both demonstrated effects of lead exposure, among the health effects (60).

Current analyses also demonstrate that primary prevention yields large economic benefits. Grosse et al. calculated that each present-day preschool child's IQ was increased by 2.2--4.7 points over what it would have been had the reduction in leaded gasoline and blood lead not taken place (61). From this, they calculated the IQ-related increase in income and estimated the economic benefit for each year's birth cohort of 3.8 million children. The benefit range for the 1998 birth cohort was between $110 billion and $319 billion (61). Landrigan et al., assuming no threshold for the lead-IQ association, estimated the loss of future earnings for the one-year cohort of children aged 5 in 1997 at $43.4 billion (62).

The evidence that lead toxicity extends down to the lowest measurable levels, that pharmacological therapies are ineffective at preventing sequelae in those with low levels, and that reduction of exposure yields huge economic as well as health benefits are strong warrants for a systematic program of abatement of lead from the single remaining major source: lead in older homes.

LITERATURE CITED

1. Gilfillan SC. 1965. Lead poisoning and the fall of Rome. *J. Occup. Med.* 7:53--60

2. Gibson JL. 1892. Notes on lead-poisoning as observed among children in Brisbane. *Proc. Intercolonial Med. Congr. Austr.* Third Session:76—83.

3. Blackfan KD. 1917. Lead poisoning in children with especial reference to lead as a cause of convulsions. *Am. J. Med. Sci.* 53:877--87

4. Byers RK, Lord EE. 1943. Late effects of lead poisoning on mental development. *Am. J. Dis. Child.* 66:471--83

5. Bailey C, Kitchen I. 1985. Ontogenesis of proenkephalin products in rat striatum and the inhibitory effects of low-level lead exposure. *Dev. Brain Res.* 22:75--79

6. Brown RS, HingertyBE, Dewan JC, et al. 1983. Pb(II)-catalysed cleavage of the sugar-phosphate backbone of yeast tRNA(Phe)---implications for lead toxicity and self-splicing RNA. *Nature* 303:543--46

7. Markovac J, Goldstein GW. 1988. Picomolar concentrations of lead stumulate brain protein kinase C. *Nature* 334:71--73

8.Simons T.1993. Lead-calcium interactions in cellular lead toxicity. *Neurotoxicology* . 14: 77-85.

9.Holtzman D, DeVries, Nguyen H, Bensch K. 1984 Maturation of resistance to lead encephalopathy: cellular and subcellular mechanisms. *Neurotoxicology*. 5: 97-124.

10. BresslerJ, Goldstein G. 1991 Mechanisms of lead neurotoxicity. *Biochem Pharnacol. 41:479-84.*

11. Krigman MR 1978. Neuropathology of heavy metal intoxication. *Environ. Health Persp.* 26:117-20.

12. . Pentschew A. 1965. Morphology and morphogenesis of lead encephalopathy. *Acta Neuropath* 5:133-60.

Krigman MR. 1978. Neuropathology of heavy metal intoxication. *Environ. Health Perspect.* 26:117—20.

10. Bressler JP, Goldstein GW. 1991.Mechanisms of lead neurotoxicity. *Biochem Pharmacol*; **41** : 479–84

13. Averill D, Needleman HL. 1980. Neonatal lead exposure retards cortical synaptogenesis in the rat. In *Low Level Lead Exposure: The Clinical Implications Of Current Research*, ed. HL Needleman pp. 201--10 New York: Raven

14. Wilson MA, Johnston MV, Goldstein GW, et al. 2000. Neonatal lead exposure impairs development of rodent barrel field cortex. *Proc. Natl. Acad. Sci. USA* 97:5540--45

15. Rice DC. 1992 Behavioral impairment produced by developmental lead exposure: evidence from primate reseach. In *Human Lead Exposure*, ed. HL Needleman, pp 137--54. Boca Raton, FL: CRC

16 Bushnell PJ, Bowman RE. 1979 Effects of chronic lead ingestion on social development in infant rhesus monkeys. *Neurobehav. Tox.* 1:207--19

17. Cory-Slechta DA. 1997. Relationships between Pb-induced changes in neurotransmitter system function and behavioral toxicity. *Neurotox.* 18:673--88

18. Nation JR, Baker DM, Taylor B, et al. 1986. Dietary lead increases ethanol consumption in the rat. *Behav. Neurosci.* 100:525--30

19.Seppalainen A, Hernberg S. 1980. Subclinical lead encephalopathy. *Am J Ind Medl* 1: 413-

20. Oliver ST 1911. A Lecture on lead poisoning and the race. *Brit Med J* 1: 1096-1098.

21. Lancrajan, Popescu JL, Gavenescu O, et al. 1975. Reproductive ability of workmen occupationally exposed to lead. *Arch. Environ. Health.* 30:396--401

22. Vuppurturi S, Muntner P, Bazzano LA, et al. 2003 Blood lead level is associated with elevated blood pessure in blacks. *Hypertension* 41:463--68

23. Landrigan PJ, Boffetta P, Apostoli P. 2000. The reproductive toxicity and carcinogenicity of lead: a critical review. *Am. J. Ind. Med.* 38:231--43

24. De la Burde B, NcLin S, Choate S, 1972. Does asymptomatic lead exposure in children have latent sequelae? *J. Pediatr.* 81:1088--91

25. Landrigan PJ, Baloh RW, Barthel WF, et al. 1975. Neuropsychological dysfunction in children with low level lead absorption. *Lancet*, pp. 708-712, March 29.

26. Perino J, Ernhart CB. 1974. The relation of subclinical lead level to cognitive and sensorimotor impairment in black preschoolers. *J. Learn. Disabil.* 7:26--30

27. Kotok D. 1972. Development of children with elevated blood lead levels: a controlled study. *J. Pediatr.* 80:57--61

28 Lansdown RG, Shepherd J, Clayton BE, Delves HT, Graham PJ, Turner WC. 1974Blood-lead levels, behaviour, and intelligence. A population study. *Lancet.* 1:538-41.

29. Needleman HL, Gunnoe C, Leviton A, et al. 1979. Deficits in psychological and classroom performance of children with elevated dentine lead levels. *N. Engl. J. Med.* 300:689--95

30. Fulton M, Raab G, Thomson G, et al. 1987. Influence of blood lead on the ability and attainment of children in Edinburgh. *Lancet* 1:1221--26

31. Lyngbye T, Hansen ON, Trillingsgaard A, et al. 1990. Learning disabilities in children: significance of low level lead exposure and confounding factors. *Acta Pedriatr. Scand.* 79:352-60.

32. Fergusson DM, Horwood J, Lynskey MT. 1997. Early dentine lead levels and educational outcomes at 18 years. *J. Child Psychol. Psychiatry* 38:471--78

33. McMichael AJ, Baghurst PA, Wigg NR, et al. 1988. Port Pirie Cohort Study: environmental exposure to lead and children's abilities at the age of four years. *N. Engl. J. Med.* 319:468--75

34. Needleman HL, Gatsonis C. 1990. Low level lead exposure and the IQ of children. *J. Am. Med. Assoc.* 263(5)673--78

35. Schwartz J. 1993. Beyond LOEL's, p values and vote counting: methods for looking at the shapes and strengths of associations. *Neurotoxicology* 237--46

36. Pocock SJ, Smith M ,Baghurst P. 1994. Environmental lead and children's intelligence: a systematic review of the epidemiological evidence. *Br. Med. J.* 309:11898--97

37. Lanphear BP, Dietrich K, Auinger P, Cox C. 2000. Cognitive deficits associated with blood lead concentrations <10 microg/dL in US children and adolescents. *Public Health Rep.* 115(6)521--29

38. Canfield RL, Henderson CR, Cory-Slechta DA, et al. 2003. Intellectual impairments in children with blood lead concentrations below 10 μg per deciliter. *N. Engl. J. Med.* 2003. 348: 1517-26.

39. Needleman HL, Schell A, Bellinger D, et al. 1990. The long-term effects of exposure to low doses of lead in childhood. An 11-year follow-up report. *N. Engl. J. Med.* 11:83--88

40. Fergusson DM, Horwood J, Lynskey MT. 1997. Early dentine lead levels and educational outcomes at 18 years. *J. Child Psychol. Psychiatry* 38:471--78

41Bellinger D, Leviton A, Waternaux C, et al. 1987. Longitudinal analyses of prenatal and postnatal lead exposure and early cognitive development. *N. Engl. J. Med.* 316:1037--43

42. Dietrich KN, Krafft KM, Bornschein RL, et al. 1987. Low-level fetal lead exposure effect on neurobehavioral development in early infancy. *Pediatrics* 80:721--30

43. Needleman HL, Riess JA, Tobin MJ, et al. 1996. Bone lead levels and delinquent behavior. *J. Am. Med. Assoc.* 275:363--69

44. Dietrich KN, Ris MD, Succop PA, et al. 2001. Early exposure to lead and juvenile delinquency. *Neurotoxicol. Teratol.* 23:511--18

45. Needleman H, McFarland CE, Ness R, et al. 2002. Bone lead levels in adjudicated delinquency: a case-control study. *Neurotoxicol. Teratol.* 24:711--17

46. Stretesky PB, Lynch MJ. 2001. The relationship between lead exposure and homicide. *Pediatr. Adolesc. Med.* 155:579--82

47. Nevin R. 2000. How lead exposure relates to temporal changes in IQ, violent crime, and unwed pregnancy. *Environ. Res.* 83:1--22

48. Silbergeld EK, Schwartz J, Mahaffey K. 1988. Lead and osteoporosis: mobilization of lead from bone in postmenopausal women. *Environ. Res.* 47:79--94

49. Muldoon SB, Cauley JA, Kuller LH, et al. 1996. Effects of blood lead levels on cognitive function of older women. *Neuroepidemiology* 15:62--72

50. Schwartz BS, Stewart WF, Bolla KI, et al. 2000. Past adult lead exposure is associated with longitudinal decline in cognitive function. *Neurology* 8:1144--50

51. Stewart WF, Schwartz BS, Simon D, et al. 2002. APOE genotype, past adult lead exposure, and neurobehavioral function. *Environ. Health Perspect.* 110:501--5

24. Tavakili-Nezhad M, Barron AJ, Pitts DK. 2001. Postnatal inorganic lead exposure decreases the number of spontaniously acive midbrain dopamine neurons in the rat. *Neurotoxicology* 22:259--69

53 Adhikari N, Sinha N, Narayan R, et al. 2001. Lead induces cell death in testes of young rats. *J. Appl. Toxicol.* 21:275--77

54. Iavocoli I, Carelli G, Sgambato A, et al. 2001. Lead inhibits growth and induces apoptosis in normal rat fibroblasts. *Altern. Lab. Anim.* 29:461--69

55. Shabani A, Rabbani A. 2000. Lead nitrate induced apoptosis in alveolar macrophages from rat lung. *Toxicology* 149:109--14

56. He L, Poblenz AT, Medrano CJ, et al. 2000. Lead and calcium produce photoreceptor cell apoptosis by opening the mitochondria permeability pore. *J. Biol. Chem.* 275:12175—84

57. Fox DA, Campbell ML, Blocker YS. 1997. Functional alterations and apoptotic cell death in the retina following developmental or adult lead exposure. *Neurotoxicology* 18:645—64

58. Lustberg M, Silbergeld. 2002. Blood lead levels and mortality. *Arch. Intern. Med.* 25:2443--

59. Rogan WJ, Dietrich KN, Ware JH, et al. 2001. Treatment of Lead-Exposed Children Trial Group. The effect of chelation therapy with succimer on neuropsychological development in children exposed to lead. *N. Engl. J. Med.* 344:1421--26

60. US Dep. Health Hum. Serv., Public Health Serv./ Cent. Dis. Control. 1991. Strategic Plan for the Elimination of Childhood Lead Poisoning. Cent. Dis. Control

61. Grosse SD, Matte TD, Schwartz J, et al. 2002. Economic gains resulting from the reduction in children's exposure to lead in the United States. *Environ. Health Perspect.* 110:563--70

62. Landrigan PJ, Schechter CB, Lipton JM, et al. 2002. Environmental pollutants and disease in American children: estimates of morbidity, mortality, and costs for lead poisoning, asthma, cancer, and developmental disabilities. *Environ. Health Perspect.* 110:721—28

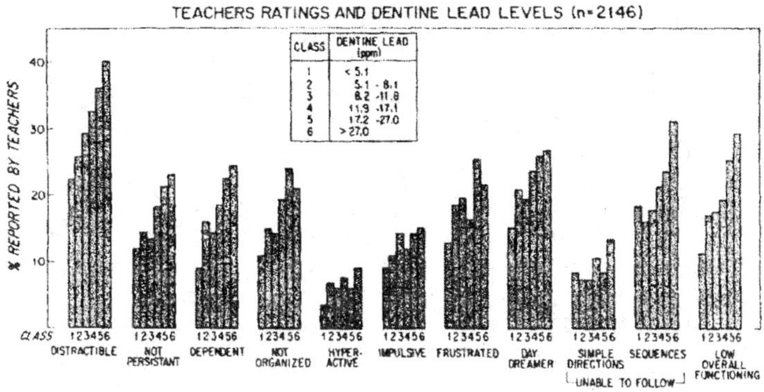
TEACHERS RATINGS AND DENTINE LEAD LEVELS (n=2146)
CLASS
DENTINE LEAD (ppm)
1 < 5.1
2 5.1 - 8.1
3 8.2 - 11.8
4 11.9 - 17.1
5 17.2 - 27.0
6 > 27.0
% REPORTED BY TEACHERS
CLASS 123456
DISTRACTIBLE
NOT PERSISTANT
DEPENDENT
NOT ORGANIZED
HYPER-ACTIVE
IMPULSIVE
FRUSTRATED
DAY DREAMER
SIMPLE DIRECTIONS
SEQUENCES
UNABLE TO FOLLOW
LOW OVERALL FUNCTIONING

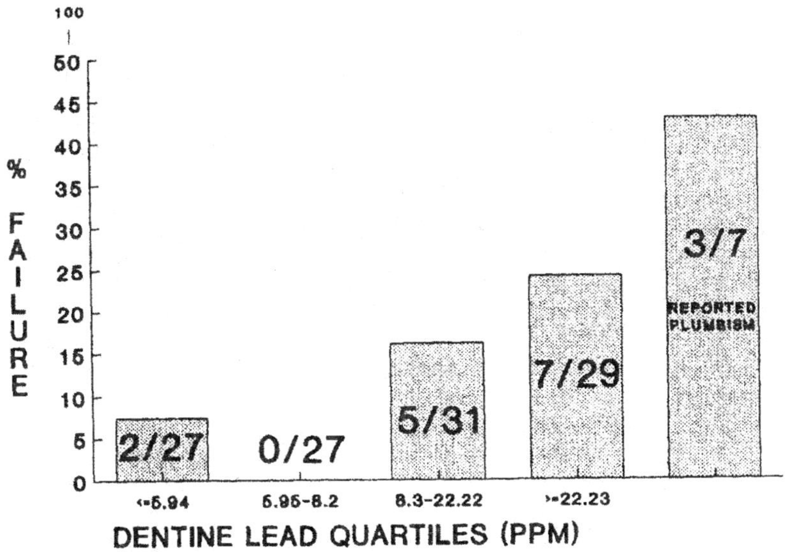
% FAILURE
2/27
0/27
5/31
7/29
3/7
REPORTED PLUMBISM
<=5.94
5.95-8.2
8.3-22.22
>=22.23
DENTINE LEAD QUARTILES (PPM)

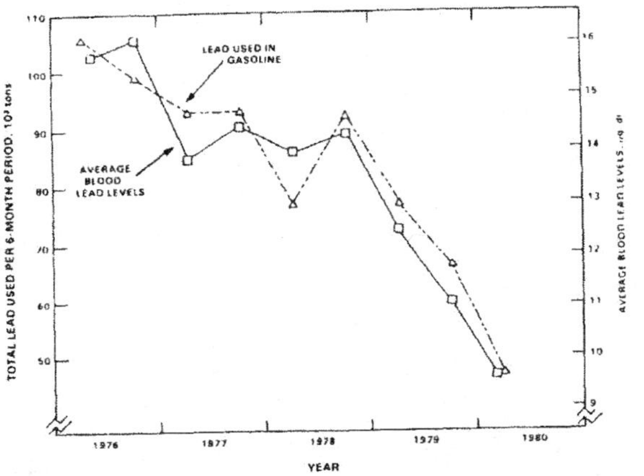
LEAD USED IN GASOLINE
AVERAGE BLOOD LEAD LEVELS
TOTAL LEAD USED PER 6-MONTH PERIOD, 10³ tons
AVERAGE BLOOD LEAD LEVELS
1976
1977
1978
1979
1980
YEAR

Senator JEFFORDS. The CDC website also states that no clinical inventions can reduce blood lead levels below 10 and that there is no evidence of a threshold below which adverse health effects are not experienced. Then the CDC draws a conclusion that any effort to reduce the standard would provide uncertain benefits, even though there appears to be ample evidence that no levels of lead in the blood is safe. This website appears to completely ignore the fact that blood lead poisoning is preventable and that in places like Washington, DC, the standard is used as a determining factor for which families receive assistance from the DC Department of Health.

The CDC identifies two focus areas to combat lead poisoning, lead paint and lead end products. There is no mention of lead drinking water. Given the apparent consensus that there is no level of lead exposure that is safe, this situation is not understandable. I can't understand it. The EPA's own website indicates that the agency estimates that 20 percent of childhood lead poisoning is due to exposure through drinking water. The agency set the maximum containment level for lead in drinking water at zero because that was the only level where no adverse effects will be experienced. I ask unanimous consent that the relevant CDC and EPA websites be included in the record also.

Senator CRAPO. Without objection.

[The referenced document follows:]

Content

- Home
- About Lead
- About the Program
- CLPP Programs
- Publications
- Partnerships
- Resources
- Data & Statistics
- Software
- Links
- Advisory Committee (ACCLPP)

- NCEH Home
- NCEH en Español
- About NCEH
- Programs
- Publications
- NCEH Topics

Why not change the blood lead level of concern at this time?

Recent studies suggest that adverse health effects exist in children at blood lead levels less than10 µg/dL. In the past the Centers for Disease Control and Prevention (CDC) has lowered the level considered elevated in response to similar reports. However, at this time the reasons not to lower the level of concern are as follows:

- No effective clinical interventions are known to lower the blood lead levels for children with levels less than 10 µg/dL or to reduce the risk for adverse developmental effects.
- Children cannot be accurately classified as having blood lead levels above or below a value less than 10 µg/dL because of the inaccuracy inherent in laboratory testing.
- Finally, there is no evidence of a threshold below which adverse effects are not experienced. Thus, any decision to establish a new level of concern would be arbitrary and provide uncertain benefits.

These studies support making primary prevention of childhood lead poisoning a high priority for health, housing, and environmental agencies at the state, local, and federal levels.

What is CDC's approach?

- The first is to focus on lead paint in housing as the most important source of lead for young children. CDC recommends designing, implementing and evaluating primary prevention strategies that prevent childhood exposure to lead. The essential elements of primary prevention for childhood lead poisoning are as follows:

  The system to identify high-risk housing and to make these units lead safe is in place. After 10 years of widespread blood lead testing and data collection by CDC-supported state and local partners, the specific addresses of housing units where children have been repeatedly poisoned are known to local officials. Systematic reduction of lead sources, particularly in old, poorly maintained housing combined with periodic maintenance monitoring will prevent children from being exposed to lead in these units in the future. Good evidence exits that those communities with the largest percent of children with very high blood lead levels, are also the communities that have the largest percent of children whose blood lead levels are lower but still well above the national average. This evidence also indicates the importance of primary prevention and the need to target those communities where the risk for exposure to lead is highest. Primary prevention in these communities would be expected to benefit all children who live in the highest risk communities.

- The second is to restrict or eliminate nonessential uses of lead particularly in toys, eating and drinking utensils, cosmetics and traditional medicines whether manufactured in the United States or imported.

  In some areas of the United States as many as 35% of children identified with elevated lead levels are reported to have been exposed to items decorated with or made of lead. In most cases, the hazardous product is only identified after a child is lead poisoned. CDC, the Environmental Protection Agency and other federal agencies are working to better identify hazardous products before they are in use.

Back to top

- Air Pollution and Respiratory Health
- Asthma
- Division of Laboratory Sciences
- Emergency and Environmental Health Services
- Environmental Hazards and Health Effects
- Environmental Public Health Tracking
- Global Health Office
- Health Studies
- Mold
- Preventing Lead Poisoning in Young Children
- Vessel Sanitation - Sanitary Inspection of International Cruise Ships

NCEH Home | Programs | Publications | Contact Us | Privacy | About NCEH
CDC Home | CDC Search | Health Topics A-Z

This page last reviewed March 23, 2004

Childhood Lead Poisoning Prevention Branch
National Center for Environmental Health
Centers for Disease Control and Prevention

U.S. Environmental Protection Agency

Ground Water & Drinking Water

Recent Additions | Contact Us | Print Version Search: GO

EPA Home > Water > Ground Water & Drinking Water > Drinking Water Priority Rulemaking: Lead and Copper Rule - Minor Revisions

Drinking Water and Health Basics
Frequently Asked Questions
Local Drinking Water Information
Drinking Water Standards
List of Contaminants & MCLs
Regulations & Guidance
Public Drinking Water Systems
Source Water Protection
Underground Injection Control
Data & Databases
Drinking Water Academy
Safe Drinking Water Act
National Drinking Water Advisory Council
Water Infrastructure Security

Drinking Water For Kids!

Drinking Water Priority Rulemaking: Lead and Copper Rule - Minor Revisions

EPA has published minor revisions to the Lead and Copper Rule (LCR) that was promulgated June 7, 1991. The revisions address implementation problems and issues arising from legal challenges to the 1991 rule. They streamline and reduce monitoring and reporting burden.

Link to fact sheet on minor revisions.

Public Health Concerns
Lead poses a significant public health threat because it builds up in the body over many years and can cause damage to the brain, red blood cells and kidneys. The greatest risk is to young children and pregnant women. EPA estimates that approximately 20% of human exposure to lead is attributable to lead in drinking water. Copper is an essential nutrient, but at high doses it can cause stomach and intestinal distress, liver and kidney damage, and anemia. The goal of the rule is to provide maximum human health protection by reducing lead and copper levels at consumers' taps to as close to the MCLGs as is feasible. The minor revisions are necessary to promote effective implementation of the LCR.

Implementation Plans
The federal effective date is 90 days after the revisions are published in the *Federal Register*. Systems should check with their state primacy agencies because not all provisions may apply in their state. The existing Lead and Copper Rule is still enforceable. EPA will post guidance documents on this site as they become available.

Related Links

- Effect of pH, DIC, Orthophosphate and Sulfate on Drinking Water Cuprosolvency [EPA600/R-95/085, June 1995] Field data from various copper monitoring studies and Lead and Copper Rule compliance data are often inappropriate and misleading for reliably determining fundamental chemical relationships behind copper corrosion control. To address this deficiency, a comprehensive solubility model for copper in drinking water has been developed, that is qualitatively and quantitatively consistent with available data for copper dissolution and precipitation.

- EPA requests comment on the Notice of Data Availability on the minor revisions to the Lead and Copper Rule that were proposed in 1996

- The law prohibits any person from introducing into commerce any pipe, or plumbing fitting or fixture that is not lead free after August 6, 1998, except for a pipe that is used in manufacturing or industrial processing. For details, read Section 1417 of the Safe Drinking Water Act and the NSF Standard: Commonly Asked Questions.

- To learn about health threats from exposure to lead in drinking water and what you can do to prevent them, read Actions You Can Take To Reduce Lead In Drinking Water.

- Back to Drinking Water Standards Program

For more information contact EPA's Safe Drinking Water Hotline, 1 (800) 426-4791.

Senator JEFFORDS. Finally, what has EPA done to coordinate with CDC on its elevation of the blood lead level standard to ensure that it is reflective of the science consensus on this issue, on the CDC's plan to combat lead poisoning and on the Healthy People 2010 goal of eliminating lead poisoning by 2010? Do you believe the lead exposure in children can be eliminated in this country without addressing lead in the drinking water problems and the plumbing factors?

Mr. GRUMBLES. Senator, I can appreciate your interest and your leadership on this issue in terms of reducing exposure to lead in drinking water, particularly to children. I would say on the first part of the question about the relationship with CDC, we have been working with CDC, we certainly have established very close working relationships with them in the context of responding to this situation here in the District.

With respect to the overall question of the No. 10 in measuring blood lead levels and the safety of that, it is one of the items that we intend to discuss and explore and review the scientific literature you have as we go about looking at what is the existing guidance and what is the existing regulation because we do share your interest in the most important bottom line and that is, what is the impact on human health. How much lead in the drinking water is too much? We welcome that and look forward to working with you on that.

Senator JEFFORDS. I look forward to working with you and communicating with you.

Mr. WELSH. May I mention something we are doing locally, not on the national scale but in Philadelphia in Region III we piloted a program called Lead Safe Babies where we worked with the National Nursing Centers Consortium to actually have health professionals who go out into homes to talk to mothers and families about the possible sources of exposure to lead that exists in their homes. We thought that was very successful in Philadelphia and we are expanding that funding and it will be called Lead Safe DC, to work with the DC Department of Health not to supplant but to augment efforts they are doing to get better information actually into the hands of people in their homes about what their exposures to lead might be so that we can take a look at the holistic issue of lead exposure, paint, dust, soil, as well as drinking water. We felt that was very successful in Philadelphia and are happy to expand that to DC to try to help improve the public's knowledge about those risks.

Senator CRAPO. Senator Clinton.

Senator CLINTON. Mr. Johnson, I want to make sure I understood your comment at the end of the previous question from the Chairman. This is, in your view, primarily a single family home problem, not a multiple unit problem?

Mr. JOHNSON. That is correct, Senator.

Senator CLINTON. One thing I think might be worth considering on the part of the DC Government, since there is some confusion about the number of service lines that are of unknown composition and I think the historical record problem is such that we may never get to the bottom of that, it might be worth considering some action that would provide filters for families in these categories of uncertainty or even go to the extent of permitting a family to purchase this filter and attach the receipt to their property tax return and deduct it from their costs. I think something needs to be done to take action more quickly and to restore confidence.

The confusion that now exists over how many lead service lines there are and how many unknown composition lines there are and whether people are going to get tested and after they get tested, how soon they can get a filter, there might be a way to short circuit that and go ahead and absorb the cost now. We can argue later about whether the EPA and the Federal Government should help to reimburse since we have primary responsibility, but at least move to get as many filters installed as quickly as possible.

I live in the District, my house was built in 1950, so I am right on the cusp and I have had an ongoing dispute about whether we do or do not have lead service lines but we have discovered that we have lead lines in the house. So even if I have an unknown composition line coming from the street, once it gets into the house, I have lead lines. So I think there are perhaps several ways you could short circuit this and also restore some confidence and provide some reassurance.

Finally, with respect to the overall review that I understand the EPA may be engaged in, Mr. Grumbles and Mr. Welsh, I would appreciate your looking at the level of testing that is required. In New York City it falls into a system serving more than 100,000 households and we have 8 million people living in New York City. In order to meet that standard, we test 100 households and that is sufficient but, there are more than 100 neighborhoods. Some are single family neighborhoods, some are duplex neighborhoods, some are apartment neighborhoods, some are public housing neighborhoods. We have as many different kinds of housing as you can find anywhere in the country.

As you are going through this review, I want you to look at the level of testing that is adequate for large systems, particularly a system as large as New York and in light of the problem we have had in DC, you think the regulations which allow systems to test every 3 years for lead are adequately protective. We only test 100 sites every 3 years. Given what we are finding out, given the fact we are changing the makeup of the chemicals put into the water, we need to be checking on this more frequently. I appreciate Mr. Jacobus' comment that we need to make it more transparent so that people know what is going on so we can immediately bring people together to respond.

This is a matter of such great concern to all of us, the quality of our drinking water, I share Senator Jeffords' concern about bottled water. Based on my review of the FDA, there is not very much testing and not very much regulation. A lot of people are spending literally hundreds of dollars a year to buy bottled water which may or may not be better or safer than what comes out of their tap.

We need to find out. People need information to make their own judgments. We need some kind of transparency that provides us information so that a concerned mother, a concerned pregnant woman, a concerned anybody can go somewhere, a website, a hotline and find out what he or she needs to find out.

Finally, Mr. Chairman, unfortunately I can't stay for the next panel buy I want to thank Gloria Borland and Jody Lanard and Dr. Best. I read their testimony. I highly recommend that WASA and everyone else read Dr. Lanard's testimony about risk communication. In fact, I think everybody in any position of public responsibility should read his testimony. It provides a very useful guideline about how to deal with crisis situation, some of which is counterintuitive. The immediate response is, we don't want to panic people, we want to give people reassurance. I ran into this with the World Trade Center. We got into a big problem because EPA and the White House didn't want to tell people that we had more particulates and problems in the air around Ground Zero than they wanted us to know, so we didn't have the urgency that would have convinced a lot of the workers and volunteers at Ground Zero to wear that protective equipment.

I think given the high level of risk that we live with in the world today, risk communication is central and it is counterintuitive because I think the natural human feeling when you are in a position of responsibility and you take it seriously is I don't want people to panic. Given the world today, it may be better to err on the side of more rather than less information. We have a lot of very mature, very active people who then can draw their own judgments.

One of the things I face all the time at Ground Zero now is construction workers, firefighters and others who say if they had told me, I would have left the mask on. Well, they were kind of told it was dangerous, they could smell it was dangerous but the EPA was saying it was OK. This is the conflict and I appreciate the conflict, but I think we need to get to a point where we level with people, we provide information transparently and I think we have to look at the laws and regulations because if we can't provide safe drinking water in our Nation's capital, that is a terrible indictment of all of us, particularly given the fact that the Federal Government oversees DC for better or worse, it is our responsibility.

Thank you so much for holding this hearing, Mr. Chairman.

Senator CRAPO. Thank you, Senator.

I would like to ask a couple more questions. Mr. Welsh, the first is for you. My question is, based on the work of the Technical Expert Working Group, how would you describe the status of the options we have for long term solutions at this point?

Mr. WELSH. The Technical Working Group is working hard to identify what actually caused the change in corrosion and to run through what the possible fixes are to that. In fact, we recently received an update of the Technical Working Group. They did a desktop analysis and got a contractor's report back just last week with a recommended treatment option. The full Technical Expert Working Group is going to be considering that report and by April 15, they will be recommending a treatment option to EPA.

When we receive that recommendation, we would like to take the opportunity to talk to the public and review that information with the public and get their input but we are trying to move as quickly as we can. It is in everyone's interest to bring about a solution as quickly as possible. So we are encouraged by the work the Technical Group has been able to do. They have forwarded this report with a recommended treatment option and we hope by May 1, we will be able to approve the selected treatment option.

The schedule we are on would call for that treatment option, once it is identified, to be tested in a limited part of the system beginning June 1. Our schedule was for full system implementation if all of the studies that are ongoing at the same time that we do that confirm that it is going to be effective and the correct solution, to go systemwide with that by September 1.

In addition to the Technical Expert Working Group, we have also had an independent peer review group of experts look at the same work and based on some of their comments, we are hopeful we may be able to accelerate the date for that full system implementation. So we are encouraged that we are moving toward the correct answer, but no final determination has been made yet as to what the correct treatment option will be, but we are getting close to that day.

Senator CRAPO. Thank you and we will look with great interest on that report as well to see what direction we should take.

I have two more quick questions. One, Dr. Lucey, is for you. I am looking at the summarized points you gave. In your summaries in the first couple of points, it indicates to me that a number of children in homes with identified elevated lead levels in their water were tested but a very low percentage of them, according to these tests, had elevated blood lead levels. Is that correct?

Dr. LUCEY. Yes, sir, that's correct.

Senator CRAPO. Then my question is I think the obvious one. Why? Wouldn't we have expected higher blood lead levels in these children?

Dr. LUCEY. Because of the lead in the water?

Senator CRAPO. Because of the lead in the water?

Dr. LUCEY. I think that is the essential question from a public health point of view. That is really why the Department of Health sent the letter on February 26 to the 23,000 homes that I signed with an abundance of caution that we should offer blood level testing because this has never been done before in the United States, it is widespread blood testing to assess the health impact of lead in the water, increased amounts of lead in the water. It is not in the textbooks. There is no clear correlation between a concentration of lead in the water and the concentration of lead in the blood.

Senator CRAPO. Is that because of a question of whether it is, I have heard the word bioavailable in the body when it is consumed by water?

Dr. LUCEY. That is part of it. It also depends on the age as Senator Clinton mentioned in terms of a child absorbing and retaining more lead than an adult. That is why children are most at risk, particularly young children with developing nervous systems where lead act to the nervous system like calcium and that is bad for the nervous system of the child.

Senator CRAPO. So we don't actually have the data foundation or whatever to make scientific conclusions at this point?

Dr. LUCEY. Scientific answers through a series of test hypotheses, get results and those studies have to be confirmed or refuted. So that is what I think we are contributing here in the District both for the District and for the country if other cities find elevated lead concentrations in the water. That is why I think it is very good the EPA is involved with the national perspective and also the Centers for Disease Control as I mentioned has been very involved with us onsite here today and multiple other times. The lead expert, Dr. Mary Jean Brown, has been up to visit with us. Dr. Gerberding, the Director of the CDC, has been very supportive.

I should say though that as I have tried to emphasize in the multiple press conferences we have had and multiple community meetings that Mr. Johnson mentioned, the way I look at this is in terms of the scientific data, these are pieces in the puzzle. One piece of the puzzle is the homes that have the highest levels of lead, more than 300 ppb. We went to those homes because we were afraid folks in those homes weren't coming to us to get their blood tested, so we didn't have the results. We found that none of 201 people had elevated blood levels.

The other very high risk group is the young kids, young children who live in day care facilities, who have lead service pipes. So we went to those facilities, provided the filter, the consent form for the parents to sign. If they signed it, we came back within a few days or a week to draw blood. Two of the children out of 280, less than 1 percent, have had elevated blood lead levels. These are two parts of the picture.

Another part of the picture is the approximately 4,500 people who have come to see us but there is another important part of the puzzle. In addition to the study that was published in the Centers for Disease Control Morbidity/Mortality Report on March 30, and that is Children's National Medical Center here in the District measures blood lead levels. They are currently in the process of doing a comparison looking at blood lead levels over the last 10 years. It is independent of the Department of Health but they have invited us to participate in the planning and we did on March 17. That is another part of the analysis that I think Dr. Best will refer to. The results aren't back yet.

I am trying to emphasize it is important to put together all the parts of the puzzle to make sure they are all consistent.

Senator CRAPO. You are working to expand the parts of the puzzle that we have to look at?

Dr. LUCEY. Yes, sir.

Senator CRAPO. The last question I have is for you, Mr. Jacobus, and that is, in addition to the changes in procedure that you described, I am aware that residents are now receiving notice of a flushing program that involves changes at the aqueduct. Could you explain how that all fits into this issue?

Mr. JACOBUS. Yes, sir. To some extent, it does not fit at all but in another way it does. Let me be specific.

The distribution system is made up of pipes of all different sizes, there are cast iron pipes, there are steel pipes, there are concrete pipes. The pipes are not sterile, there can be sediment in them from lime that settles out, the pipes can have rust on the inside of them, oxidation from that, so it is good management practice every spring to open fire hydrants in the distribution system and push water through the lines and essentially clean and blow out the lines. That removes the debris and sort of cleans up the inside of the pipe.

If you were to go inside a pipe and run your hand along it, you might feel what we would call biofilm. That could be a harboring location for bacterial growth in the pipe. The disinfectant in the pipe in the water, the chlorine-based disinfectant that is put into the distribution system to protect the water all the way to the tap, and the water can stay in the distribution system 3 or 4 days before it is consumed, so it is important to have that disinfectant residual there, the bacteria that might be out there, and we are measuring for those all the time, but the bacteria we might find out there if you are using a chloramine-based disinfectant as your secondary disinfectant, it is standard and common industry practice in the springtime after the roads would not freeze, to switch the disinfectant for a few weeks back to free corine rather than the corine ammonia complex. That kind of confuses the bacteria, shocks the system and that in conjunction with the flushing process cleans the system and gives good distribution system maintenance for the rest of the year, especially for the summer months when bacterial growth could be prompted by the temperature. That would be a normal practice.

Since we changed to chloramine, we have been doing that. We did that in 2002 and 2003, are doing it again in 2004. I say we, I mean we in conjunction with all of our wholesale customer partners.

When we go to a phosphate-based inhibitor as an additional chemical to go after the lead leaching problems in the lead service pipes, we know from other cities' experience that when you change the chemistry on the distribution system, you might get some reaction inside the pipes. With the pipes being as clean and as blown out if you will as possible, that will help make the corrosion inhibitor not have any secondary effects like creating red water which would be rusty effects.

So it is the chloramine disinfectant change and back and forth and the flushing as a standard practice but it blends very nicely into what we are about to do in June and then the full system in September.

Senator CRAPO. Thank you.

Senator Jeffords.

Senator JEFFORDS. Dr. Lucey, using 10 mpd as a standard, your testimony provides several data points indicating relatively small numbers of children had what you define as "elevated" blood lead levels. How do you define "elevated" and how did you select that number?

Dr. LUCEY. As you noted, the Centers for Disease Control has used that number of 10 mpd for children under the age of 6 but also children from the ages of 6 to 15 and also for women who are pregnant and nursing because the primary concern is with the unborn baby or the newborn baby. The CDC refers to that level as a level of concern or the definition of an elevated blood level. For an adult, it is a higher value, 25 mpd. So it is really a national guideline.

Senator JEFFORDS. I have already mentioned the April 17, 2003 study which shows harmful effect from blood levels well below 10. Based on that information, do you feel it is appropriate to ignore children and families with blood level test results above zero but below 10?

Dr. LUCEY. I am aware of that paper. Dr. Lamphere is in Cincinnati and was the senior author. He is a well-respected person in the research community with regard to lead. I think that was a very important paper last April in the New England Journal of Medicine. I think it needs to be corroborated like anything in medicine. Its findings need to be duplicated to show they are reproducible but I know that is a very important finding.

At this time, I have been in contact regularly with the Centers for Disease Control with regard to whether that level is going to be changed—the 10 mpd—or not anytime soon. To the best of my knowledge it is not at this time. I think, as in most things in medicine, there is a state of knowledge today and then there is what the state of knowledge or the standard practice might be in the future. Sometimes it is clear which way things are going and sometimes it is not so clear.

If I could mention briefly, perhaps to illustrate even better your point, I mentioned in point No. 5 that in the United States 1976–1980, in Attachment 5, 88.2 percent, 9 out of 10 children in the United States who were then 1–5 years of age, now 24–28 years of age, had blood lead levels of 10 mpd or higher. How could that be acceptable? At that time the level of concern was much higher, 40 mpd. I think that provides some historical context. I am not saying what happened then was good or not compared to now but it is some historical context.

Senator JEFFORDS. Mr. Johnson, I have a couple questions for you.

In placing lead service lines in WASA and moving the location of the meter, and if so, how is it affecting the link to the service line that you are replacing and the associated costs?

Mr. JOHNSON. If I understand your question, you are asking if we are removing the service line—what is the differential in the cost of moving the service line for the meter versus another location?

Senator JEFFORDS. Yes.

Mr. JOHNSON. Currently the cost or the EPA requirement has us to remove the service line that we have control or responsibility for. The District of Columbia law defines that as being that portion of the service line in public space. Frequently the terminus of where public space might exist where the meter is may be a very different point. So the Health Department expressed some concern about cutting the line at that point and then adding the copper line to it and suggested that we go directly to the first joint which would be the meter. We agreed and think that is a much better practice because you don't get particulate lead in that system.

The cost of doing a service line in public space, we have estimated at an average of about somewhere between $10,000–$14,000 per service line. That is because we have to do substantial rebuilding of the street and the roadway system when we do the excavation to meet the District's standards. The total cost we have estimated for doing all the service lines is about $350 million in public space and we believe and estimate for the cost of going on the other side of that meter and getting the portion that is in the private space which would normally be the responsibility of the homeowner to average about $2,000–$3,000. Because we think the preference again is to go to that threaded joint which would frequently be inside the foundation wall or in the building itself, we think that cost is probably going to average around $3,500.

Senator JEFFORDS. Recently WASA undertook a program to replace aging water meters. Can you describe the program for me? Indicate if you installed lead-free water meters and indicate if WASA uses lead-free parts when replacing parts through its system.

Mr. JOHNSON. That is somewhat of a technical question, Senator but I will attempt to answer it on the basis of what I understand the case to be.

The water meters we installed are considered by EPA and are characterized as lead-free meters. As I understand it, most metal components of things in this universe have some small portion of lead in them. You will find some of your metal faucets and brass parts and the like will have some metal in them but this classification of lead-free takes it down to a level where if water is moving through it, typically it does not leach and doesn't create a problem. So we replaced all the meters in the system.

Meters have not changed a great deal over time, so the technology is basically the same. We added a piece of technology to it so that we could read the meters remotely using a cell phone technology so we can read the meters without having someone to physically go on the street every day.

The question has been raised in removing those meters and changing them, why didn't you look down in the hole and see whether you had a lead service line. The answer to that is the meters are mounted on something called a riser. That is a device that comes from the service line, coming from the main as well as the line coming from the house and literally is a riser the meter is mounted in, so you don't really have a chance to expose the full line.

In response to one of your earlier questions, we are doing a series of dig ups in cases where we have undetermined service lines and trying to explore those and see what is actually there when we don't have good information.

Senator JEFFORDS. I would join the Chairman's request that you grant us the privilege of having you sitting here while we have the next panel.

Mr. JOHNSON. I will certainly do that, sir.

Senator CRAPO. With that, we would like to excuse this panel. We want to thank you all for your attendance today and for your attention to this issue. There very well may be questions from Senators who were not able to get here or further questions from those of us who were here. We would ask you to be very responsive to us if we forward those questions to you in writing.

Thank you very much. This panel is excused and we will call our second panel.

Our second panel consists of Gloria Borland, who is a member of the Dupont Circle Parents; Jody Lanard who is a Risk Communication consultant; and Dana Best, director, Smoke Free Homes Project, medical director, Healthy Generations Program, and assistant professor, George Washington University School of Medicine and Health Sciences with the Children's National Medical Center.

Senator JEFFORDS. If I can take a moment to give Dr. Dana Best from the Children's National Medical Center a special welcome. During my time as Chairman of the Health, Education, Labor and Pensions Committee, Dr. Best worked for me as a Fellow. It is nice to see you here again. She did her residency there and is from my home State. I am so pleased to have you here now.

Dr. BEST. Thank you very much. I am thrilled to be here. To bad it is about such a nasty topic.

Senator CRAPO. Thank you. Ladies, we appreciate your coming and participating with us in our second panel. Were all of you here when I gave my strong lecture about paying attention to the clock. We do appreciate your doing that because it does give us the opportunity with the limited time we have to engage in some dialog. Again, I encourage you to try to pay attention to that clock and stick to the 5 minutes to summarize your written testimony. Your written testimony is a part of the full record. We have already read it and I believe every Senator will read it before the week is out.

Ms. Borland, are you ready to start?

STATEMENT OF GLORIA BORLAND, DUPONT CIRCLE PARENTS

Ms. BORLAND. I want to thank all the Senators on the committee and also all the parents that are here today. I am here representing Dupont Circle Parents. I am a mom to a 22 month-old girl who has lead in her blood at twice the national average.

We parents are angry and full of anxiety because we don't know what the long term implications of lead poisoning in our water will have on our children, brain damage, lower IQ, behavioral problems, and I will defer to the experts who will be speaking next on that topic.

There are three points I want to make but in the interest of time, I would like to start with the third point first. The problems with our water here in DC are so huge, the cost to solve the problem is very expensive, the organization managing it right now is so dysfunctional that the only answer is to put WASA under Federal leadership in our opinion. Only under Federal Government control will you be able to restore the trust we parents need in our water system.

Drastic action? Yes, but look at the victims. Look at all the young children here in the audience and out in the hallway. See our babies, see their faces and that is why we are asking for Federal receivership to make sure they are safe.

I want to give you a couple of examples of communication. First of all, communication and trust must go hand in hand. I am sure if the Washington Post's David Nakamura had not exposed this scandal, our young children today on April 7 would still be drinking leaded water and WASA would still be hiding this crisis from us.

Communicating is not rocket science. It is the easiest and simplest thing to do when there is leadership and an organization willing to do it. The problem is not in the process of communication, the problem lies in deception. They deceived us. They tried to hide extraordinarily high levels of lead poison in our water supply thus putting our young children at risk. The EPA and the Army Corps of Engineers went along with this deception in violation of their Federal oversight responsibilities.

How was this crisis communicated to us? For most DC parents of young children, our day of infamy was Saturday, January 31, 2004 when we read the headline of the *Washington Post* that morning and we were hit with the shocking bomb that our infants, toddlers and young children had been secretly poisoned by the lead in the drinking water in our homes.

Lead in young children lowers their IQs and the experts can go into all that. I was pregnant in 2001 and 2002 when the high lead levels first became noticed by WASA. The right and legal thing for WASA to do was to issue an emergency warning to the public and to obstetricians and pediatricians to warn their patients not to drink tap water. It does not matter that WASA hadn't identified the source of the problem; that kind of research could take months, even years. In the interest of public safety, you issue the warning to the public to take the precaution first and take the time and the money and the resources to figure out the cause.

It is just like when a fireman sees a house burning, their first priority is to save lives. Put the fire out, then they figure out what caused the blaze. WASA got it backward. Because they got it backward, they wanted to figure out the cause of the lead crisis first and that was putting our babies' lives at risk. All WASA had to do was warn me and other mothers, don't drink your water without a filter or buy bottled water. All they had to do was warn me. When I learned about the lead in the water, I wanted to cry. I had been so careful. I even gave up coffee, for God's sake and I hear about lead in our water.

My daughter attends a very good pre-school day care center in Dupont Circle. There are 77 students and the building was built in 1989 so it is a modern building. For their national accreditation, there was no problem when the water was tested. Lat year, WASA dug up the streets and it caused $2,000 worth of damage in the center but more importantly, when this crisis broke, the executive director tested the water and in the building some of the fountains had 3,100 ppb and 5,900 ppb.

When I saw the letter she sent to the parents, I thought it was a typo, thousands not like 15 ppb which is the threshold but thousands, almost 6,000 ppb in the drinking fountain and 77 students at one of the best day care centers in the city? As of today, no one from WASA has contacted the director of our day care center. She has not received a phone call or letter. All this talk about reaching out to day care centers is baloney. The day care center at their own expense immediately went to bottle water and you can see.

Most parents have been perplexed as to why an entity entrusted with public safety would lie and then cover up their lies. What is their motive? When I spoke over the weekend to my very wise friend, Joe Louis Ruffin III, father of a 3-year-old boy living in Chevy Chase, I asked, how could this happen. Joe said, "WASA wanted to protect their bond ratings. Communicating the truth would have brought their assets down".

So this is like Enron. Enron was only about money, this is about money, bond ratings, but the consequences here, the innocent victims are the lives of our next generation.

I want you, Senators, to see the victims—our young children. When bond ratings get in the way of public safety, look into the faces of our babies, see the photos on display, look around the room. When bond ratings get more important than our babies' brain development, their nervous system and IQ, we have a serious problems.

Here are some stories from parents all over the city. This is Paul McKay, co-founder of www.purewaterdc.com and his son. He is the one that launched the website.

This is Theresa Brown who lives in LeDroit Park. This is her daughter, London, who is 1 years old. Theresa told me on Friday, I feel completely and utterly betrayed. They have a responsibility to the citizens of the city, especially to those young babies and children who are completely defenseless. You cannot put a price on brain damage. How about if we deliberately caused brain impairment to their kids or grandkids and see how they liked it.

This is from Denise Senecal, a Dupont Circle mom. Can anyone at WASA assure me that my son will not suffer harmful effects from this exposure of lead?

This is from Desa Sealy Ruffin, wife of Joe who I mentioned earlier. Desa told me on Friday, I can only say I think the District, WASA and EPA have all broken a real fundamental covenant with the citizens in the District of Columbia and I no longer trust them to do anything. I think the three agencies conspired to keep us in the dark. They should be facing criminal charges.

This is from Valerie Jablow, a mother on Capitol Hill. She says her son likes to go to the libraries, the public pool, stores, restaurants in our neighborhood and she is not sure how the water will be when she goes out with her child in the neighborhood.

This is Parker. Parker is 16 months old. His father, Terrance sent me this on Sunday. He said, his son, Parker, was adopted and thus was bottle fed water and formula as a toddler. Early this year they learned that his first lead screening resulted in very high elevated levels of lead in his blood.

This is Ronnette Bristol who lives in northeast DC in an apartment building. She has four kids and says, "We are buying lots of bottled water until someone can come out and test our apartment building."

This is Lyubov Gurjeva originally from Russia. She told me, I never believed DC water was safe to drink. This is from someone from Russia.

This is Xin Chen and they were notified by WASA that they had lead service lines. She has an infant, a newborn, and a 3-year old. She says, "I don't trust them, I don't trust their test results. No trust at all with every parent I spoke to."

Same thing with Maria DePaul and her husband, Ethan, who live on the Hill, the same thing. Her husband said, "WASA will find legal loopholes so they do not have to help you out".

Many parents complained about the cost of buying expensive filters and bottled water. People don't mind buying bottled water in an emergency for a couple of days or couple of weeks, but when a couple of weeks turns into a couple of months with still no end in sight, people are beginning to feel the financial burden of buying bottled water for every day use.

You talk about how to restore trust. My feeling and that of other parents is that day of infamy is outrage. We expected our elected leaders in the District to quickly step in, fire the managers at WASA and respond with swift action to fix our water crisis. Instead, they didn't. I hope the EPA and Army Corps leaders remember the faces of our young babies when they carry out their daily oversight duties from now on.

The seat of the problem here is management culture instilled by Jerry Johnson, Mike Marcotte and board chairman, Glenn Gersten. To restore trust, do what is done in the private sector, remove managers for extremely poor judgment and failed performance and put in new management. These are the necessary first steps toward restoring trust with the parents here in the District of Columbia. New managers are now in place like Enron, Adelphia and Worldcom and Gersten is a Wall Street attorney so he understands why you need to clean house in order to restore public confidence in an institution.

We know our Mayor is asking the Federal Government for more money to solve this problem. I don't think it makes sense to put good money in the hands of bad managers. The problems and dysfunction at WASA are so huge, they require Federal intervention. We parents encourage the U.S. Senate to institute its powers to begin the process of putting WASA under Federal control, Federal receivership. A new management team and new board of directors needs to be put in place to work on solving this lead crisis. We need to make sure the Army Corps and the EPA are listening to our demands for better communication, a two-way dialog with the public they are entrusted to serve.

Yes, we need Federal help and Federal dollars to solve this crisis, but that goes hand in hand with new management and Federal control of the system until our water is deemed drinkable again.

Thank you.

Senator CRAPO. Thank you, Ms. Borland.

Before we go to Dr. Lanard, I want to correct an oversight of mine. Earlier I should have noted that we have with us Mr. Paul Strauss, who is the U.S. Shadow Senator for the District of Columbia who has also submitted testimony and was also with us yesterday as we met with local residents. Mr. Strauss, I apologize for my oversight at the beginning in acknowledging your presence here.

Dr. Lanard.

STATEMENT OF JODY LANARD, M.D., RISK COMMUNICATION CONSULTANT

Dr. LANARD. I will use part of my 5 minutes to make one comment about Ms. Borland's magnificent statement on behalf of the stakeholders. The desire to fire everybody and start anew is very understandable but I have seen in my work with other officials who have really screwed up communication that sometimes the reformed sinner who has learned the hard way becomes one of the best managers and officials I have ever seen. Their attention is focused on the issue they have screwed up more than anybody who is going to come in and start anew. So I hope maybe you will cut them a little slack and notice if they ever start to learn to do it the way you hope they will. I don't know whether they will or not but I am hopeful they could learn.

I am Jody Lanard, a risk communication specialist from Princeton, NJ. Thank you, Senator Crapo, for inviting me here today.

My written testimony includes a list of 25 communication strategies that underlie my critique of WASA's handling of this issue. If I run out of time, I invite you to ask me during the question period to give you some examples of really wonderful risk communication practices from other officials and other issues.

Some of the communication strategies that WASA should pay attention to are, and these are very counter intuitive, as Senator Clinton said, "Don't over reassure people", err on the alarming side (which in this case would have meant informing early, not waiting for a red flag but hoisting the yellow flag, giving people a heads up even before you know what is going on), acknowledging uncertainty. The general public, and even I when I am outside my own field, think other people know much more in their field than they actually do. The public thinks doctors know much more than they do. We all think officials know much more than they do, and officials collude with this by being so paternal sometimes and by over-reassuring us,

Go out of your way to acknowledge uncertainty and break the cycle of being so over reassuring and then having us be shocked when you tell us later, "We are learning new and interesting things every day."

I am going to cut to the chase and tell you the whole list of 25 is in my written testimony. The two other most important things are: No. 1, not to aim for zero fear. The public is much more resilient than you think. I am working on this with several different groups and trying to persuade them that panic is very rare. Anxiety happens, even a little bit of hysteria happens, but we are very resilient. Look at how the people in Washington reacted after the Pentagon was attacked. People were not panicking, people were bearing it. They felt panicky but they were not actually panicking. And, No. 2, in this case, most of all, if any of these officials want to be rehabilitated in the eyes of their citizens, they have to acknowledge all the errors, deficiencies, mistakes and misjudgments they made and they have to apologize for them a hundred times more than they think they need to. It is not for them to say when it is time to put this behind us. So they should wallow in their apologies until people get sick of hearing them. First they have to understand more about what they need to apologize for.

I am as upset as everybody else about WASA not notifying people early when those first 53 houses were found to have high lead levels, surprising high lead levels. Fifty percent of those houses had high lead levels, but in the brochure where WASA tried to tell everybody about this, or they think they tried to tell everybody about this, they say, "Some houses had high lead levels." They didn't say, "Fifty percent of the sample that year had high lead levels." So it came across as minimizing.

I am going to give you an analysis of this brochure for as long as I can get it in to explain how they worked really hard not to scare people, but unfortunately they were allying with peoples' apathy instead of trying to find a way to get their attention.

The brochure WASA put out in October 2002, which they cite to say they were not trying to hide the lead problem from the public, was entitled, "The District of Columbia Water and Sewer Authority and the District of Columbia Department of Health Acknowledge Lead Awareness Week and Its Impacts on Your Health." The purpose of the brochure is to acknowledge Lead Awareness Week. There is this awareness week, that awareness week, every week there is some awareness week. This did not look like a warning, even a very subtle warning.

The brochure has absolutely excellent educational content but it only weakly signals to the public that there are new reasons to take this issue seriously. The title makes it sound like the PR Department decided to use National Lead Awareness Week as a news peg for sending out good information about what to do about lead. It has a pretty picture of water on the cover, it has a smiling pregnant woman on the second page and it is not until you get to the third page that they say really low down, some homes in the community have lead levels above the EPA action levels. By the time a reader gets to that sentence, the context of the brochure suggests that "some homes" are very few and "above the EPA action level" is only a little above. The cheerful informative tone of the preceding pages in the context of the celebratory title of the brochure does not signal "Do something, this is a surprising change in our findings. Take this seriously." It is as if Paul Revere, and I will tell you this in risk communication terms, announced in celebration of National Freedom Awareness Week. "There is no need to panic, but some British are coming: Hey, meet me at the old North Bridge." I hope that is not too irreverent for a committee hearing.

Senator CRAPO. I think we can handle it.

Dr. LANARD. Good. You are resilient, I know that.

The two main things that WASA did wrong, one before the story broke, the other after the story broke. Before the story broke, other than not announcing the story themselves and helping the public get ready for this problem, they tried to use facts to attack apathy. Using facts alone is not a good way to attack apathy. I think I will go on to the next point because I think I have made it there.

I want to give you an example of what WASA could have said instead of this. I would like someone at WASA to say,

> "I am so sorry to tell you that we are finding a lot of unexpected high lead levels in water coming out of the taps in our 53 sample homes this year, 26 out of 53 is half. We don't know yet why this is happening; we don't know yet whether any people, especially children have increased blood levels because of this; we don't even know all the recommendations we want to make to you but because we feel you deserve to get this information quickly, we will give you some preliminary, precautionary recommendations. We will be learning things over the next weeks that we will wish we had known months ago. We may make mistakes or retract things we have already said."

This is called anticipatory guidance, warning people about what might happen.

> "We may make mistakes. New information is going to come in but we are committed to sharing this with you early. We know you will be worried. We share that worry and we will bear this together and get through it."

This would have expressed confidence in the public and that is a compliment the public might well have returned along with its appropriate anger at WASA and its anxiety. Telling the public you don't think they can handle bad news is insulting, it is patronizing and it generates mistrust. Now, I hope in the questions you will ask me for some really good examples.

Senator CRAPO. Thank you, Dr. Lanard.

Dr. Best.

STATEMENT OF DANA BEST, M.D., DIRECTOR, SMOKE FREE HOMES PROJECT; MEDICAL DIRECTOR, HEALTHY GENERATIONS PROGRAM; ASSISTANT PROFESSOR, GEORGE WASHINGTON UNIVERSITY SCHOOL OF MEDICINE AND HEALTH SCIENCES; AND CHILDREN'S NATIONAL MEDICAL CENTER

Dr. BEST. Thank you for providing this opportunity to discuss the harms of lead poisoning in children.

I am a pediatrician and preventive medicine physician with expertise in pediatric environmental health from Children's Hospital. The Children's system provides primary care for thousands of DC children, particularly those of low socioeconomic status.

The history of lead provides some interesting background for today's hearing. Lead's utility has been recognized for thousands of years. The dangers of exposure to lead have been recognized almost as long. Unfortunately, the two primary sources of lead in our environment, leaded gas and lead in paint, were not banned until decades after reports of harm from their use. While the impact of leaded gas has declined significantly since its banning, lead paint continues to be the primary source of lead poisoning today.

Because of the number of homes that still contain lead paint, discriminating between lead poisoning from lead paint and from lead contaminated water is difficult, particularly since many of the homes with lead pipes also have lead paint.

The focus on children and lead poisoning is because children are more likely to ingest lead than adults and because they are undergoing critical periods of development at the same time they are ingesting lead. Toddlers put everything in their mouths, including lead laden soil and paint chips. They live closer to the floor where the lead dust settles. They breathe faster, eat and drink more per body weight and absorb lead more efficiently than adults. Their rapid growth means their bones absorb calcium at a higher rate than adults and since lead is similar enough to calcium it can be stored in bone resulting in lifelong stores of lead in some cases. These stores can contribute to the lead poisoning of the next generation when pregnant and breast feeding women release lead into their blood stream during their pregnancy or the nursing period. Lead crosses the placenta and is readily incorporated into breast milk.

There are many effects of lead poisoning from death to subtle but significant changes in cognition and behavior. No study has determined a lead level below which an effect is not seen. Even at levels under 10 mcg/dl, the current action level in children, IQ scores have been shown to decline four to five points. That loss of four to five points can mean the difference between normal and subnormal intelligence and the ability to function independently. Other studies have demonstrated similar effects, some even under 5 mcg/dl.

The behavior changes associated with lead exposure include increased distractibility, decreased reaction time, poor organizational skills, Attention Deficit Hyperactivity Disorder and poor classroom performance. Unfortunately no treatment has been shown to reverse the lifelong effects of lead poisoning and the primary treatment for severe lead poisoning has been implicated in decreasing IQ itself. Many studies have shown persistent cognitive and behavioral effects long after lead levels have dropped to those considered low. Prevention, not treatment, is the only solution to lead poisoning.

In the District, lead poisoning is part of the larger picture of children's health risks. Many of the lead poisoned children are the same children living in poverty, exposed to tobacco smoke and without health insurance, all conditions which can add to the impact of lead poisoning. The recent CDC report of lead levels in DC residents indicated that the long term decline in children's lead levels halted in the year 2000 when chloramines were added to the water. This disturbing indication needs to be confirmed. At Children's we have begun an analysis of the last 10 years of lead tests performed in our laboratory. We will look at the average lead level, noting any changes, and look for associations between lead levels and household water supply, lead paint in the home, insurance status and other potential influences. This study is an extremely high priority and we will inform Congress and the District of Columbia of our results as soon as they are available.

There is no safe level of lead. Prevention is the only solution. There is no way to place a dollar value on the harm from lead poisoning of DC children, no matter the source, water, paint or otherwise. The harms of lead poisoning have been known for thousands of years, with many missed opportunities to remove lead from the environment in a timely fashion. We should not add the District of Columbia to this history.

All potential lead sources should be eliminated including reduction of lead concentrations in drinking water to below the EPA action level. This is the law. The children of the District of Columbia deserve this and nothing less.

Thank you.

Senator CRAPO. Thank you very much.

Ms. Borland, I would like to start with you. I was very interested in the example you gave of the day care center which if I understand you correctly, it was constructed in 1989 and in the initial tests of the water, there was no problem.

Ms. BORLAND. Right.

Senator CRAPO. And then there was some kind of construction in the streets?

Ms. BORLAND. Last year is what the director said. My daughter just started in June, so I don't know all the history but it was last year.

Senator CRAPO. There was some kind of construction. Presumably it had to do with the water delivery system and then very high levels of lead were found in the water. You may not know the answer to this but I was curious about that because of the earlier testimony, that we tend to think that the earlier dates when we stopped using lead pipes were safe for buildings. Do you know whether that level of high lead in the water has maintained and whether they have continued testing and it stayed high or did it spike?

Ms. BORLAND. They had a private firm test it about 2 or 3 days after the story broke. I don't know if they have tested it since but they immediately went to bottled water.

Senator CRAPO. I am sure we can have our authorities check that facility because obviously if that kind of dynamic can happen at a facility constructed in 1989, that indicates there may be a potential risk issue there that we are not paying attention to which gets back to some of the questions I was asking earlier to the first panel about whether anybody can ask for a test to be made and if that shows a high lead level in the water regardless whether the home fits into a category that would be considered at risk, then they can get the necessary response to address the issue.

You also indicate in your testimony that you feel the Federal regulations seem to be in place for communicating with the public but the regulations were simply not followed and the protections in place were ignored by WASA, correct?

Ms. BORLAND. I am not an expert on legislation and you can probably put in new legislation that you have to have a public press release immediately but if the managers don't want to follow that, it is the leadership integrity, if it is not there, it is not going to happen, the final loophole. It is our opinion that it is a management structure problem, not the process but a management problem of the people involved and lack of integrity.

Senator CRAPO. Thank you.

Dr. Lanard, you asked me to ask you for some examples. Do you want to give us a few?

Dr. LANARD. These are examples I love. I much prefer to teach by good examples than by bad examples. These are some imitatable examples from real live experts. Jerry Johnson and Glen Gerstell had been quoted as saying they believe in using facts to overcome fears to educate the public. I hope some of these examples will convince them to use even the scary facts and to go beyond the facts to help people bear their fears, because that is part of the job as Rudy Giuliani showed on September 11.

The first two examples are from State epidemiologist Jeff Engel in North Carolina. In June 2003, North Carolina had its only confirmed SARS patient, one of only eight confirmed SARS patient in the United States. Dr. Engel responded with a series of news conferences. At one, a reporter asked if all the news coverage had the potential to cause more hysteria and fear. The reporter asked this kind of hopefully; reporters like hysteria.

Dr. Engel replied,

> "We need to involve our community in all aspects of public health. Certainly a disease like SARS, so new, so frightening, should instill fear. Fear is an appropriate response, for me as a public health physician, for everyone in the community. We need to transfer that fear into positive energy and keep the facts in front of the hysteria. SARS is a new disease, it spreads person to person, it can kill. That is newsworthy."

I am trying to prove that is not an accidental kind of statement. That is a decision he made to make that kind of statement.

Two months later, he made essentially the same kind of statement about Eastern Equine Encephalitis of which there had been less than one case a year in North Carolina. He said,

> "The State has only documented 12 or 13 infections since 1964, the most in one year was 3 in 1989. Nevertheless, fear is appropriate. My God, here you have a mosquito that can kill. What we are trying to do through you guys, the media, is to use that fear in a positive way."

Dr. Engel told me 2 days later that the local Wal-Mart sold out of insect repellant right after that news conference. So he generated preparedness, not panic.

My final two stories are from Julie Gerberding, the master of the universe at risk communication. She did not start out as the master of the universe during anthrax. In fact, I am not sure she was allowed to say very much during anthrax. She learned risk communication by the time SARS came along. This is also to prove it is learnable. She was asked early on whether SARS could be bioterrorism. She answered, "while we have lots of reasons to think the SARS outbreaks are not due to terrorism, we are keeping an open mind and being vigilant." Many other officials were asked the exact same question and answered, "There is no evidence of a terrorist attack." They didn't say the other half. So Dr. Gerberding's version is paradoxically more reassuring. We know she is actually paying attention to the possibility of terrorism. Later in the SARS outbreak, she reassured us and cautioned us at the same time. "Although we haven't seen community transmission of SARS, we are not out of the woods yet." So she reassured them in the first half and cautioned them in the second half.

When people hear these kinds of examples, these real life ones or the ones I make up when I am trying to tell officials how they should have said it, they say to me, "aren't some people naturally inclined to do good crisis communication or is this something you can really learn?" I usually answer by telling them this final story.

One day during SARS, there was a really weird newspaper article quoting an astrobiologist from Wales that SARS and other viruses might come from outer space on meteor dust. It had to be a very quiet day on the SARS front when the newspapers had space for this. At a CDC telebriefing, CNN's reporter Miriam Falco, a very professional reporter, said, "Dr. Gerberding, I just have to ask you about this outer space thing. I am embarrassed but what do you think?" Dr. Gerberding answered with a wicked twinkle in her eye but an absolutely straight face, "Although we have no evidence that SARS is from outer space, we are keeping an open mind."

[Laughter.]

Dr. LANARD. So crisis communication is learnable. That is one of my main messages to you.

Senator CRAPO. Thank you very much, Doctor. My time is up so I will turn to Senator Jeffords for his questions.

Senator JEFFORDS. I don't mean to start with you right away.

Dr. LANARD. I gave away all my good examples.

Senator CRAPO. And they were good.

Senator JEFFORDS. In 1986, lead was discovered in drinking water in the Palisades section of Washington, DC. Residents were quoted as saying "The runaround has been unbelievable. No one in the bureaucracy has even begun to take this seriously." The Director of Water for the city stated, "Premature to contact residents throughout the city before the city developed a plan to handle and finance free testing."

I ask unanimous consent to insert several newspaper articles on this topic into the record.

Senator CRAPO. Without objection.

[The referenced document follows:]

The Washington Post, November 3, 1986

The Washington Post

November 3, 1986, Monday, Final Edition

SECTION: METRO; PAGE D1

LENGTH: 1309 words

HEADLINE: City Officials Say **Lead** in Water Poses Problem in Palisades Section of NW

BYLINE: Margaret Engel, Washington Post Staff Writer

BODY:
District government water safety officials say that **lead** contamination in the water of older houses in the Palisades section of Northwest is an "urgent" problem and that they will begin systematic water testing of the area this week.

The city's decision to begin testing follows months of complaints from families in the neighborhood and a case of 21-month-old **twins** whose development was affected by excessive **lead.**

"The runaround has been unbelievable," said Bob Wenger, an environmental health lawyer whose two children have joined many neighborhood children in drinking bottled water in recent months. "No one in the bureaucracy has even begun to take this seriously."

City water officials said they have relied on neighbors to spread the word about possible **lead** contamination but now are planning to meet with Palisades residents and federal environmental experts.

The Palisades area is bounded on the north by Loughboro Road, on the south by Chain Bridge Road and on the west by the Potomac River.

The officials said the problem undoubtedly exists in other sections of the city in older houses and buildings containing **lead** pipes -- especially structures recently disturbed by construction.

However, Lacy Streeter, director of water administration for the D.C. Housing and Environmental Regulation Administration, said that "it is premature to contact residents throughout the city" before the city develops a plan to handle and finance increased testing.

Lead is a potentially lethal toxin that accumulates in the body, commonly through exposure to **lead**-based gasoline or paint. At low levels, it causes fatigue, stomach pain, appetite loss and irritability. Large or accumulated doses poison the nervous system and can cause kidney problems and stunted growth.

Airborne and ingested **lead** has long been a problem in the **District.** In the 1970s, nearly one-third of children tested had what officials termed unacceptably high levels of **lead** in their systems. A **District lead** screening program and a fed- eral ban on **lead**-based paints have helped improve children's health. Fewer than 2 percent of children screened now are determined to have had excessive **lead** exposure.

The problem in the Palisades first surfaced in October 1985 when Judy Southerland and her husband Maurice Sanders, who live in the 5600 block of Sherrier Place, could not understand why their infant **twins** were not growing as fast as other babies. The development of one daughter dropped to 50 percent of what was typical of her age group. She had chronic constipation, lost interest in playing and wanted to be held constantly, according to her **parents.**

Tests by a pediatrician showed alarmingly high **lead** levels in the **twins'** blood -- 37 and 34 micrograms per liter. The federally accepted maximum for children is 20 micrograms per liter.

For the following year, the family was frustrated by delays, inaccurate tests and false cures as they worked with **District** government officials in trying to find the cause of their childrens' problem and correct it.

The pediatrician in October 1985 sent them to the **District's lead** detection program within the Commission on Public Health. "They advised us to repaint the interior of the house, which we did at a cost of $ 2,000," Southerland said. The effort later was found to be unnecessary. City **lead** detection workers took water samples that same month, which they reported were negative for **lead.**

Three months after the repainting, officials from the city **lead** program tested the **twins** and found lower **lead** levels -- 27 and 19 micrograms per liter. These tests, and the earlier water tests, now are believed by the family and others to have been inaccurate.

A third round of blood tests, performed by the city in April, found high levels again, 26 and 25 micrograms. But because of a mixup, the family did not get the results until July.

"By this time, we were just sick of all the delays and errors," Southerland said. The family hired a private testing lab to check their water and found that it contains three times the 50 parts per billion of **lead** currently allowed by the U.S. Environmental Protection Agency. The limit is soon to be cut to 20 ppb under a pending safety standard.

District officials theorize that Southerland and Sanders' water is contaminated because the **lead** pipes leading to their house, built about 1921, probably were disturbed when the city constructed a new water main in their neighborhood in 1984. The construction may have jarred the protective carbonate scale that grows around **lead** pipe after its first year, a protection that reduces the ability of water to leach **lead** from the pipe.

In recent years many builders have switched to copper or plastic pipes, although newly constructed houses can have contamination problems for several years if **lead** solder and other **lead**-based plumbing substances are used. The federal Safe Drinking Water Act, enacted June 19, bans **lead** solder. Virginia and Maryland took similar actions this spring. A ban has been proposed in the **District,** but no action has been taken.

"I'm furious that it had to happen," said Southerland, whose family switched to bottled water two months ago after receiving the private test results. "We cannot understand why the city doesn't alert the rest of the people who could have been affected by this."

Streeter said the city has tested a few other nearby houses. "We have not gotten to the point of finding the other homes with **lead** pipes," he said. "There are probably others we'll find in a systematic search."

The city's water mains are constructed of concrete, not **lead,** he said, but the builders of many older houses used **lead** pipe to connect to the water mains. Records in the city's Department of Public Works show which houses have **lead** service pipes, but the city has not considered making a search of the records because of the scope of the project.

Jim Collier, chief of the water hygiene branch in the D.C. Department of Consumer and Regulatory Affairs, has told residents who contact the city with concerns about **lead** to let the water run in the pipes each morning for a minute before using the water.

Lucenia Dunn, a spokeswoman for the agency, said officials will meet with residents of any older neighborhood who have concerns about **lead** contamination.

But Palisades residents are upset about the time it took to get action.

Marsha Verville, a resident whose water tested above the EPA's proposed **lead** limit, finally took her complaint to Mayor Marion Barry at an election function in September. The mayor called her back and directed her to the consumer affairs department's water hygiene branch, which tests city water.

"We made 15 phone calls to get through to them," said Wenger, who was trying to get his family's water tested. "In frustration, we talked to someone in the mayor's office. He told us someone will call within three days. No one did, and when we called him back he blamed the bureaucracy."

In exasperation, Wenger and others hired private labs to run water tests, but they note that most of the houses near the construction have not had their water checked.

Collier and Streeter said the city is prepared to expand its testing of the Palisades area and encourages residents to call the Department of Housing and Environmental Regulation at 727-7395 to schedule a free water analysis.

The Southerland-Sanders **twins** are among the few children who have been tested. Their switch to bottled water has lowered their **lead** levels to below 16 micrograms per liter and caused immediate, positive changes in their behavior, their mother reports.

"We're thrilled the **lead** has dropped, but we know it's probably in their bones," she said. "We don't know what the long-term effects will be, but we know that all the delays cost us another year of drinking **lead** water."

GRAPHIC: Map, no caption, TWP

The Washington Post, December 6, 1986

The Washington Post

December 6, 1986, Saturday, Final Edition

SECTION: METRO; PAGE B3

LENGTH: 622 words

HEADLINE: Fear of **Lead** in D.C. Water Spurs Requests for Tests

BYLINE: Margaret Engel, Washington Post Staff Writer

BODY:
District officials have been flooded with requests by 883 households, schools and businesses to sample their water for **lead** after newspaper reports were published about **lead** problems in the drinking water of the Palisades area of Northwest Washington.

Officials of the D.C. Department of Consumer and Regulatory Affairs said yesterday that the problem could extend to thousands of houses and other structures built between 1890 and 1930 with **lead** water pipes. A meeting of city officials is planned this month to determine the magnitude of the problem.

Department officials said they plan to tell hundreds of families in the next few weeks whether their tap water is contaminated by dangerous amounts of **lead.**

According to Jacqueline Davison, administrator for the department's Housing and Environmental Regulation Administration, the mayor must decide ultimately whether the city bears responsibility for the health problem and whether the **District** can afford to replace the pipes or correct the problem by softening the water.

In the meantime, hundreds of **District** families are forgoing city tap water and paying up to $ 20 a week for bottled water until the results of city tests are known.

"Just about everyone we know is drinking bottled water at considerable expense," said Kate Shafer, who lives in the Palisades area. A private lab found her tap water slightly exceeds the proposed federal limit of 20 parts per billion of **lead.**

Lead is a potentially lethal toxin that accumulates in the body, particularly in the bones and kidneys. Large or accumulated doses can cause stunted growth and kidney problems.

A recent Environmental Protection Agency study in support of changing the federal **lead** standard from 50 parts per billion to 20 parts per billion found that excessive **lead** in public water supplies is a health hazard throughout the nation, particularly in older cities with **lead** water pipes and newer plumbing soldered with **lead.**

District water officials say that the city's water mains are constructed of concrete, but that builders of older homes used **lead** pipe to connect to the mains.

In the past, the city budgeted less than $ 2,000 a year for water testing because it received only 10 to 30 requests annually for the free service.

Last month, however, **parents** of 22-month-old **twins** publicly discussed the excessive **lead** in their daughters' blood that stunted their growth and described the family's yearlong effort to get city agencies to diagnose the problem. After a private lab found the water in the family's house contained nine times the proposed federal **lead** standard, city officials declared the problem "urgent" and deposited sample bottles on the doorsteps of 83 houses in the Palisades neighborhood.

The response of anxious residents throughout the city is causing the city to boost its water testing budget to at least $ 50,000, Davison said. A private lab in Rockville has been hired to test the 1,219 samples from the 883 homes and establishments.

Results from about half of the houses are complete, said James Collier, chief of the agency's water hygiene branch. Of the 83 homes tested on Sherier Place NW, the street where the **twins** live, **lead** levels in 15 percent exceed the proposed federal limit.

Collier said, however, that it may never be possible to determine what caused the **twins'** **lead** problems. "We can't determine what there was a year ago, two years ago," he said.

The **twins'** father, Maurice Sanders, said the city is "attempting to deny responsibility for anything." His wife, Judy Southerland, added, "The city persists in talking about **lead** paint as a problem. We don't believe they could breathe enough [**lead** paint] dust. We know they imbibed the water."

GRAPHIC: Photo, Judy Southerland and **twins** Abigail and Olivia, right, drink bottled water now. (Judy Sutherland and **twin** daughters Abigail and Olivia), Harry Naltchayan

Source: News & Business > News > **Major Newspapers**
Terms: **parents and lead and twin and district and date(geq (01/01/1983) and leq (12/31/1987))** (Edit Search)
View: Full
Date/Time: Friday, April 2, 2004 - 4:14 PM EST

The Washington Post, December 21, 1986

The **Washington** Post

December 21, 1986, Sunday, Final Edition

SECTION: METRO; PAGE B3

LENGTH: 592 words

HEADLINE: Tests on Lead in D.C. Water to Take 3 Months

BYLINE: Margaret Engel, **Washington** Post Staff Writer

BODY:
The extent of lead contamination of District drinking water will not be determined for another three months, according to the acting city Public Health Commissioner, Dr. Reed Tuckson.

High lead levels caused by aged lead pipes and lead solder is rapidly becoming one of the District's thorniest public health problems. Nearly 1,000 households have swamped the city's water testing program with requests for tests after residents of the Palisades neighborhood, in lower Northwest along the C&O canal, learned that several older houses had lead-contaminated water.

The issue became public last month after a yearlong effort by the parents of twin infants to find the source of their daughters' serious **lead poisoning.** After repainting their home at the city's suggestion with little improvement in the children's health, the couple hired a private laboratory to test their water.

The laboratory discovered abnormal lead levels and pinpointed the source as the lead service pipes leading to their turn-of-the-century house.

Tuckson said the city must analyze the water samples, which are being tested by a private lab in Rockville, before it can tell what the health risks from city water may be. The analysis will take at least three months, he said.

"We're trying to figure out how many parts of the city need to be tested," Tuckson said. "For the last couple of weeks we've been trying to design a study." He said the city will hire an outside contractor to perform the water survey.

Thousands of homes in the District are believed to be serviced by lead pipes leading from concrete water mains. The city's Department of Public Works has not done an inventory of affected homes because records do not exist from some builders, officials said, and because of the enormity of the project.

"We don't know yet whether it's a problem of older neighborhoods or one that's citywide," Tuckson said.

Because of residents' concerns, the city is recommending that pregnant women and children under the age of 6 in homes where the water has not been tested drink bottled water as a safety precaution. In all other homes where lead pipes might exist, the city is suggesting that residents let the water run three to four minutes each morning before using it for drinking.

Also, all children under 4 should have their blood tested twice a year for lead levels, said Dr. Martin Levy, director of the District's Preventative Health Services Administration. Private pediatricians can perform the test or the city will do it for free, he said. If their children's blood lead levels are low, parents should not be concerned about the content of their water, he said.

Levy said that the federal Centers for Disease Control's Center for Environmental Health doubts that the lead problem in the District will require replacing all existing lead pipes.

"No one has found water as the major source of lead in children," he said, conceding that the District's **lead poisoning** project has never tested water. "The CDC doubts that it will come to replacing pipes."

The city of Boston, which has similar lead pipe problems, reduced much of the high levels by diluting the acidity of the water. The acid ate away the protective calcium carbonate that forms inside pipes to keep lead from leaching into the water.

However, the District's water already is low in acid, which may eliminate a chemical cure, Levy said.

"This is just a situation in flux," he said. "It certainly is becoming a major health problem, but we haven't reached the answers we need yet."

The Washington Post, January 23, 1987

January 23, 1987, Friday, Final Edition

SECTION: METRO; PAGE C1

LENGTH: 661 words

HEADLINE: Lead Found in Water Of Many City Homes;
Contamination May Affect 56,000 Houses

BYLINE: Margaret Engel, Washington Post Staff Writer

BODY:
A large number of the nearly 1,000 District households whose water was tested recently have excessive lead contamination, according to the Environmental Protection Agency's director of the Office of Drinking Water.

The director, Michael Cook, said the District's growing problem of lead contamination of tap water may affect up to 56,000 houses in the city.

"There were high enough percentages [of elevated levels in samples] that anybody with a lead service pipe should be concerned," said Cook, who is familiar with the contents of a draft report on the problem.

In a related development, residents in the Palisades area of Northwest Washington, where the problem first surfaced, said they will ask the Army Corps of Engineers to open a tap at the Dalecarlia Reservoir for families whose water is contaminated by lead.

Hundreds of District residents received letters from city officials this week informing them of the lead levels in their tap water. Those whose water exceeded the proposed federal standard of 20 parts per billion were told not to drink or cook with water from the "hot" side of the tap and to run the "cold" side of the kitchen tap for three to four minutes each morning and after several hours of nonuse.

The residents requested the tests this fall after the parents of 22-month-old twins publicly discussed the excessive lead in their daughters' blood that they believe stunted their growth and described the family's year-long effort to get city agencies to diagnose the problem. In several areas of the city, particularly in Northwest Washington where the twins live, families have been drinking bottled water since November.

Dr. Reed Tuckson, acting D.C. public health commissioner, said yesterday the city was conducting a "massive coordinated effort" on the lead problem and he, along with federal health and environmental specialists, will brief city administrator Thomas Downs on the situation this afternoon.

"We understand residents' anxiety," Tuckson said. "We want to hurry and come to the city in an organized, sensible way."

Tuckson said city officials will discuss the extent of the problem publicly on Monday. He estimated that about 10 percent of the results exceed the current federal lead standard of 50 parts per billion. Neither he or Cook had exact figures yesterday.

Families with high readings should switch to bottled water immediately, Tuckson said, and all families with lead service pipes should have the blood lead levels of their children under 6 years old tested. The city's **lead poisoning** prevention program, at 727-9870, conducts free tests. Pregnant women in homes with high readings also should be tested, he said.

"We will speak to our citizens on Monday," Tuckson said. "You can't go out and talk to people about a confusing issue when the state of art is not decided."

Several of the residents receiving the city's letter were confused because the federal lead standards were expressed as parts per billion and their household test results were measured in micrograms per liter. The two measures are the same, but this was not spelled out.

"People couldn't deciper what the results meant," said Robert Archer, an Advisory Neighborhood Commissioner for Ward 3. His family began buying bottled water at a local grocery store this week after receiving one reading of 89 micrograms per liter, well above both the current and proposed federal standards.

"We're asking the Army Corps of Engineers [which operates the city's reservoirs] to open up a tap so we can get the water we need rather than spending $ 2.68 for each 2 1/2 gallon Safeway jug," Archer said.

Officials of the Corps could not be reached yesterday.

Lead is a potentially lethal toxin that accumulates in the body, particularly in the bones and kidneys.

In the past, the city's lead prevention efforts have dealt solely with lead in paint, soil and air.Staff writer Michael Weisskopf also contributed to this report.

Source: News & Business > News > **Major Newspapers**
Terms: **lead poisoning and date(geq (1/1/1984) and leq (9/1/1987))** (Edit Search)
View: Full
Date/Time: Friday, April 2, 2004 - 3:59 PM EST

Senator JEFFORDS. Given this repetitive failure, what recommendations do you have for how to improve the situation?

Dr. LANARD. I have one very concrete recommendation I am very pleased to pass on to you. After the CDC had its communication problems during anthrax, that focused their minds and they developed an extremely intense risk communication training program which they have now packaged into a CD-ROM and a series of roaming 3-, 4-, and 5-day trainings. People can use this training program, local agencies can use it. It is called "Crisis and Emergency Risk Communication." I helped my husband, Peter Sandman, work on it for 18 months. They also hired some of the other top risk communication people in the world. They paid us good money, it is all there now and belongs to the people of the United States. We don't get any more money if people use it. We just get the satisfaction of seeing people learn it. "Crisis and Emergency Risk Communication, CD Cynergy" is available from the CDC. They will also come and give seminars at agencies. The actual CD-ROM costs like nothing and the seminars are very cheap. They are run by extremely experienced, unfortunately experienced because of hard lessons, CDC personnel.

One of my main lessons is to take a little training. It is the cheapest training you can get because it has already been paid for.

Senator JEFFORDS. Ms. Borland, I want to thank you again for testifying before the committee. I want to thank Senator Crapo for agreeing to allow a DC resident to speak today.

Gloria, if there were three or four actions you could have the government agencies immediately take that were represented here today, what would they be?

Ms. BORLAND. Again, because the problem is so huge, people still don't know where the lead pipes are. It is a huge problem. The costs are going to be enormous, the receivership. You need to really clean house and start over. It really needs to have that outside management come in, outside people and just that drastic action. That will instill the trust that is needed here in the District of Columbia from parents.

Senator JEFFORDS. Dr. Best, Dr. Lucey implies in his testimony that there is not a problem with the elevated blood levels citywide based on the voluntary testing program that has been conducted. Can you give me your evaluation of the adequacy of the set of self-selected samples to determine the extent of citywide problems?

Dr. BEST. I want to first emphasize that Children's is working closely with the Department of Health and would like to continue to do so.

From statistical soundness, however, a self-selected sample is not representative of any population, including the sample that Children's has which goes back for 10 years. That is not a representative sample. The only way to do a representative sample is to pick out a grid and use your random number generator to pull out an appropriate number of houses and find an appropriate number of children.

On the other hand that is very expensive and these are data we have here in our pockets so to speak, which we can produce quickly. To generate the data that would be truly representative we would need several years and probably $1 million.

Senator JEFFORDS. Dr. Best, what other factors, environmental or social for example, might compound the negative health effects of lead exposure in children?

Dr. BEST. If you are trying to overcome a small drop in IQ, outside of preventing the drop, the best thing you can do is provide the most enriching environment you can possibly provide. Unfortunately this is where one of those socioeconomic strata define themselves. Those of us who are educated and are more affluent can provide those environments without having to completely demolish our lives.

Many of our children in DC, however, do not have those advantages. Sixty percent of children in DC live in poverty or some huge number. Those children do not have the opportunity to go to the best day care center in the city. They often have to attend poorly maintained, poorly funded schools. Their parents may need to work two or three jobs if they are lucky to have two parents at home. Those parents are not reading to them every day, those are the advantages that we as more affluent people have over the other families. You can't buy that.

Senator JEFFORDS. Is there an acceptable level for lead concentrations?

Dr. BEST. No.

Senator JEFFORDS. Is there a safe level?

Dr. BEST. No.

Senator JEFFORDS. CDC notes on their website that there is no level at which adverse effects are not expected and that aiming for lower standards would be arbitrary and provide uncertain benefits. Can you comment on that assessment?

Dr. BEST. Since the first article that noted a loss of IQ in children subjected to levels lower than 10 which was the first really strong article in 2000, there has been a big discussion about whether or not the CDC should lower their level of 10. Even the CDC doesn't call that level a threshold level. It is a level above which you need to take some action to educate the family, to make sure they understand the harms and the need for action, good risk communication.

CDC's point is that when they wrote that, testing is not accurate below 10 mcg/dl. I suggest strongly now that is no longer true. We have instruments at Children's Hospital that can measure with a great deal of specificity levels below 5 and in fact, we would consider a level of 1 or higher to be an accurate level.

However, that still doesn't excuse this misleading indicator. When I tell parents your child has a lead level of 3 and they say, "Oh, is that OK?" I have to say we are not going to take any action because there is nothing we can do. We don't have a treatment for low lead levels. we don't have a treatment until your lead level gets up to 45 mcg/dl. Then that treatment, as I stated in my testimony, can actually reduce your IQ as well.

The only thing we can do is prevent and the CDC argued in that article that they are advocating prevention. I am not sure that is the right approach. I think we should lower the level at which we start doing this increased education. I think that should be zero.

Senator JEFFORDS. I guess I don't quite understand why?

Dr. BEST. Because if your child has a measurable lead level, you should be doing what you can to reduce exposure to that child.

Senator JEFFORDS. I think that is what ought to happen.

Dr. BEST. You should be figuring out whether you have lead paint in your house, doing the appropriate abatement procedures for lead paint which is not necessarily removing the paint. It is covering up paint that is chipping, wiping down surfaces so your toddler doesn't go eating all the dust in the house, checking the water supply so you can make sure that is not contributing to the source and checking the dirt around your house where children play.

Senator JEFFORDS. Thank you all. I am sure the Chairman agrees with me we have had wonderful testimony and very helpful to us.

Senator CRAPO. Thank you, Senator Jeffords.

I have one more question and Dr. Best, it is for you as well and kind of follows the same line that Senator Jeffords has been pursuing just now. That is, I understand the point about self-selected testing and so forth. However, I am still perplexed a little as I look at the numbers that were reported here by Dr. Lucey—that of the 201 persons who were measured, who lived in homes with the highest levels of lead in the drinking water, which was 300 ppb or more, none had elevated blood levels and the same kind of things came through in a number of other different ways of singling out even though they were self-selected, singling out those who had been drinking water.

Dr. BEST. That 300 sample would not be self-selected because they looked at the water first and then went back and looked.

Senator CRAPO. That is the point I was getting at. When you analyze it that way, they wouldn't really be self-selected, would they?

Dr. BEST. Not that sample but it is still not representative of the DC population as a whole.

Senator CRAPO. I understand that.

Dr. BEST. It is only representative of the houses tested.

Senator CRAPO. Clearly if you are trying to get a sample of the DC population as a whole, but if you are trying to look at the question of what is the impact on the blood level of drinking water that has lead in it, wouldn't those numbers tell us something?

Dr. BEST. What they tell us is that in a sample of 300 cases where children—how old were the children?

Senator CRAPO. It doesn't say in this report here.

Dr. BEST. They were under six I think.

Dr. LUCEY. Can I comment?

Senator CRAPO. Certainly.

Dr. LUCEY. This is everyone. We went out to their homes.

Senator CRAPO. This is not just children?

Dr. BEST. Is this adults as well?

Dr. LUCEY. Yes.

Dr. BEST. That is even a more important point because children absorb lead more efficiently than adults, so if you have a sample with any adults in it, then you skew it to the lower end of the lead levels because they are not as efficient.

Senator CRAPO. When we talk about issues like paint or fuel or dust or water, have there been sufficient studies in the country to rank those risks?

Dr. BEST. No. That is only because we haven't had a history like we have in DC. One of the good outcomes of this I hope is a better understanding of lead and water and how it is ingested and absorbed by children. Hopefully we will find that it does not make a large contribution to lead poisoning of children. It is still not safe. There is no way I can say that it is safe and I am sure there is no one else who will admit they think it is safe.

Senator CRAPO. Certainly.

First of all, we have gone over some so both Senator Jeffords and myself are late to other important responsibilities, so we are going to have to wrap up now. Unfortunately I don't know if you will be able to stick around afterwards, Senator Jeffords. I am going to have to rush to get to another responsibility.

I did want to thank all of you for coming. Ms. Borland, I met with some of the residents whose pictures you have shown there today and listened to their stories in person. You raise very, very important points and actually put a very human face on this issue. I think that is very helpful.

Dr. Best, your scientific knowledge and Dr. Lanard, your scientific knowledge and your suggestions are both very helpful to us as well as the information we received from the other panelists. I did note, Dr. Lanard, when somebody says to me, "I have no evidence of such and such", I am left wondering whether they have studied it or whether they are going to. If somebody says, "we have no evidence but it is a concern and we will be looking at it", I can see how that leaves you with a much more comforting feeling. Those very simple and sometimes counter intuitive points are very helpful for us in communication. We appreciate that.

At this point, I am going to conclude the hearing. Again, I want to thank all our witnesses today. I want to thank all the people of the community here in Washington, DC and others involved who have brought this to our attention and who are continuing to work with us. This panel will continue its oversight responsibilities and will continue to work with all the appropriate authorities to make sure we identify the problem, get immediate solutions put into place for those at risk and then move forward to deal with the question of what caused it, how can we solve that and resolve the issue so we can safely drink our water here in Washington, DC.

Both Senator Jeffords and I have noted to each other as we have talked, we both live here in Washington, DC.

Senator JEFFORDS. I got my home in 1970. It was constructed in like 1812.

Senator CRAPO. My home was constructed in 1890. So we are way on the other side of that scale.

We really do appreciate your attention and concern today, all of you.

This hearing will be adjourned but our focus on this issue is certainly not ended.

Thank you.

[Whereupon, at 5:23 p.m., the subcommittee was adjourned, to reconvene at the call of the chair.]

STATEMENT OF HON. BOB GRAHAM, U.S. SENATOR FROM THE STATE OF FLORIDA

Mr. Chairman, thank you for calling this hearing today. I share the concerns of the families here today, as well as everyone who lives in DC, about the condition of the water supply in our Nation's capital and throughout the United States. We have all observed the recent finger pointing of the relevant agencies, but today is about sorting through the rhetoric, getting some answers, and finding some real solutions for DC residents. The dangerous effects of lead are well documented. Major government agencies from the Centers for Disease Control and Prevention (CDC) to the Environmental Protection Agency (EPA) agree that lead in drinking water can cause a variety of adverse health effects, particularly dangerous for children under 6.

Despite this risk, the responsible agencies failed to inform the public about the lead problem in a timely or appropriate manner. The District of Columbia's Water and Sewer Authority (WASA) and the EPA first became aware of the lead problem in 2002. WASA, however, failed to notify homeowners of the problem until more than a year later, at a public meeting in November 2003. Even when announcing the meeting, WASA officials neglected to tell citizens the gathering would be discussing the lead issue. This long-delayed public education program has essentially failed failed to provide district residents with timely or thorough information.

Another issue is the role of the EPA in this situation. EPA has a primary oversight role over WASA and is responsible for the lead testing programs. We must ask, is there a problem with the testing standards issued by EPA that delayed this result? Additionally, are the current EPA water composition standards safe for the varying water infrastructure of this Nation? Is it possible that these standards need revisiting?

The first panel of this hearing, comprised of EPA and WASA officials, will have an opportunity to address these issues. I look forward to hearing their suggestions for next steps for information gathering, providing that information to the public, and most importantly, a strategy to solve the lead problem.

Today, we will also hear from witnesses who live in the district. I am interested to get their perspectives on how the exchange of information can be improved. What were the mistakes and how are they going to be fixed?

It seems incredible during this day and age, with all the technologies available, that we are having problems communicating. WASA must improve its public communication program. How should this be done? What role will the EPA and the Department of Health play in this process? We need to know what the new strategy will be.

The health and safety of the community is second only to its faith in those in power to provide them with useful and updated information. I want to make sure that we learn from these mistakes and not repeat them during the next phases of fixing the problem.

Thank you again for conducting this hearing. I look forward to hearing the testimony from our witnesses.

STATEMENT OF PAUL STRAUSS, U.S. SENATOR (SHADOW), DISTRICT OF COLUMBIA

Subcommittee Chairman Crapo, Ranking Member Jeffords, and others on this subcommittee, as the elected United States Senator for the District of Columbia, and the father of two little girls, I would like to express my deepest concern about the safety of our region's drinking water. Because of the potential health problems for local residents, it is imperative that this issue is promptly addressed.

I particularly want to thank you Chairman Crapo, and Senator Jeffords for taking the time yesterday, to join me and my Mayor at a very moving meeting with local residents right here in the community. It was extraordinary that some of the most influential members of the U.S. Senate sat down in a living room with ordinary residents of the District of Columbia to hear their concerns. This gesture has set a productive tone for these hearings, and demonstrated a level of personal concern on your part, which makes me hopeful that meaningful solutions are truly on the way.

District of Columbia's problems have had an anti-home rule sentiment. They tend to have a limited focus that examines only the failures of our local officials. This is not what this situation calls for. It is instead time for full cooperation and support to appropriately address this emerging problem. I want to emphasize that there is a need for real solutions, not finger pointing.

When it is appropriate to do so, DC residents are more than capable of criticizing the response of their own government. We do not need the assistance of Congress when it comes to complaining, we need your help to solve this problem. Gratuitous

District bashing will not help, and neither will new restrictions on our local autonomy.

The DC Water and Sewer Authority has estimated that 23,000 homes have lead lines and more than 5,000 homes have water with lead levels exceeding the Federal limit. As it is well known, lead disrupts production of hemoglobin, which leads to anemia, can cause cognitive problems, affects kidneys, which can lead to hypertension and even renal failure. Lead is considered to be one of the top environmental health dangers for children 6 and under due to the future health risks that can arise from its exposure. In pregnant women, lead crosses into the placenta and is absorbed by the fetus.

While the District's response to this problem has been far from perfect, the Mayor is quickly moving in a new direction. It is important to remember that the majority of these service lines were installed not by our elected home-rule government, but placed at a time when it was Congress itself that had actual control over the District of Columbia. A more significant reality is that DC is not unique when it comes to this problem. This is a national problem, a problem that we have to find real solutions for. The unfortunate truth is that the municipal drinking water in DC is probably not appreciably worse than the water in many other areas of the United States. It has been estimated that the costs for the needed improvements on a national level could exceed one trillion dollars. None of our States and localities can afford this kind of effort without Federal financial assistance. I urge the members of this committee to work with your colleagues on the Appropriation's committees to see that the resources needed will be made available.

In conclusion, I would like to thank the subcommittee for holding this important hearing. I particularly would like to thank Senator Crapo for his commitment to leave this hearing record open so that DC residents, despite their lack of equal representation in the U.S. Senate, can at least place written statements in the record. This will at a minimum permit their concerns to be documented and hopefully addressed. If you have not already done so, I would like to include with this statement a petition signed by over 600 of my constituents who are urging actions on this issue. My office intends to followup with those other constituents of mine who wanted to attend today's hearing, but are not able to. Finally, I would like to thank Regina Szymanska and Paola Nava, of my staff for their assistance in preparing this statement. I look forward to further hearings on this topic, and I'm happy to respond to any requests for additional information that you or any member of this subcommittee may have.

Statement of Benjamin H. Grumbles, Acting Assistant Administrator, Office of Water, Environmental Protection Agency

Good afternoon, Mr. Chairman and Members of the Committee. I am Benjamin Grumbles, Acting Assistant Administrator for Water at the United States Environmental Protection Agency (EPA). I welcome this opportunity to discuss the issue of lead in drinking water, the specific situation related to elevated lead levels in the District of Columbia's (DC's) drinking water, and actions that EPA is taking at the national level to address the matter.

LEAD AS A PUBLIC HEALTH CONCERN

EPA places a high priority on reducing exposure to lead. This contaminant has been found to have serious health effects, particularly for children. Health effects may include delays in normal physical and mental development in infants and young children; slight deficits in the attention span hearing, and learning abilities of children; and, high blood pressure in some adults (which may lead to kidney disease and increased chance of stroke). But pregnant women and children are our primary concern. The Centers for Disease Control and Prevention (CDC) has identified a blood lead level of 10 micrograms per deciliter as the level of concern for lead in children.

Lead exposure in young children has been dramatically reduced over the last two decades. According to a 2003 CDC report [Surveillance for Elevated Blood Lead Levels Among Children—United States, 1997–2001. Centers for Disease Control and Prevention. *Surveillance Summaries*, September 12, 2003. MMWR 2003:52 (No. SS–10)], 88 percent of children between the ages of 1 to 5 were estimated to have blood lead levels that exceeded 10 μg/dl for the period between 1976–1980. By 1999–2000, this estimate had decreased to approximately 2 percent. This decrease is largely due to the 1973 EPA regulation to phaseout lead in gasoline between 1973 and 1995, and to the reduction in the number of homes with lead-based paint from 64 million in 1990 to 38 million in 2000. Some decline was also a result of EPA regulations reducing lead levels in drinking water and banning lead from paint. Other reasons include bans on lead in food and beverage containers and reductions in lead in industrial emissions, consumer goods, hazardous waste, and other sources. There are several EPA programs that continue to be successful in reducing the public's exposure to lead in the environment.

The most common source of lead exposure for children today is lead in paint in older housing and the contaminated dust and soil it generates. [see Risk Analysis to Support Standards for Lead in Paint, Dust and Soil (EPA 747–R–97–006, June 1998] This is primarily from housing built in the 1950s and homes with pre-1978 paint. Several Federal programs and surveillance and prevention programs at the State and local level continue to work toward reducing exposure to lead. In addition, EPA works with Federal agencies—mainly the Departments of Housing and Urban Development, Health and Human Services, and Justice through the President's Task Force on Environmental Health Risks and Safety Risks to Children—on implementing a Federal strategy to virtually eliminate childhood lead poisoning.

LEAD IN DRINKING WATER

Although the greatest risks are related to paint, lead in drinking water can also pose a risk to human health. As indicated in EPA's public education language for the Lead and Copper Rule, approximately 20 percent of a person's exposure to lead can come from drinking water. The level of exposure can be greater for children and infants, particularly when tap water is used to mix juices and formula. EPA has set a maximum contaminant level goal of zero for lead in drinking water and has taken several actions over the last 20 years to reduce lead in drinking water. The 1986 Amendments to the Safe Drinking Water Act (SDWA) effectively banned the new use of lead solder and leaded pipes from public water supply systems and plumbing, and limited faucets and other brass plumbing components to no more than 8 percent lead. To address lead in schools, the Lead Contamination Control Act (LCCA) of 1988 recalled drinking water coolers with lead-lined water reservoir tanks, and banned new drinking water coolers with lead parts. The 1986 SDWA Amendments also directed EPA to revise its regulations for lead and copper in drinking water.

An interim standard for lead in drinking water of 50 micrograms per liter, or parts per billion (ppb), had been established in 1975. Sampling of customer taps was not required to demonstrate compliance with this standard. In 1988, the Agency proposed revisions to the standard and issued a final standard in 1991. The revised standard significantly changed the regulatory framework. Unlike most contaminants, lead is not generally introduced to drinking water supplies from the source water. The primary sources of lead in drinking water are from lead pipe, lead-based solder used to connect pipe in plumbing systems, and brass plumbing fixtures that contain lead. Setting a standard for water leaving the treatment plant fails to capture the extent of lead leaching in the distribution system and household plumbing.

EPA requires public water suppliers to meet the regulations governing treated water quality distributed via the public water system. The regulations do not require homeowners to replace their plumbing systems if they contain lead. To reduce consumers' lead exposure from tap water, EPA used its available authorities to require public water suppliers to treat their water to make it as non-corrosive as possible to metals in their customers' plumbing systems. These treatment requirements were issued in EPA's Lead and Copper Rule (LCR) on June 7, 1991.

The rule requires systems to optimize corrosion control to prevent lead and copper from leaching into drinking water. Large systems serving more than 50,000 people were required to conduct studies of corrosion control and to install the State-approved optimal corrosion control treatment by January 1, 1997. Small and medium sized systems are required to optimize corrosion control when monitoring at the consumer taps shows action is necessary.

To assure corrosion control treatment technique requirements are effective in protecting public health, the rule also established an Action Level (AL) of 15 ppb for lead in drinking water. Systems are required to monitor a specific number of customer taps, according to the size of the system. If lead concentrations exceed 15 ppb in more than 10 percent of the taps sampled, the system must undertake a number of additional actions to control corrosion and to inform the public about steps they should take to protect their health. The rule was subsequently revised in 2000 to modify monitoring, reporting and public education requirements, but the basic framework, including the action level, was not changed.

The LCR has four main functions: (1) require water suppliers to optimize their treatment system to control corrosion in customers' plumbing; (2) determine tap water levels of lead and copper for customers who have lead service lines or lead-based solder in their plumbing system; (3) rule out the source water as a source of significant lead levels; and, (4) if action levels are exceeded, require the suppliers to educate their customers about lead and suggest actions they can take to reduce their exposure to lead through public notices and public education programs. If a water system, after installing and optimizing corrosion control treatment, continues to fail to meet the lead action level, it must begin replacing the lead service lines under its ownership.

Although we are currently seeing problems in the District, the LCR has proven to be successful in reducing levels of lead in drinking water. Following issuance of the rule in 1991, EPA required medium and large systems to conduct initial rounds of monitoring by December 1992. The results from the first round of sampling for large systems (serving more than 50,000) conducted in 1991 showed that 130 of 660 exceeded the action level of 15 ppb. We recently reviewed Consumer Confidence Reports for 109 of the systems that were on the list and found that only 9 were above the action level within the last few years (one of which was DC). Thus it would appear that the actions taken by systems to reduce corrosion through appropriate treatment have significantly reduced the public's exposure to lead in drinking water. However, even though we have had success in reducing exposure, we must remain vigilant to ensure that treatment continues to control corrosion and that information on potential risks is communicated to the public.

THE CURRENT SITUATION IN DC

In the District of Columbia, WASA failed to achieve the intended objectives of the the regulatory framework that EPA established. Within the last few years lead concentrations have increased significantly. Public education efforts were not effective in reaching the people who needed to know about the problem or in conveying the risks posed to some customers by elevated levels of lead in the water.

The provision of safe drinking water is not an easy task. Treatment processes must be balanced to address multiple risks. EPA has developed guidance to assist systems in achieving simultaneous compliance with different standards to, for example, balance treatment processes between the need to control corrosion within a system and also avoid harmful byproducts that can result from disinfection processes. As Regional Administrator Welsh will describe, EPA is working with WASA and the Washington Aqueduct, managed by the U.S. Army Corps of Engineers, which supplies water to WASA, to determine if changes in treatment processes to reduce disinfection byproducts resulted in elevated lead levels. The situation in DC appears to be unique. In surveying States and regions, we have not identified a systemic problem of increasing lead concentrations in tap monitoring conducted by public water systems. However, we will continue to investigate this matter in the weeks ahead.

ACTIONS UNDERTAKEN BY EPA HEADQUARTERS

This event is a reminder of what we take for granted—that we can turn on our faucets, whenever we want, to draw a glass of clean, safe water. I also see it as indicative of the challenges in managing the Nation's water infrastructure. We face the possibility of interruptions in service quality and public health protection as a result of deterioration of aging infrastructure or outdated components, such as the lead service lines serving older homes in the District. This will require significant levels of coordination on the part of local, State and Federal Governments, and an understanding of the true investment needs on the part of customers.

With respect to the situation here in DC, I fully understand the concerns that congressional Members and committees and city leaders have regarding timely and effective public notification. EPA is reviewing the actions taken by all parties to ensure that we use the lessons learned to prevent such an event from taking place in the future—here in DC and in other communities across the Nation.

Staff from my program and EPA's Office of Research and Development are currently working closely with the Region to provide technical assistance and are participating on the Technical Expert Working Group (TEWG) evaluating potential technical solutions to elevated lead levels. I directed staff to convene a peer review panel that is conducting an independent review of the TEWG's Action Plan released on March 10 and which will also review subsequent reports. Staff are also participating in a review of WASA's public education material and are working with the Region on communication issues in the District.

As the head of the national water program, I have undertaken a number of actions to address the specific issue of lead in drinking water from a national perspective.

All of us want to ensure that the Nation's school children are not exposed to elevated lead levels in their drinking water. While States and schools may have acted immediately to remove harmful lead lined coolers in accordance with the 1988 Lead Contamination Control Act, lead solder and plumbing fixtures can still contain low levels of lead. States and schools should continue to monitor their water outlets to ensure that children are protected using EPA's recommended protocol for testing water in schools for lead. On March 18 I sent letters to every State's Director of Health and Environmental Agencies seeking their help in understanding and facilitating State and local efforts to monitor for lead in school drinking water. We want to know if additional guidance might help States and local governments conduct more comprehensive monitoring in schools and day care facilities.

I am also working with our enforcement and regional drinking water program managers to embark on a thorough review of compliance with and implementation of the Lead and Copper Rule. Our initial focus is to ensure that EPA has complete and accurate information on the Lead and Copper Rule in its Safe Drinking Water Information System. States were required to report specific results of monitoring (i.e., 90th percentile lead levels) to EPA for systems serving populations greater than 3,300 people beginning in 2002. On March 25, I sent a memorandum to Regional Administrators asking them to work with the States to ensure that all available information is loaded into the data system by the end of June.

With more complete information, we will be able to work in consultation with enforcement and regional staff to assess national compliance and implementation. Our review will attempt to answer three questions: (1) Is this a national problem? Does a large percent of the population receive water that exceeds the lead action level? Do a large number of systems fail to meet the lead action level? (2) How well has the rule worked to reduce lead levels in systems over the past 12 years, particularly in systems that had demonstrated high lead levels in the initial rounds of sampling? (3) Is the rule being effectively implemented today, particularly with respect to monitoring and public education requirements?

We expect this analysis to continue throughout the year and will release interim reports as results become available. Part of the analysis will include a review of the existing requirements of the rule and associated guidance. To help the Agency obtain additional information from experts, we have scheduled workshops in May to discuss sampling protocols for the rule and utility experiences in managing simultaneous compliance with multiple drinking water rules. Workshops on additional subjects may be scheduled later in the year.

The compliance review, expert workshops and other efforts underway will help us to determine whether it is appropriate to develop additional training or guidance or make changes as part of our review of existing regulations. Our immediate goal is to ensure that systems and States have the information they need *today* to fully and effectively implement the rule and minimize risks to public health.

Mr. Chairman, this reminds us all of the importance of communication—especially with the public. To maintain public health and confidence, information communicated to the public must not only be accurate, but timely, relevant and understandable. While I believe that communication efforts on the part of the Region, the District's Department of Health and WASA have improved, there is still much to be done to ensure that the city's residents are aware of the steps they can take to protect their health. As you will hear from Mr. Welsh, the Region is completing a thorough review of WASA's and the Aqueduct's activities to determine if any violations of environmental law have occurred. EPA will take the appropriate steps to protect public health, both by enforcing existing regulations and by using its additional authorities to address imminent and substantial threats to public health, as appropriate.

We will work closely with the Region, our public service partners and concerned citizens to investigate the situation in DC and to review implementation of the rule nationwide. EPA wants to ensure that citizens across the country are confident in the safety of their drinking water.

Thank you for the opportunity to testify this afternoon. I am pleased to answer any questions you may have.

STATEMENT OF DONALD WELSH, DIRECTOR, REGION III, ENVIRONMENTAL PROTECTION AGENCY, PHILADELPHIA, PA

Good afternoon, Mr. Chairman and Members of the Subcommittee. I am Donald Welsh, Regional Administrator for Region III of the United States Environmental Protection Agency (EPA). Thank you for the opportunity to appear before you today to discuss the important issue of lead in the tap water of District of Columbia residents, and the steps EPA and other agencies are taking to resolve the problem on a short-and long-term basis.

Elevated levels of lead in the environment, whether in drinking water or lead paint, can pose significant risks to health—particularly to pregnant women and young children. Reducing exposure to all sources of lead is vital to protecting the health of our citizens.

It is unacceptable to us that many families in the District continue to live with fear and uncertainty over the quality of the water they drink. At EPA, we will not be satisfied until all aspects of this problem are resolved. There is no higher priority for my office than to work with the city to protect residents.

To that end, EPA and the District of Columbia have directed—and are closely monitoring—a series of interim measures to ensure that residents have safe drinking water and proper precautionary guidance. At the same time, a multi-agency Technical Expert Working Group is acting as quickly as possible to identify and correct the cause of the elevated lead levels.

The city and EPA have had regular meetings and conversations to monitor progress and to ensure necessary actions are being taken. While we are satisfied at this point that the District of Columbia Water and Sewer Authority (WASA) is taking the required steps to deal with the immediate threat to public health, EPA reserves the right to use enforcement authorities to compel action if necessary.

Among a host of immediate steps being taken to safeguard public health, WASA is delivering certified water filters to occupants in the estimated 23,000 homes and businesses with lead service lines, and conducting additional tap water sampling to fully identify the extent of the problem in the District. There also are expanded outreach efforts underway to ensure residents have essential information to protect themselves and their families. Details of these and other actions will be provided later in this testimony.

HISTORY AND EXTENT OF THE PROBLEM IN DC

In the District of Columbia, the regulatory framework established in EPA's 1991 Lead and Copper Rule (LCR) did not achieve key aspects of its intended objectives. The LCR requires systems to optimize corrosion control to prevent lead and copper from leaching into drinking water. To assure corrosion control is effective, the rule establishes an action level of 15 parts per billion for lead in drinking water. If lead concentrations exceed the action level in more than 10 percent of the taps sampled, the system must intensify tap water sampling and undertake a number of additional actions to control corrosion and to educate the public about steps they should take to protect their health. If the problem is not abated, the system must also begin a lead service line replacement program.

Within the last couple of years in the District of Columbia, lead concentrations in tap water in many homes increased well above the 15 parts per billion action level. In addition, public education efforts were ineffective and, we believe, not fully compliant in all instances with EPA rules.

By way of background, two public water systems are responsible for complying with provisions of the LCR in the District. The Washington Aqueduct owns and operates two water treatment plants which provide finished drinking water to WASA, as well as to Arlington County and the city of Falls Church in Virginia. The Aqueduct is responsible for all corrosion control treatment for its three customer systems. WASA, which distributes water from the Aqueduct to customers in the District, is responsible for monitoring lead and copper at its retail customers' taps. EPA's Region III office in Philadelphia has primary oversight and enforcement responsibility for public water systems in the District.

The results of DC's required tap monitoring exceeded the 15 ppb action level for 10 percent of taps monitored during six of 15 reporting periods since January 1992—three times prior to 1994 and three times since 2002.

During the 1990s, several studies were conducted by WASA, the Aqueduct, and EPA to support identification of an optimal corrosion control treatment (OCCT) for the drinking water supplied by the Aqueduct. EPA gave the Aqueduct conditional OCCT approval in 1997 and, after reviewing results from several required reports, provided final approval in 2000. Later that year, the Aqueduct replaced its secondary disinfection treatment by converting from free chlorine to chloramines, primarily to ensure compliance with EPA's more stringent requirements to combat serious health consequences related to disinfection byproducts.

The OCCT implemented by the Aqueduct appeared to be effective in minimizing lead levels until the sampling period between July 2001 and June 30, 2002. EPA received a final report from WASA on August 27, 2002 indicating that the 90th percentile value had increased to 75 ppb during that period. The high level required that WASA conduct more frequent monitoring and carry out public education. The lead action level was also exceeded for subsequent monitoring periods in 2003, with 90th percentile values at 40 ppb (January 1 to June 30, 2003) and 63 ppb (July 1 to December 31, 2003).

In 2003, EPA Region III worked through an EPA Headquarters' contractor to hire Professor Marc Edwards of Virginia Tech to help analyze the situation and make recommendations to assist the region in working with WASA. Professor Edwards' draft report delivered in October 2003 provided key input to the evaluation of the problem during the fall of 2003 and continues to aid the expert technical team convened by EPA to identify the underlying causes for elevated lead levels.

In addition to the stepped up monitoring, when WASA exceeded the action level, the authority was required to undertake a lead service line replacement program. The LCR requires that a system replace at least 7 percent of the lead service lines the system owns annually until all of the lines have been replaced, or until tap water monitoring indicates that its 90th percentile lead level is equal to or less than 15 ppb. If the sample for a service line shows a concentration below the action level, the line serving that house is considered to be replaced.

Starting in March 2003, WASA began an expanded lead service line sampling program to determine the concentrations of lead contributed by individual lines. Region III received sampling results from this program on October 27, 2003. The information was reviewed by our technical staff with an eye toward determining how to address the underlying cause of the corrosion problem. The results indicated that 385 lead service lines had been physically replaced and that an additional 1,241 lines were considered replaced because they had monitoring results below the 15 ppb action level. The report also indicated that 3,372 of 4,613 lead service lines tested through September 30, 2003 had lead levels that exceeded the lead action level. In many cases, lead levels from customer taps were very high, with levels above 300 ppb in nearly 3 percent of lines and above 100 ppb in 18.5 percent of lines. Frequently, several months passed between when a sample was collected and information was provided to homeowners who participated in the expanded sampling program. In addition, the notifications were not effective in relaying to the customers the significance of the problem.

INTERIM STEPS BEING TAKEN TO PROTECT RESIDENTS

As indicated earlier, WASA and the District of Columbia government are undertaking a series of actions outlined by EPA to address the immediate public health threat posed by lead in drinking water. The actions include:

- WASA will ensure delivery by April 10 of the NSF International-certified water filters and consumer instructions to occupants in homes and buildings with lead service lines. Periodic replacement of the filters also will be ensured.
- Additional tap water sampling has begun in schools as part of a program to test a representative group of homes and other buildings that are not served by lead service lines to help determine the full scope of the problem. The city began the school testing program on March 27 based upon EPA's approval of the sampling protocol, and as of last Friday (April 2) WASA had anticipated the completion of sampling at approximately 150 schools. WASA has agreed by today (April 7) to address EPA questions on the sampling plan for other facilities in the city so that we can authorize WASA to move forward on that initiative.
- WASA has committed to an accelerated schedule for physically replacing lead service lines in the District. WASA has agreed to complete 1,615 actual service line replacements during the compliance period that ends September 30—a far greater pace than was met during the prior compliance period. As part of EPA's annual grants to the District, the agency is providing a total of $11.3 million to WASA for lead line replacement this year.
- WASA is expediting notification to customers of the results of water sampling at their residences, committing to providing results in 30 days or less. WASA has stated that residents now receive a letter that provides more detail about their sampling results, and those with high lead levels are referred to the DC Department of Health.
- As EPA, the District and WASA continue to expand outreach efforts to provide important information to consumers, WASA will provide to EPA for our review no later than today (April 7) an enhanced public education plan to satisfy a directive that communications on the lead issue convey the proper sense of urgency and concern for public health. The goal is to reach all sectors of the population in an effective way. WASA already has committed to a series of activities, including placing advertisements in ethnic and foreign language newspapers, developing and distributing public service announcements to radio stations, and taping a program this

week on the District's cable channel that will be shown in a variety of venues. WASA also will continue to meet with various community groups and organizations.

EPA is undertaking a compliance audit of WASA's lead service line and public education actions. In letters to WASA last week, EPA identifies instances in which requirements were apparently not met and, as part of the enforcement process, requires WASA to provide information and documentation to EPA responding to those findings. Once EPA receives the requested information, we will officially determine whether there have been violations and take the appropriate action.

In a separate initiative, an internal EPA team is evaluating WASA's prior outreach efforts—a process to be completed by month's end that involves a review of materials, interviews with residents and public officials, and a survey of best practices from public water systems around the country.

It is clear that WASA was ineffective in informing the public of the magnitude of the problem of lead in drinking water and in conveying the steps families and individuals should take to protect themselves. The spirit of the LCR encourages robust communication focused on the public's right to know. Mass media tools, including direct contact with media representatives, as is recommended in EPA guidance, were not used effectively.

The Region is taking a critical look back at how it could have done a better job in its oversight of WASA. There will continue to be lessons learned that will benefit the agency in the future. We have revised our oversight procedures to assure that shortcomings in public outreach are identified earlier and corrected. We are more closely monitoring WASA's activities to ensure that system-wide notices effectively inform customers about the lead risk and we will ensure that information provided in WASA's next Consumer Confidence Report to customers is clear with respect to information about lead levels in drinking water.

In addition to our collaborative efforts with the city, EPA has taken a number of actions to provide information to residents and others on the issue of lead in the District's drinking water.

The Region has created a new program with the National Nursing Centers Consortium, called Lead Safe DC, at an initial cost of $100,000, to bring lead education information, home visits and blood level testing to District neighborhoods. The consortium is the Nation's only network of nurse-managed community healthcare centers, and has enjoyed great success with a similar lead information program with EPA in the city of Philadelphia.

The Region has created a comprehensive Web site that includes advice for consumers, frequently asked questions, health effects information, and links to informational hotlines, WASA and the DC government. It can be accessed at *www.epa.gov//dclead.*

EPA has dispatched community outreach specialists to provide information and get input on the lead issue from community groups and individual residents in the District. EPA has a National Safe Drinking Water Hotline, and the region is proactively providing consumer information to radio stations for use in the District. Nearly a dozen regional employees have volunteered to assist with translation to Spanish of written and broadcast materials.

ACTIONS TO IDENTIFY AND CORRECT SOURCE OF HIGH LEAD LEVELS

The Technical Expert Working Group from the public and private sectors is in the midst of its investigation to identify the cause of the elevated lead levels and present a solution as quickly as possible. The team already has met early milestones in the process. EPA is scheduled to receive a proposal by April 15 from the technical team for a water chemistry change to reduce corrosion while maintaining the optimum protection against other harmful contaminants that can be found in drinking water. The preliminary recommendation is to conduct a partial system test using orthophosphates at WASA's Fort Reno pumping station and thereafter, a full system test by feeding orthophosphates at the Dalecarlia and McMillan water treatment plants. Under the proposed timetable, the partial system test is currently planned for June 1, followed by full system implementation on or about September 1. To review the team's findings, EPA has formed an independent peer review group made up of experts who are not involved in the planning.

Until the revised treatment process is in place and lead levels in tap water are reduced, all consumers are advised to follow appropriate tap flushing recommendations and to heed the District Department of Health's advisory that pregnant or nursing women and children under 6 years of age who live in homes that have a lead service line should not drink unfiltered tap water.

CONCLUSION

In closing, working closely with the District of Columbia, our public service partners and concerned citizens, we will continue to aggressively act to protect residents and resolve the lead problem. We are taking action to hasten the day when the citizens of the District of Columbia can once again be confident in the safety of their drinking water.

Thank you for the opportunity to present this information this afternoon. I am pleased to answer any questions you may have.

RESPONSES BY DONALD WELSH TO ADDITIONAL QUESTIONS FROM SENATOR INHOFE

Question 1. When did EPA first know of elevated lead levels in DC tap water?

Response. DC exceeded the 15 ppb action level during three reporting periods between 1992 and 1994 before it installed corrosion control treatment (CCT). The CCT implemented by the U.S. Army Corps of Engineers Washington Aqueduct (Aqueduct), and given interim approval by EPA Region III in 1997 and final approval in 2000, appeared to be effective in minimizing lead levels until the reporting period between July l, 2001 and June 30, 2002. EPA received a final report from the DC Water and Sewer Authority (WASA) on August 27, 2002 indicating that the 90th percentile value had increased to 75 ppb during that period. The high level required that WASA conduct more frequent monitoring every 6 months. The lead action level was also exceeded for subsequent monitoring periods in 2003, with 90th percentile values of 40 ppb (January 1 to June 30, 2003) and 63 ppb (July 1 to December 31, 2003).

Question 2. Has EPA identified any specific events or system changes that would have caused these high lead levels?

Response. EPA is reviewing past reports and performance data related to how the water was treated and how water chemistry changed in the Aqueduct wholesale customer's distribution systems. The review will investigate several factors which could have affected pH within the distribution system or otherwise affected chemistry in such a manner so as to increase corrosion. The areas that will be reviewed include, but are not limited to:

- the process change from chlorine to chloramine disinfection in November 2000 and potential impact of nitrification,
- the potential impact of drought during the period 2001–2002 on pH, comparing to changes observed in chemistry during the drought of 1998–99,
- the effects of water distribution system maintenance and operation practices, and
- the effects of corrosion control treatment management practices.

The Technical Expert Working Group (TEWG), which includes contractors for EPA, WASA and the Aqueduct and staff from WASA, the Aqueduct and EPA's Region III Headquarters and Cincinnati offices is assessing the possibility that some or all of the above factors contributed to the failure of the approved optimal corrosion control treatment to keep lead concentrations below acceptable levels.

Question 3. What actions did EPA initiate to determine the proximate cause of the higher lead levels?

Response. Following notification of WASA's exceedance of the lead action level in August 2002, EPA met with WASA staff in September 2002 to discuss activities that WASA needed to undertake to comply with the LCR. The main requirements were implementation of public education and lead service line replacement programs. WASA also agreed to resume full lead and copper tap sampling. EPA, WASA, and the Washington Aqueduct also discussed potential causes of the lead level changes and the need for a review of the corrosion control treatment process to determine what could have caused the lead action level exceedance and what could be done to correct it.

In January 2003, EPA Region III began the process of conducting the review on the behalf of WASA and the Aqueduct. WASA later decided to engage its own contractor to conduct a similar review. EPA developed a work assignment to conduct the corrosion control review under the scope of an existing contract and circulated it to WASA, the Aqueduct, Arlington County and the city of Falls Church to ensure that the concerns of all parties involved would be addressed. Work began in May 2003. WASA's contractor began working at about the same time. The Aqueduct assisted in this process by providing data and technical information required by the reviewers. In the fall of 2003, the contractors recommended that additional studies be conducted to better identify the cause of the lead action level exceedance so that the proper corrosion control treatment could be applied. WASA began implementing these studies in December 2003. Analyses of WASA's lead service line testing results in December 2003 found high lead levels which appear to be caused by the lead service lines.

EPA and WASA continue to use their contractors to implement the planned studies, review the results, and recommend new corrosion control treatment. A Technical Expert Working Group (TEWG), which includes contractors for EPA, WASA and the Aqueduct and staff from WASA, the Aqueduct, EPA's Region III, Headquarters and Cincinnati offices, the District of Columbia Department of Health, and the Centers for Disease Control and Prevention has also been formed to more fully investigate the matter.

Questions 4 and 5. What recommendations did Region 3 make to WASA or the Corps in terms of changes to the operations of the water system to mitigate for high lead levels?

Did EPA recommend adding zinc orthophosphate, or another phosphate additive to the water?

Response. In June 1994, the Aqueduct submitted a corrosion control treatment (CCT) study to EPA, in compliance with the requirements of the LCR for all large public water systems (serving more than 50,000 people) which recommended that pH adjustment be used to control corrosion. In 1995, EPA engaged a contractor to conduct a sanitary survey of the District's drinking water storage and distribution system. EPA included a task to evaluate the CCT recommendation in the Aqueduct's corrosion treatment study. The sanitary survey recommended that additional consideration be given to the use of a phosphate corrosion inhibitor.

In 1996, EPA engaged a contractor to review: (1) the Aqueduct's CCT study and recommendation, (2) observations and recommendations of the sanitary survey, and (3) additional data about the District's distribution system water quality, particularly with respect to the coliform bacteria problem the District had experienced over the previous few years. Because the contractor had concerns that a phosphate corrosion inhibitor might have adverse effects on water quality in an older water system such the District's, he recommended against use of a phosphate corrosion inhibitor. The contractor believed that it would be more appropriate for the Aqueduct to refine its pH adjustment so that it could maintain pH at a higher level.

In 1997, EPA conditionally designated pH adjustment as the optimal corrosion control treatment (OCCT) for the Aqueduct. WASA conducted tap sampling during 1997 to 2001 and sample results did not exceed the lead action level. In 2000, EPA designated pH adjustment as the OCCT for the Aqueduct.

With respect to the current elevated levels, EPA has been working with the Technical Expert Working Group to determine what changes should be made to the operation of the water system to reduce lead levels. Based on the recommendations of that group, Region III has approved an interim modification of the approved OCCT for the Aqueduct and WASA to allow for an initial application of a phosphate-based corrosion inhibitor. The April 30, 2004 approval letter from the Region to WASA and the Aqueduct, which outlines the process is available on the EPA web site at *http://www.epa:gov/dclead/corrosion*_letter_4-30.pdf. If the trial application is successful and does not cause any other unanticipated treatment concerns, full system application could begin as early as mid-July.

RESPONSES BY DONALD WELSH TO ADDITIONAL QUESTIONS FROM SENATOR JEFFORDS

Question 1. What explanation are you giving to parents whose children were totally unnecessarily exposed to lead in their drinking water and what steps are you taking to regain the trust of the citizens of Washington, DC?

Response. As Regional Administrator Welsh noted in testimony before the subcommittee, it is unacceptable to EPA that many families in the District are living with fear and uncertainty over the quality of the water they drink. The goals of the Federal requirements for the lead and copper rule are to ensure that the public is informed about potential risks and the steps they can take to minimize exposure. In its outreach efforts, WASA did not fulfill its responsibility to effectively inform all affected parties about a problem with their water. Notifications to individual residents were often not timely and did not achieve the goal of getting information to those who needed to know. Mass media tools were not used as effectively as they could have been. There should have been more widespread and urgent communication of the problem District-wide.

In hindsight, EPA could have more quickly assessed the timeliness, effectiveness and impact of WASA's public notification program and more promptly directed WASA to correct its deficiencies. We have revised our oversight procedures to assure that any shortcomings in public outreach are identified early and corrected. We are also investigating potential violations of the regulatory requirements relating to public education.

An EPA team has completed a report that reviewed the effectiveness of WASA's public education efforts. The report contains recommendations to enhance the effectiveness and delivery of public education under the lead and copper rule. It was transmitted to WASA on May 16, 2004. This report also included recommendations for EPA Region III to improve its oversight of WASA's public education program. We are more closely monitoring WASA's activities' to ensure that system-wide notices effectively inform customers about the lead risk and we will ensure that information provided in WASA's next Consumer Confidence Report to customers is clear with respect to information about lead levels in drinking water. WASA has been directed to submit an Enhanced Public Education Plan by May 21, 2004.

We are also taking our own actions to provide information to the public. EPA Region III dispatched community outreach specialists to provide information and get input on the lead issue from community groups and individual residents in the District. We continue to provide information on our Regional and National web sites and the National Safe Drinking Water Hotline. We have provided public service announcements, in English and Spanish, to media outlets in the Washington, DC area and have held or participated in 10 public meetings since early February. We have also been meeting with a coalition of environmental and consumer groups—the Lead Emergency Action for the District (LEAD) coalition, to both hear their concerns and to identify how to better communicate with the general public.

Question 2a. How was it that the EPA was able to identify these incidents of noncompliance [with respect to public education] 2 years after the fact but failed to identify them at the time they mattered most—when the public needed to have accurate information about the potential risks associated with lead contaminated drinking water?

Response. The technical staff in EPA Region III were focused on solving the problem of elevated lead levels. Staff noted that, public education activities were undertaken by WASA, but did not conduct a detailed review of the public service announcement. After the lead problem was observed to be more widespread, additional staff were assigned to perform an audit of all of WASA's public education materials and actions. This group conducted an onsite file review of WASA's records in March, 2004. The review team identified several potential violations which were outlined in a March 31, 2004 "show cause" letter from the Region to WASA and an information request letter to determine if there are any additional violations. The audit is ongoing.

As Regional Administrator Welsh mentioned in his testimony, standard procedures for handling and reviewing WASA's, as well as the Washington Aqueduct's compliance documentation have been revised. This was done to ensure that more detailed reviews of relevant materials are conducted by staff from the program office as well as the enforcement office. Public education materials will also be reviewed by those with experience in public outreach, writing and education to help ensure that WASA's outreach and education materials' are sufficient to gain the attention of customers so that they may take appropriate actions to reduce their exposure risks to lead or other contaminants.

Question 2b. How can you provide this Committee and the American people with any true assurances that our environmental laws are being enforced given the Agency's record on enforcement of environmental laws under this Administration, and the Agency's performance on this DC lead issue?

Response. EPA continues to work with our State partners to take appropriate steps to ensure that public health is protected. With respect to the provision of safe drinking water, this is accomplished by enforcing the Safe Drinking Water Act and its implementing regulations; including the use of EPA's authority under the Act to address situations that may present an imminent and substantial endangerment to public health. In fiscal year 2003 alone, EPA took 419 formal enforcement actions to address noncompliance by public water systems. Over the past 6 years, EPA has taken more than 4,000 formal enforcement actions under the Safe Drinking Water Act (to address public water systems and underground injection control). EPA also uses compliance assistance where appropriate as a tool to address drinking water noncompliance. EPA remains committed to ensuring that clean safe drinking water is available to every person, every day, no matter where they are in the Nation.

Regarding lead issues in particular, EPA's enforcement and compliance assurance program, in Headquarters and in the Regions, is actively participating in ongoing review of compliance with the Lead and Copper Rule. EPA Region III is working closely with the District of Columbia government to ensure that the Water and Sewer Authority takes appropriate actions to protect public health immediately and to ensure that their future actions are effective and meet both the intent and the letter of the regulations. If, at any time, EPA feels that the current intervention efforts are not working, Region III will issue an administrative action or take other appropriate action to enforce public health protections provided by our laws and regulations.

Question 3. Can you describe how the WASA action plan was developed and is that a public document? What actions will EPA take if WASA diverts from the contents of the plan and what oversight is the Agency conducting to ensure that the plan is executed?

Response. On March 4, 2004, EPA Region III issued a letter to the District of Columbia government listing ten actions that the Region believed were necessary to (1) reduce the public's risk of lead exposure, (2) improve the knowledge base on lead levels in tap water by conducting widespread testing and (3) improve the effectiveness of public education. The District's City Administrator's office transmitted a letter to WASA ordering that these ten actions be met and requiring that WASA submit plans to address each of the ten areas. Action plans were required to be submitted at different times depending on the action item to be addressed. WASA has developed and submitted each of the required plans. The Region and the District of Columbia have requested further clarification or more details on some of these plans. The letters and the plans are considered public documents and are available from EPA Region III's DC lead in drinking water website located at *www.epa.gov/delead.*

Question 4. [Re: inaccurate testing instructions from WASA]. Can you comment on your assessment of the importance of providing residents with accurate testing instructions and a description of how you have corrected this problem?

Response. EPA believes that it is critical that WASA provide accurate sampling instructions to residents. The data generated by this sampling is used to prioritize lead service line replacements, identify the people in the "unknown service line" category who should receive water filters and help inform residents about potential exposure. The data obtained from monitoring conducted using faulty sampling instructions distributed by WASA in early March 2004 is nearly useless for the purposes outlined above.

Upon being alerted to errors in the testing instructions, EPA Region III staff sent e-mail messages on March 11, 2004 to the Deputy General Manager and to the water quality manager of WASA describing our concerns. These e-mail messages requested that WASA immediately correct the instructions, identify who received the incorrect instructions with their sampling kit and describe a plan to correct the problem. EPA Region III staff held a conference call with WASA's water quality manager the following day to discuss the instruction sheet and data generated with its use. Although the water quality manager was not involved in this effort, which was managed by a WASA contractor, he confirmed to EPA Region III staff that WASA would not use these data for any compliance purposes and that he would work with others within WASA to identify which residents received sampling kits with the faulty instructions and send out new kits. The water quality manager sent, via e-mail, the corrected version of the instruction sheet to review by the end of the day on March 12, 2004.

On March 16, the Regional Administrator sent a letter documenting the previous correspondence between his staff and WASA and required that WASA report back to him on the number of people who had received sampling kits with faulty instructions. He also requested that WASA send out new kits to all persons so identified. WASA, in a March 24th letter to the Region, responded that more than 2,000 residents had received the faulty instructions and that new sampling kits with the corrected instructions would be shipped to all of them via UPS by March 27th, with delivery to be completed by the end of the following week. WASA has since confirmed to the Region that this schedule was met.

Question 5. [Re: Lead service line replacement only required for public portion]. I am interested in your views on whether public water systems or private homeowners should bear the responsibility for lead service line replacement?

Response. Where corrosion control is effective, removal of a lead service line may be unnecessary. However, when removal of the line does become necessary, EPA believes that removing the entire service line is the most prudent approach to assure public health protection. However, current legal authorities are limited to requiring water utilities that enter into the lead service line replacement phase of the Lead and Copper Rule to replace only the portion of the line that is under their control. The rule does not, however, preclude a utility from replacing the private portion at no expense to the homeowners if they determine it is appropriate. Our regulations require a utility to notify property owners at least 45 days prior to partial line replacement and to offer to the homeowner the opportunity to replace the private portion of the line at the same time.

The decision on who should pay for the private portion of the line replacement is a local, legal and policy decision. The Region knows of one instance where residents were required to replace their privately owned portion of the lead service line; at property owners' own expense. Because its source water chemistry poses challenges in implementing corrosion control treatment, Madison, Wisconsin began a program to replace all of the lead service lines in the city. The local government passed an ordinance that requires the property owner to replace their privately-owned portion of the service line, at the property owner's expense, when the water utility replaces the publically owned portion to ensure that all of the lead service lines are replaced on an accelerated schedule.

Question 6. In 1986, lead was discovered in drinking water in the Palisades section of Washington, DC. Did anyone here refer to previous instances of lead contamination in the District when formulating a response plan? If so, please describe how you used this information, and if not, why not?

Response. None of the staff working on this issue in EPA Region III were in the drinking water program in 1986. Our records for that time period have been archived or destroyed according to EPA records management protocols. As a result, current staff had no knowledge of the 1986 incidents related to lead in the tap water within the Palisades area of the District. The Regional Administrator has asked his staff to research this incident to determine if there are any lessons to be learned from it.

RESPONSES BY DONALD WELSH TO ADDITIONAL QUESTIONS FROM SENATOR CRAPO

Question 1. In testimony, you suggested that EPA would like to judge more accurately whether public messages are received about the risks of lead contamination in drinking water. You alluded to professional marketing practices for measuring the effectiveness of commercials and other advertisements. What measures such as this have you identified?

Response. At the request of the Regional Administrator, EPA staff conducted a review of the education and outreach activities which occurred in response to elevated levels of lead in the District's drinking water during 2002 and 2003. The report is entitled "Recommendations for Improving the Washington DC Water and Sewer Authority Lead in Drinking Water Public Education Program." It was transmitted to WASA on May 6, 2004 and made available to the public through our Internet site.

The review was undertaken to advise the Regional Administrator of potential areas for improvement by both WASA and EPA so that our agencies may implement the most effective outreach and education programs on the important issue of safe drinking water. The report was prepared by a team from various offices of EPA with expertise in drinking water on technical and regulatory issues, as well as outreach, education and risk communication. The report identified several steps that WASA can take to move toward more effective public education and outreach on the subject of lead in drinking water. In carrying out research for this report, we gathered input through interviews conducted with District of Columbia residents on their impressions of WASA's public education efforts and how best to reach them through a revamped education program.

A major recommendation in this report was that, in addition to following mandatory requirements and making use of EPA guidance, WASA should hire internal or consultant expertise in the areas of marketing research and risk communications. This expertise is needed to help the utility in assessing the audience to be reached, making recommendations for design and content of materials to be used, as well as delivery methods. The authors also suggested that these consultants assist with conducting a communications audit, developing a strategic communications plan and creating a tracking/measurement tool to assess the effectiveness of education efforts. Additionally, use of an outside consultant may help at least partially address the trust factor mentioned by some of those interviewed.

These recommendations should be viewed as a key input to WASA's continuing efforts to plan and carry out enhancements to drinking water education efforts both for regulatory compliance and "beyond compliance" efforts.

In order to have a clear path for the future, we requested that WASA prepare an Enhanced Public Education Plan in one central document to be submitted to EPA by May 21, 2004. We suggested that the Enhanced Plan incorporate the input of the report, as well as other recommendations that have been made as an outcome of other reviews and internal WASA reviews.

Question 2. What is the status of the Lead Safety DC program, which you mentioned in testimony? You described the program as an outgrowth of a pilot project in Philadelphia called Lead Safe Babies.

Response. The Region has created a new program with the National Nursing Centers Consortium, called Lead Safe D.C., at an initial cost of $100,000, to bring lead education information, home visits and blood level testing to District neighborhoods. The consortium is the Nation's only network of nurse-managed community healthcare centers, and has enjoyed great success with a similar lead information program with EPA in the city of Philadelphia.

The Lead Safe Babies (LSB) pilot served approximately 100 new/pregnant mothers in North Philadelphia on the issues of lead poisoning prevention. The program consisted of an initial home visit where the care taker is given a pre-test knowledge questionnaire, and clarification and/or additional education is provided regarding lead poisoning prevention. The visit includes detailed education about preventing lead poisoning; including but not limited to, the importance of hand washing, washing toys, house cleaning for lead dust, avoiding peeling paint, and good nutrition. When the child is approximately 8 months old, a second home visit is conducted where a post-test knowledge questionnaire is given to determine the knowledge retained by the care taker and to encourage initial blood lead testing for the child and each year subsequently until their sixth birthday. Due to the success of the pilot, another grant was awarded that expanded LSB to all of Philadelphia with an emphasis on high risk zip codes.

In 2002, the LSB program was expanded to five additional counties in Pennsylvania and the NNCC conducted an analysis of the effectiveness of the LSB program so that the program could be improved and easily adopted by community organizations. The analysis of the program showed that the average blood lead levels of children in the LSB program were much lower than compared to the average blood lead level of children in the same census tracts. An analysis of the LSB Program shows that in the four participating health care centers, the average LSB blood lead level ranged from 3.7 micrograms per deciliter (ug/dL) to 9.5 ug/dL on the first test (at approximately 8 months of age) as compared to neighborhood averages (neighborhood where the centers are located) of 16.56 ug/dL to 24.10 ug/dL. The second test is conducted at 2 years of age when children are more mobile and are at increased risk for exposure to lead. There were only two centers that had enough data from the second tests to show results which indicated that there was an average of 10 ug/dL to 10.75 ug/dL for LSB program children versus 14.72 ug/dL to 15.42 ug/dl for the neighborhood.

To date the LSB program has served over 1,100 at-risk infants. Last year, the Centers for Disease Control awarded the city of Philadelphia along with NNCC and its member nurse-managed facilities funding to continue in home visits to approximately 1250 mothers in the Philadelphia area. The LSB materials have also been translated into Spanish. We look forward to bringing the program to the District and hope to achieve similar levels of success.

Question 3. You testified to the need for better teamwork if the Lead and Copper Rule is to be implemented effectively. Now that your agency and the other authorized agencies are working very closely in what has become a closely-scrutinized effort, what improvements in teamwork have you leanred? Also, how will you change routine procedures for working together to implement the Rule after the DC drinking water system returns below the action levels?

Response. We learned that, internally, EPA can improve our oversight of the District's water supplies by better utilization of our resources and by involving expertise from elsewhere within the Agency. This expertise should be tapped even when monitored parameters are within acceptable ranges. As mentioned in previous testimony, internal procedures in EPA Region III have been changed so that data and compliance reports are reviewed and tracked by several technical staff rather than one program manager. Education and outreach materials are being reviewed by communications and public relations staff members.

WASA has agreed to share their materials in draft form so that EPA can review outreach documents and offer suggestions not only on compliance with the regulations, but on clear, concise messages related to obtaining the public's attention. My staff have already begun the process of gathering input from State agency staff who deal with many public water supplies and have decades of experience in monitoring normal water supply operation parameters. We have improved our coordination and communication on drinking water issues with the District Department of Health, WASA, and the Aqueduct, and have committed to maintain this closer relationship.

We have also learned how important it is to get local community groups involved early in any issue affecting the general public, whether it is EPA or WASA garnering their involvement.

STATEMENT OF JERRY N. JOHNSON, GENERAL MANAGER, DISTRICT OF COLUMBIA WATER AND SEWER AUTHORITY, WASHINGTON, DC

Good afternoon, Chairman Crapo, Ranking Member Graham and other members of the Committee. Good afternoon, Chairman Crapo, Ranking Member Graham and other members of the Committee.

Thank you for your invitation to the Authority to provide information to the Subcommittee on the exceedance in the District of Columbia of the action level under the Lead and Copper Rule.

I am Jerry N. Johnson, General Manager of the District of Columbia Water and Sewer Authority, and I am accompanied by the Authority Chief Engineer and Deputy General Manager, Michael Marcotte.

I know that you have many questions, so I will be very brief. Providing high quality services under the Safe Drinking Water Act for people who live in, work in and visit this city, has been, and continues to be our top priority.

WASA'S MISSION

The Board of Directors only recently updated WASA's Strategic Plan for 2003 through 2005.

WASA's vision of its future is to be the industry leader and environmental steward in providing excellent water service and wastewater collection and treatment services for all customers.

The mission of WASA is to serve all it regional customers with superior service by operating reliable and cost effective water and wastewater services in accordance with best practices.

Among our values are to be respectful and sensitive to the needs of our customers, ethical in professional and personal conduct, and committed to equity, trust and integrity in all that we do.

In facing the current challenge, we pledge every effort to: (1) understand the phenomenon of increased levels of lead concentration in certain households; (2) allocate the necessary resources and work with the District government and our partner agencies to address the problem; (3) be candid with the public and with you about the information we have and about our plans.

WASA'S HISTORY, GOVERNANCE, AND OPERATIONS

The District of Columbia Water and Sewer Authority, or "WASA", was created as an independent agency of the District of Columbia in 1996 following an extended period of disinvestments in the critical infrastructure serving the District and the diversion of over $80 million in water and sewer ratepayer enterprise funds for unrelated expenditures.

WASA is an independent, quasi-governmental regional entity that is governed by an eleven-member board of directors that is appointed by the Mayor of Washington, DC and confirmed by the District of Columbia Council. Six of the Board members, including its chairman, are District residents chosen by the Mayor. Five members, though named by the Mayor, are selected by the county executives from surrounding jurisdictions.

WASA's Board has fiduciary responsibility for the Authority, which has independent bond authority and a "double A" bond rating. In fiscal year 2003, operating revenues totaled approximately $255 million. Capital expenditures reached approximately $200 million for the year, and are part of a $1.8 billion 10-year capital reinvestment infrastructure program.

WASA provides wastewater treatment services for the District of Columbia, and Montgomery and Prince George's Counties in Maryland, as well as portions of Loudon and Fairfax Counties in Virginia at Blue Plains, the largest advanced wastewater treatment plant in the world. WASA also operates the District of Columbia's 1800-mile storm water and wastewater collection systems.

As you know, WASA purchases treated drinking water from the U.S. Army Corps of Engineers Washington Aqueduct for delivery through the District's 1300 mile distribution system to retail customers in the District of Columbia.

DISTRICT HISTORY IN EXCEEDING THE LEAD & COPPER RULE

Since 1991 when EPA promulgated the current Lead and Copper Rule, the District of Columbia, like other municipalities, routinely tested water for lead concentrations in a small number of homes.

The District of Columbia exceeded the action level in the period 1993–1994, and WASA's predecessor agency began taking steps to address the chemical makeup of the drinking water.

The Washington Aqueduct implemented a corrosion control regime that alleviated the problem of lead leaching in 1994, and lead levels remained below the action level for several years. The EPA conditionally approved the corrosion control approach in 1997.

When WASA began its operations in 1997, annual compliance testing from 1996 through 2001 indicated that DC did not exceed the action level. In fact, in 1999, the EPA offered, and WASA accepted, a reduced regular sampling program. As a consequence, the number of samples required for compliance monitoring was reduced to 50 households sampled annually.

THE RECENT EXCEEDANCE

In May 2002, EPA approved the revised optimum corrosion control program.

In August 2002, WASA provided the official notice to the EPA that for the first time since WASA was created, but the second time since 1993–1994, water in the District exceeded what is now well known, but still sometimes misunderstood, the "action level".

The action level is a regulatory trigger of 15 ppb that no more than 10 percent of the samples can exceed. The 50 samples taken in the District for that compliance period included 23 samples over 15 ppb.

EPA then requires that a utility take three basic steps until there is a reduction in the number of samples that test above 15 ppb to fewer than 10 percent of the total: (1) begin a public education campaign to inform the public about lead as an environmental contaminant; (2) begin a program to replace or test 7 percent of the public section of the known inventory of lead service line pipes that take water from the public water mains to the individual properties; and (3) undertake immediate steps to achieve optimum corrosion control of the treated water.

STEPS TAKEN TO COMPLY WITH THE LEAD AND COPPER RULE (2002–2003)

WASA is a regulated utility, and it is accountable not only to the customers and broader public that we serve, but to the Board of Directors, and the Environmental Protection Agency. The District of Columbia Council also maintains legislative oversight over WASA.

I believe the culture of this organization is one that is consistent with the mission statement I shared with you earlier. Clearly, there are also a number of entities to which we are accountable and which share this common interest.

In this regulated environment, as in all others, every locality is in some respects unique. We've worked hard to make sure our regulators understand the specific operational challenges we confront. We share information, we ask questions, we seek guidance and instruction, and then we take what we understand to be appropriate action with the full knowledge of the regulator. Again, regulatory compliance is of paramount concern. We undertook a serious effort to ensure compliance under the Lead and Copper Rule in 2002. As a consequence, in:

- August 2002—WASA determined/reports to EPA that action level was exceeded;
- October 2002—DCWASA/DCDOH jointly release "Living Lead Free in DC;" Brochure is delivered to customers and editorial departments of *Washington Post* and *Washington Times*, as required (at this time, WASA was not issuing monthly bills);
- October 2002—WASA releases Public Service Announcement raising awareness and encouraging testing;
- June 2003—EPA approves first year of replacement program, but changes compliance date from 12/31/03 to 9/30/03;
- June 2003—WASA's 2002 Drinking Water Quality Report includes information on lead monitoring and exceedance (language approved by EPA);
- July–October 2003—WASA begins intensive sampling program to accompany the first 400 physical replacements undertaken within timeframe that was shortened by EPA;
- August 2003—"What's on Tap" Customer Newsletter contains focus on lead;
- September 2003—Initial program year completed with 1615 services replaced or cleared by sampling;

• September 2003—"An Information Guide on Lead in Drinking Water" developed by WASA and EPA and distributed by WASA in two languages (schools, clinics, libraries, ANC Chairs, DCDOH);
• September 2003—WASA again releases public service announcement encouraging testing;
• October 2003—Washington Post advertisement: "WASA and EPA recommend that you have your water tested for lead"; and
• December 2003—Meeting held by WASA to discuss projects funded by Safe Drinking Water Act, including lead replacement program.

OPTIMAL CORROSION CONTROL

Apart from the gradual replacement program, and the public education program, reducing the corrosivity of treated water is a principal objective of a water utility once the action level has been exceeded. WASA, in conjunction with the Washington Aqueduct and the EPA has pursued this goal.

Optimizing corrosion control in the treatment process has, and continues to be the critical next step in addressing this issue.

I am very pleased to report that there has been measurable progress on that front. There is now a draft plan that is being circulated for a 15-day comment period. Consensus on this draft plan will lead to concrete steps that can be taken in the treatment process within the next few months and into the fall.

STEPS TAKEN BEYOND THE REQUIREMENTS OF THE LEAD AND COPPER RULE

The Water and Sewer Authority, however, went beyond the requirements of the regulation, principally by working directly with customers. This effort resulted in our responding openly to many individual inquiries, but also to our responding to media inquiries, requests for participation in community meetings and to participation in some of the active community listserves. Specifically, in:

• October 2002—Responded to media inquiries on the exceedance (*Washington City Paper*, article, "The District Line—Plumbing the Depths");
• October 2002—WASA notifies DCDOH that action level was exceeded;
• January 2003—Lead Services Hotline begins—providing specific information to customers;
• February 2003—Written communication to Mayor and all DC Council members advising of likelihood of constituent calls that result from lead replacements;
• March 2003—WASA sends letter of notice of replacement program and an invitation to meet and discus with WASA (transportation provided) to Advisory Neighborhood Commissioner chairs and civic association leaders;
• May 2003—WASA holds two community meetings on lead replacement program;
• November/December 2003—Three neighborhood meetings held by DCWASA to discuss replacement program;
• December 2003—Meeting held by WASA to discuss projects funded by Safe Drinking Water Act, including lead replacement program;
• December 2003—Full summary of 2003 test results compiled by DCWASA; and
• December 2003—WASA requests that DCDOH provide additional assistance.

Starting in April 2003, between 14,000 and 15,000 WASA customers were contacted and solicited to participate in the sampling program to test the concentration of lead in the water at the tap. Sampling was undertaken by volunteers (residents), and no customers were forced to participate in the program.

Initially the customers were contacted by mail. After an initial low response, customers were offered a $25 incentive to participate. From April 2003 through September 30, 2003, about 3200 customers were contacted by telephone. Approximately 11,000 sampling kits were sent to customers by Federal Express through September 2003.

Lead Services Sampling Program 2003—Direct WASA Customer Contacts

[Direct WASA Customer Contacts]

Sample Letters/No Incentive	8,000
Sample Letter/Incentive ($25–$50)	6,000
Phone Solicitation	3,200*
Sample Kits Shipped by Fed Ex	11,000
Approx. Total Contacted	14,800
Total Samples	6,131

* Partial overlap w/letters.

By the end of 2003, WASA had also created a customer e-mail account, *wqp2003@dcwasa.com*, for customers to make inquiries and express concerns. WASA also implemented extended call center hours in September 2003 to include Saturdays.

MULTIPLE LEAD SERVICES REPLACEMENT PROGRAM INQUIRIES

On March 4, 2004, the DCWASA Board of Directors announced that it has retained a law firm to investigate WASA's management of elevated lead level sampling and notification. The investigation will be conducted by Covington & Burling, an internationally recognized law firm headquartered in Washington, DC. Heading the investigation is Covington partner Eric H. Holder, Jr., the former U. S. Attorney for the District of Columbia and former Deputy Attorney General of the United States. The investigation is expected to be completed and results published in the first weeks of May.

The other inquiries of which we are aware, include the:

- U.S. General Accounting Office;
- District of Columbia Office of the Inspector General;
- District of Columbia Council Committee on Public Works and the Environment;
- House Committee on Government Reform; and
- House Energy and Commerce Subcommittee on Environment and Hazardous Materials.

The Authority continues to believe that it took appropriate steps to comply with the Lead and Copper Rule. Our efforts took place as we continued our efforts to consult with the Environmental Protection Agency and the District of Columbia Department of Health.

We have, none-the-less, been severely criticized by some public official and citizens. We acknowledge this criticism, and it is our obligation to listen and to understand.

We await the outcome of each of these inquiries, and we are prepared, speaking on behalf of the management and the Board of Directors in this instance, take whatever actions are appropriate.

CONTINUING COMMUNITY OUTREACH

Let me assure you, Mr. Chairman and every other member of this committee that with this, as on any other challenge this relatively young agency has been confronted with, we seek to learn from the past and continually improve our services.

Building and maintaining public confidence in this vital service with which we have been entrusted on a continual basis must be an integral part of what we do.

In the past 10 weeks:

WASA has shipped filters to every residence that is believed based on our records to have a lead service line pipe. This latest step was undertaken in conjunction with the Environmental Protection Agency and Mayor Anthony Williams.

WASA will also supply replacement cartridges.

On Saturday, February 14th, working closely with the District of Columbia public schools we tested all of the District's public schools—an extra precaution even though we believe the pipes leading into the schools are lead free. WASA has trained DC Public School staff to conduct another round of public school tests which is underway.

I recommended and the Board of Directors supported a decision to increase the number of lead service pipes in public space that we replace this year by more than 500. These physical replacements will be at properties with the highest concentrations and where a pregnant woman or where a child under the age of six lives.

The Board approved a resolution and is distributing for public comment new steps it may take to address this issue in the long term. Two examples include the question of replacing lead service lines in public space with a timetable that goes beyond the requirements of the EPA's Lead and Copper Rule, and the difficult challenge of financing the replacement of service lines in private space.

The Lead Services Hotline, a program that EPA did not require, was initiated in January 2003 to facilitate direct communications with our customers. Since February 5, we expanded with more personnel allowing us to staff the operation for 12 hours Monday through Friday and for 9 hours on weekends. Since February 4, 2004 the Hotline received 45,746 calls, and 6,233 e-mail messages. We have shipped over 19,000 test kits.

A summary of the 2004 Sampling Program results is attached, for your information. We have now conducted a total of over 11,000 tests of water provided by our customers, and we are processing several thousand more results in a much-improved process that minimizes customer inconvenience.

As you may know, Mayor Williams established and co-chairs with the DC Council Committee on Public Works and the Environment Chairman, Carol Schwartz, an Interagency Task Force. This body has been enormously helpful in coordinating the efforts of District agencies in response to this challenge.

We have worked closely with the District of Columbia Emergency Management Agency and Department of Health. WASA is, for example, providing DC DOH with $1.5 million in expenses for the DOH blood lead-level testing program and associated activities to improve their data processing systems. WASA has budgeted $1.7 million for WASA, DOH and other joint outreach/communications initiatives, excluding $1.5 million for expanded Lead Services Hotline command center operations.

WASA is continuing our efforts to communicate effectively with our customers and the general public. We continue to update our web site, *www.dcwasa.com*, which is averaging over 1,700 visits daily. Our April customer newsletter, *What's On Tap*, includes information on lead, our flushing advice, as well as the annual distribution system citywide flushing program. This newsletter is distributed to between 125,00 and 130,000 customers, and the March and April editions both focus on the lead issue.

Since February, we sponsored about ten joint meetings with the Department of Health and the Washington Aqueduct all across the city. These meetings have been advertised and nearly 1,000 residents have attended these WASA sponsored events. WASA has also participated in many civic group meetings to discuss the lead issue.

In early February, WASA sent over 300,000 letters in English and Spanish with information to every address in the District of Columbia. This letter included a DOH Fact Sheet, again in Spanish and English, general information on the subject of lead in water, as well as precautions for potentially affected properties. This letter was mailed in a specially printed envelope with a large letter message printed on the front ("Please Read: Important Lead Information").

RECENT OUTREACH EFFORTS

Our work continues as we speak, Mr. Chairman. The last two editions of the WASA customer newsletter, "What's On Tap" (March and April), have also been devoted to this issue.

WASA is currently contacting by mail the residents that reside in homes that are believed served by a lead service line pipe. We are providing up to date information on service line pipe flushing. The letter also urges those that have not yet had their tap water tested to use our testing process at no cost to the individual customer.

Similarly, we are contacting the residents who live in properties for which the WASA customer information system has no record for service line pipe material. As you may know, the only way to determine the service line pipe material with certainty at this time is to dig it up.

Finally, WASA has also reached a preliminary agreement with the George Washington University School of Public Health Department of Environmental and Occupational Health to provide the Authority with assistance and advice on lead in drinking water issues.

ADDRESSES WITH NO RECORD OF SERVICE LINE PIPE MATERIAL

The Authority has been working to more carefully define and fine-tune our initial inventory of properties that rely upon a lead service line pipe. You may know of the Weston Study, undertaken in 1990 by the District, and which was used by WASA and EPA to establish the initial inventory of addresses with likely lead service line pipes. WASA has accelerated efforts to gather all the information available to us on addresses that may be served by a lead pipe, but the Weston Study provides the only estimate of the overall number of lead service lines in the District.

It is our continuing challenge to refine our information and the specific addresses that are most likely among those with lead services.

There have been media reports that there are as many as 40,000 to 50,000 addresses with no record of a pipe material. Those numbers are greatly exaggerated.

WASA is now contacting by mail the approximately 21,000 customers at addresses for which we have no record of a pipe material. We are urging them to use our testing program to sample their tap water, recognizing that a number have already been sent test kits. We are also urging those residents to take the same precautions as residents with known lead service line pipes—following the flushing recommendations.

WASA is also undertaking some test "dig-ups" where test results suggest the presence of a lead service line, and we are now developing an appropriate plan to provide filters to additional properties that are likely relying upon lead services, and we will work with these customers and the Task Force in the coming weeks on any related issues.

On Friday, April 2, 2004, the Board of Directors approved a resolution that addresses future WASA policy with respect to the replacement of lead service lines, separate and apart from the EPA requirements. The Board has planned three community meetings in the next few weeks in order to obtain community comment on this topic.

Earlier I mentioned the number of public meetings we have participated in and hosted. Since February, the Board has also conducted four media briefings. We appeared at the House Government Reform Committee hearing, four DC Council hearings, twice weekly Mayoral Press Briefings and three congressional staff briefings. As I hope you can see, the Water and Sewer Authority is actively engaged, we are continuing to learn, we are sharing information, and working hard to openly communicate in an environment that has been especially challenging.

Each one of these appearances is another opportunity to share facts and improve public understanding of a fairly complex set of health, regulatory, engineering, chemistry and policy issues.

On behalf of the Authority and its Board of Directors, I wish to express our appreciation for this Committee's interest and contribution to this effort.

We will be happy to respond to any questions.

RESPONSES BY JERRY N. JOHNSON TO ADDITIONAL QUESTIONS FROM SENATOR JEFFORDS

Question 1. I have been frequently told by parents—If I had only known, I could have taken precautions. I want to ask each of the witnesses at the table to tell me how you are responding to this question—what explanation are you giving to parents whose children who were totally unnecessarily exposed to lead in their drinking water and what steps are you taking to regain the trust of the citizens of Washington, DC?

Response. The Water and Sewer Authority's highest priority has been and continues to be to ensure that it delivers safe and high quality drinking water to residents and visitors to the District of Columbia. The Lead and Copper Rule of the Safe Drinking Water Act sets forth, in detail, the type and form of information that the EPA deemed appropriate to disseminate to the public. WASA followed the guidelines of the public education program set forth in the Lead and Copper Rule. In February of 2004 it became clear that the proscribed public education program did not reach all residents.

Following is a summary of the type of information we continue to share with the public, particularly populations believed to be most susceptible to harmful lead exposure.

In 2002, the results of EPA required annual compliance sample testing, indicated that the "action level"—a regulatory trigger that informs water authorities that optimum corrosion control is not being achieved—had been exceeded. The data was provided in a timely manner to the Environmental Protection Agency as required under the Lead and Copper Rule. In fact, the information was provided informally 3 months prior to the end of the compliance reporting period. The DC Department of Health was also notified. The exceedance of the action level indicated that the Water and Sewer Authority should begin:

- working with the EPA and the Washington Aqueduct to achieve optimum corrosion control (WASA is a customer of the Aqueduct—buying water for retail distribution to consumers in the District of Columbia);
- a program of public education about the hazards of environmental lead exposure from water and other sources;

• a program to replace (or test to clear) 7 percent of the lead service line pipes in public space yearly until the action level was no longer exceeded, and;

• implement an expanded compliance monitoring program.

WASA worked with the District of Columbia Department of Health to develop and widely distribute a brochure that provided information to residents about the risks associated with environmental lead exposure, including lead concentrations in water. WASA also distributed public service announcements and prepared pamphlets in English and Spanish providing information to populations at greatest risk, including information on flushing, tap water testing, and other information; all in accordance with the proscribed EPA public education program. WASA conducted several informational meetings throughout 2003 designed to increase WASA's outreach.

In 2004, WASA significantly expanded its outreach program. High-level managers have attended countless public meetings, answering questions and disseminating information. WASA has been working with the media (including seeking many corrections) in an attempt to make sure correct and helpful information is published. WASA has written and submitted for publication articles and letters to the editor.

WASA has continued its free testing program—any single-family residence in the District may have its water tested for lead concentrations. WASA has sent letters to all addresses believed to have a lead service line, encouraging participation in the sample testing program, and alerting them to the flushing procedures. Property owners for which we have no information on service line pipe material have also been contacted and provided the same information.

WASA has distributed water filters to residences that we believe are likely to have a lead service line pipe. Any residence that participates in the sample test program with a test result that exceeds 15 ppb will be sent a filter. WASA intends to continue its filter program until at least next spring.

WASA has retained and is working with a national expert on corrosion control. WASA has participated in the Aqueduct's planning for a treatment approach to reduce the corrosivity of the water.

Also in 2004, out of an abundance of caution, and before the blood level testing had yielded significant data, the Department of Health advised pregnant or nursing women and children under six with a lead service line to avoid drinking unfiltered water. The District of Columbia Department of Health has conducted over 5000 blood level tests, including over 1,500 tests of the targeted population. Thus far, the test results indicate that there has been no general increase in elevated blood levels in the population. Also, for those households that have undergone environmental assessments where a very limited number of individuals have been found to have registered elevated blood lead levels, other significant environmental exposures (paint and dust) exist.

Consistent with Department of Health findings and conclusions announced to date, Dr. Tee Guidotti, Professor and Chair of the Department of Environmental and Occupational Health, School of Public Health and Health Services, and Director, Division of Occupational Medicine and Toxicology School of Medicine and Health Sciences, The George Washington University Medical Center, has advised the Water and Sewer Authority that children who already have a body burden of, BPb >10 μg/dL are most susceptible to harm from heightened levels of lead. Dr. Guidotti has also indicated that:

• Drinking water is at most a minor source of lead for children (seven percent of total exposure for toddlers, children aged 2 years);

• Concentration in water does not translate directly in BPb;

• EPA's Integrated Exposure Uptake and BioKinetic Model predicted that BPb in infants exceeded 10 μg/dL only when 100 percent of water consumed contained 100 ppb on a sustained basis; and

• A discernable effect on BPb of children requires at least sustained levels of 300 ppb.

WASA is moving forward with its lead line replacement program. By September 30, 2004, WASA will have physically replaced over 1,600 lead service line pipes in public space in the District. WASA's Board of Directors is considering a plan to increase the pace of the service line replacement—more than doubling the amount that will be replaced yearly.

Question 2. I am pleased that WASA has hired a George Washington University toxicology team to advise the agency. Does this team contain anyone with pediatric experience on this team or anyone with risk communication experience and if not, why not? Will this group be a permanent part of WASA and what role with this group play in your day-to-day operations?

Response. Dr. Tee Guidotti is the Co-Director of the Mid-Atlantic Center for Children's Health and the Environment, a pediatric environmental health specialty unit. The team of professionals working with WASA from the George Washington University team also includes individuals with experience in risk communication. Members of the team participate in weekly meetings with WASA executive management. They also participate in select community and interagency meetings involving the local and Federal Government officials. The partnership with the University is on a contractual basis, and the Board of Director's has not yet determined that a permanent engagement is required.

Question 3. WASA recently undertook a program to replace aging water meters. Can you describe this program for me, indicate if you installed lead-free water meters, and indicate if WASA uses lead-free parts when replacing parts throughout its system? In addition, have you cross-referenced your data for those homes with high lead levels and those that have received a new meter, and if so, what are your conclusions?

Response. The meter replacement program installations total 110,047 of 124,298 meters through March 2004. The meters are certified lead free by the manufacturer, consistent with EPA requirements. Any contact with the service line during the installation would be unusual. The installation of the meters does not require direct contact with the service line pipe because the old meters are removed from and the new meters are placed on a parabola-shaped meter "setter" that ties into either end of the service line pipe. Also, the fact that ten of the 25 compliance samples taken in the first half of the compliance year 2001–2002 exceeded 15 ppb clearly indicates that the meter replacement program is not related to the exceedance because this sampling was completed about 6 months before the meter replacement program began.

Question 4. I know people who have homes with elevated lead levels, but not above 15 parts per billion. I know people who have homes with elevated lead levels well above 15 parts per billion who are not on your inventory of "known lead service lines." Given what we know about the harmful effects of lead, what is WASA and the DC Department of Health doing to help those residents take appropriate health precautions?

Response. The EPA has indicated that 15 ppb is not a maximum contaminant level (MCL) indicating the highest level of an contaminant that is allowed in drinking water, nor is it a maximum contaminant level goal (MCLG), the level of a contaminant in water below which there is no known or expected risk to health. Rather, it is a an "action level". An action level is a concentration of a contaminant that, if exceeded, triggers actions a water system must undertake.

Sample test results that exceed 15 ppb do not indicate that a service line pipe is necessarily made of lead, just as a test result that is below 15 ppb does not necessarily mean that a service line pipe is made of a non-lead material. WASA continues to work to provide accurate information, address public concern and to correct misinformation. For example:

- Any resident who is concerned about lead water concentrations is encouraged to use a flushing protocol, and this information has been repeated on a monthly basis in the customer newsletters, frequent media briefings, and dozens of community meetings, hearings, and community newspaper ads.
- Pregnant and nursing women and children under six have been advised to avoid drinking unfiltered water if they have reason to believe that they may have a lead service line.
- WASA has directly and strongly encouraged households with a lead service line to participate in the sampling program.
- Residences for which we have no record of a pipe material have been contacted directly, and encouraged to participate in the sampling program.
- Despite the fact that 15 ppb is not a health standard, WASA is providing a filter and replacement cartridges to any residence that participates in the sampling program and tests above 15 ppb.
- WASA has participated in over 24 community meetings.
- WASA continues to share information with the public by seeking major electronic media interviews.

Question 5. Under current regulations, public water systems are only responsible for replacing the portion of lead service lines in public ownership. In replacing lead service lines, is WASA moving the location of the meter, and if so, how is this affecting the length of the service line that you are replacing and the associated costs?

Response. WASA is responsible for the maintenance, and replacement as necessary, of a service line pipe in public space. WASA is not routinely moving water meters as a part of the physical replacement of lead service line pipes, and the position of a meter does not affect our responsibility for maintaining the service line pipe that rests in public space.

Occasionally, while undertaking other work, WASA determines that a meter is located in private space and some distance away from the property line. WASA repositions such meters from a private yard to an area closer to the property line (usually the sidewalk or "tree box") because it is a sound business practice and a matter of efficiency. This action helps avoid any potential conflict with a homeowner with respect to determining responsibility for any water usage (leaks) that occur on private property, regardless of the location of the meter. Repositioning the meter, when necessary, also allows WASA to conduct routine maintenance without entering private property.

Through 2003, WASA's physical replacement program resulted in the removal of a lead service line pipe in public space, specifically from the water main in the street to the property line as is consistent with Federal and local requirements. However, recently WASA has replaced approximately 400 service lines from the water main up to the next threaded joint. This practice was adopted in conjunction with the Department of Health and the EPA, based on the known likelihood of a temporary spike in lead concentrations in tap water following the cutting and reattachment of an existing lead pipe at the property line and its reattachment to a copper pipe in public space. In order to avoid creating this temporary spike, WASA, pending a new finding by the Department of Health, will replace a service line from the water main to the next threaded joint, regardless of the placement of the meter.

Question 6. In your cooperation with the other municipalities you serve or during the Washington Aqueduct Wholesale Customer Board Meetings, did you ever discuss with or notify those communities of the results of your lead testing?

Response. The Water and Sewer Authority is the retail distributor of water to residents of the District of Columbia. WASA does not provide retail water services to suburban jurisdictions. The suburban jurisdictions, like WASA, are direct wholesale customers of the Washington Aqueduct. In the fall of 2002, following the trigger of the action level, there were discussions with EPA and the Aqueduct regarding a study of the optimal corrosion control methodology and why it was no longer being achieved.

Question 7. Several weeks ago, I notified the EPA that several residents of the District had received inaccurate testing instructions from WASA. It seems to me that incorrect testing instructions could invalidate the entire sampling plan that WASA had in place before the lead was identified and the entire sampling plan that is ongoing to further evaluate the severity of the lead crisis in DC. Can each of you comment on your assessment of the importance of providing residents with accurate testing instructions and a description of how you have corrected this problem?

Response. Clearly, it is important to provide clear instructions to ensure adherence to any testing protocol. The Lead and Copper Rule does not, and prior to your inquiry EPA did not provide specific direction with respect to the protocol or the instructions for the non-compliance sampling program. Because this is an important issue, WASA has modified these instructions in response to questions or concerns, including concerns expressed by non-compliance sampling program participants, in order to make them clearer and more understandable.

Specifically, we have modified in conjunction with the EPA the instruction that you believe may have caused confusion has been modified. Participants who received that instruction sheet have been asked to submit new samples.

With respect to the validity of the compliance sample tests, the instruction item about which your office inquired is relevant only to the second draw sample that is part of the free sampling program offered to District residents. It was not pertinent to the Lead and Copper Rule compliance sampling program in effect prior to or following the District's exceedance of the action level.

In the current compliance year, the test results have been used to help determine the potential need to take steps to limit potential exposure. For example, any residence with a test result that exceeds the regulatory action level of 15 ppb receives a water filter and replacement cartridges.

Question 8. Can you describe exactly why you believe that lead contamination in drinking water in apartment buildings is not a problem and what evidence you have to support that claim?

Response. WASA's policy is to replace service lines for multi-family units that are known to be lead as a high priority. However, WASA's best information is that larger multi-family properties are served by service line pipes that exceed 2 inches in diameter. Lead pipes are very malleable but have very thick-walled cylinders with small diameter interiors, and are therefore, a poor choice for uses that require the deliver of large amounts of water to a large multi-family dwelling or commercial property.

The principal concern for larger properties (those bigger than a single-family sized dwelling) are the internal plumbing components (lead solder in the plumbing or brass fixtures). WASA proposed a test plan to EPA that will test these assumptions. EPA approved the plan and WASA has begun implementation of the test plan.

WASA discourages individual tenants from seeking to participate in the sampling program, but encourages owners or managers of multi-family units to have their water tested. When an investigation (CIS, Lead Information System, as-built plans or direct observation) indicates that the line is likely lead or was unknown, additional steps are taken. EPA protocols for water sample testing in such facilities are not the same as those for water utilities under the Lead and Copper Rule. However, WASA will provide a limited number of sample test kits in coordination with a building owner or property manager.

Question 9. How is the Action Plan being made available to the public?

Response. WASA has held 12 community meetings since February, and participated in as many civic, ANC and other community group meetings. The District of Columbia Department of Health and the Washington Aqueduct are usually also represented. WASA also attends Mayor Anthony Williams' press briefings that have routinely occurred twice weekly. The steps WASA is taking on the lead services program have also been noted in the General Manager's Monthly Report to the Board of Directors and Water and Sewer Authority press notices. Information is routinely updated on the Authority web site *www.dcwasa.com*. The lead services program is routinely on the agendas of the monthly meetings of the Board of Directors and the Board committee meetings, including operations, finance and budget and customer. These meetings are open to the public (calendars and agendas are published). WASA has issued public service announcements, and is preparing ads for publication in local community newspapers.

Question 10. I know people who live on Capitol Hill in houses with elevated lead level, but not above 15 parts-per-billion. The 15 parts per billion is not a health-based standard. Are you providing filters to those homes with lead service lines with test results below 15, but above zero, and if not, why not?

Response. WASA is not providing filters to residences with sample test results that do exceed 15 ppb. As you may know, lead concentrations below 2 ppb are not easily measured (non-detectable), and as you state, 15 ppb is a regulatory trigger, not a health standard. Fifteen (15) ppb is, therefore, not directly tied to a specific measure of exposure that is linked directly to health risk. Also, the second draw service line concentrations suggested in sample test results may not easily extrapolate into sustained rates of exposure noted by some public health experts when discussing risks of environmental lead exposure.

WASA is providing a filter and replacement cartridges to sample program participants whose tap water sample tests exceed 15 ppb out of an abundance of caution and an effort to address public concerns.

Question 11. I understand that you are also not providing filters to those residents whose homes have tested high for lead that are not in your data base of homes with a confirmed lead service line. Are you providing filters to those homes that are outside of this data base but have tested high for lead?

Response. WASA is providing water filters and replacement cartridges to residences that participate in the sampling program whose test results exceed 15 ppb, regardless of the service line pipe material composition.

Question 12. I understand that there is a lack of certainty about whether or not about 37,000 service lines are lead or not. What are you doing to eliminate this uncertainty and what are you doing to ensure that those residents take appropriate health precautions?

Response. The 37,000 estimate reported in the media is inaccurate. WASA has no information on service line pipe material for approximately 21,000 addresses. Based on experience gained from test pits and physical public space replacements that allowed actual physical observation of service line pipes in 2003, the initial inventory, based principally on the Weston Report, of likely lead service lines is accurate approximately 80 percent of the time (generally, of 100 service lines identified as lead, 80 of the service lines are actually made of lead.) WASA does have other sources of information that provide some data on pipe material. The WASA customer information system includes a record of pipe material on most but not all of the customer accounts. This information, again based upon actual physical observation following an excavation, is accurate about 60 percent of the time.

At this time, the only means of identifying service line pipe material with certainty remains direct observation. A WASA consultant is currently working to develop a technology that will permit WASA to more accurately identify/confirm service line pipe material without an excavation. This technology is being field tested over the next 6–12 months. WASA is also investigating other approaches to help identify service line pipe material more efficiently.

WASA has mailed letters to each of the properties for which we have no information on service line pipe material. Eve though we expect that a very small proportion of these addresses actually have lead service line pipes, we have provided to them the same precautionary information provided to residences with a lead service line. The letters invited these residences to participate in the water lead sampling program (post card requesting UPS delivery of the test kit). These residences will be sent a filter and replacement cartridges if test results exceed 15 ppb.

Question 13. Can you describe the progress WASA has made sending out water filters to lead-affected homes? How effective does WASA judge these filters to actually be in controlling lead exposure for its customers?

Response. WASA completed the major program to distribute filters to all households believed to have a lead service line on April 6, 2004. About 300 of the filters could not be delivered despite a requirement that UPS attempt three deliveries at each address and obtain a signature receipt. Filter and replacement cartridge distribution continue consistent with a distribution plan already discussed.

WASA provides filters that are certified by the National Sanitation Foundation to adequately address lead water concentrations of 150 ppb, as noted in the manufacturers' information accompanying the filters. WASA has provided additional instructions with the filters that reiterate the flushing instructions to which the Department of Health and the Environmental Protection Agency have agreed. The filters together with the proscribed flushing instructions that WASA includes in the packaging with the filters substantially reduce lead level concentrations.

Question 14. One of the issues here is who bears the responsibility to replace lead service lines. Under current regulations, the homeowner bears the responsibility for the lead service lines between the meter and the house. This troubles me because it does not seem appropriate for the ability to pay for this pipe replacement to be the determining factor in who gets lead-free water. In addition, I know that the city bore some responsibility in knowingly selecting leaded pipes—I would like to submit *Washington Post* stories from 1893 and 1895 discussing the concerns over lead pipes and their impact on public health. It appears that over 100 years ago this city knew of the danger and continued to use lead pipes. I am interested in your views on whether public water systems or private homeowners should bear the responsibility for lead service line replacement?

Response. WASA did not exist before 1996, and although its immediate predecessor, WASUA, was responsible for water distribution for many years, the division of responsibility between Federal and local authorities for this function has evolved over time.

Today, WASA is responsible for maintaining the portion of a service line pipe that rests in public space. The Lead and Copper Rule requires that WASA replace only the portion of the service line pipe that rests in public space (specifically addressed in relatively recent local law). It is important to remember, however, that it is a property owner that constructs the water service line pipe that must pass through public space in order to tie a private residence to a public water main. There are local jurisdictions in the United States that do not require that the local public water utility exercise responsibility for this privately constructed and financed component of the infrastructure that serves only an individual property.

WASA is an independent agency of the District of Columbia. The Mayor and Council of the District of Columbia exercise legislative oversight of the District of Columbia Water and Sewer Authority, and have established a very clear policy in this matter. Legislation recently introduced by District Council member Harold Brazil, would if approved, provide governmental assistance (general fund) to some homeowners in replacing lead service line pipes that rest on private property. The legislation includes income eligibility criteria that suggest that questions of equity in the allocation of public resources may be relevant to the discussion.

With respect to the Lead and Copper Rule, careful and complete implementation of its provisions, regardless of whether the action level has been exceeded, will not guarantee lead free tap water. Full compliance with the provisions of the rule, regardless of whether the action level has been exceeded, should prompt a water system to seek and maintain optimum corrosion control treatment, or "OCCT." OCCT is intended to minimize corrosion, and subsequent leaching into tap water from any pipe material.

Question 15. In 1986, lead was discovered in drinking water in the Palisades section of Washington, DC. Residents were quoted as saying, "The runaround has been unbelievable. . . . No one in the bureaucracy has even begun to take this seriously." The Director of water for the city stated that it was, ". . . premature to contact residents throughout the city" before the city developed a plan to handle and finance increased testing. I ask unanimous consent to insert several newspaper articles on this topic into the record. I find it unbelievable that no one at this witness table learned anything from this previous experience. Did anyone here refer to previous instances of lead contamination in the District when formulating a response plan? If so, please describe how you used this information, and if not, why not?

Response. After the fact, WASA management was aware of the issues that arose in 1986 (WASA did not exist until 1996.)

In 2002, the Water and Sewer Authority and the EPA determined that for the compliance period that ended in 2002, the data from the compliance samples indicated that the "action level"—a regulatory trigger that optimum corrosion control is not being achieved—had been exceeded. The data was provided in a timely manner to the Environmental Protection Agency as required under the Lead and Copper Rule. The DC Department of Health was also notified. The material provisions of the regulation, and the response by the EPA and local public health authorities did not indicate that the steps WASA had taken in implementing a lead services program under the Lead and Copper Rule were inappropriate. Specifically, the exceedance of the action level indicated that the Water and Sewer Authority should begin:

- working with the EPA and the Washington Aqueduct to achieve optimum corrosion control;
- a program of public education about the hazards of environmental lead exposure from water and other sources;
- a program to replace (or test to clear) 7 percent of the lead service line pipes in public space yearly until the action level was no longer exceeded.

Since January 2004, WASA's public education efforts have vastly expanded, but the newspaper articles to which you refer suggest that WASA's response to the exceedance in 2003 was very aggressive relative to those efforts undertaken in 1986. WASA's response was also undertaken in a different environment (OCCT plan approved by EPA in 2002). The public education materials were shared with regulators in advance of publication, and in one instance, *Living Lead Free In DC*, was prepared in collaboration with the District Department of Health. WASA also, for example, responded to media inquiries (*Washington City Paper*, 10/18–24/2004), published a newspaper ad, participated in community meetings, and contacted several thousand residents by mail to solicit their participation in an expanded sampling program.

Question 16. During the hearing, you mentioned that residents who receive filters would be provided with a 6-month supply. How many replacement filters does that include and what is the average cost of a 6-month supply?

Response. Two filter cartridges provide a 6-month supply (in excess of 80 gallons). The cost for two filter cartridges, including shipping, is approximately $22.

Question 17. What priority system is WASA using to determine which homes will have lead service lines replaced first, and does that give consideration to the presence or absence of vulnerable populations such as pregnant women, infants, and children?

Response. The physical replacement schedule of addresses in 2004 was established in calendar 2003, and was based upon the initial inventory of lead service line pipes submitted to EPA. The sample data collected in 2003 had not been analyzed when the schedule for replacements in 2004 was developed. Among the important factors the Authority considered included the number of services per block in order to maximize the number of replacements that could be undertaken while causing as little disruption of residential traffic and parking patterns as possible.

WASA has made provision for 500 "priority" physical replacements in 2004. These priority replacements of lead service line pipes in public space will target day care centers, residences with people having elevated blood lead levels, and also be based on the presence in a household of a member of the at risk populations (children under the age of six and women who are pregnant or nursing). This schedule of replacements will be selected in coordination with the Department of Health.

Question 18. What priority system is WASA using to determine which homes will receive filters, and does that give consideration to the presence of absence of vulnerable populations such as pregnant women, infants, and children?

Response. WASA is providing a filter and replacement cartridges to sample program participants whose tap water sample tests exceed 15 ppb out of an abundance of caution and an effort to address public concerns. Recall that the action level of 15 ppb is not directly tied to a specific measure of exposure that is linked directly to health risk. Also, the second draw service line concentrations suggested in sample test results may not easily extrapolate into sustained rates of exposure noted by some public health experts when discussing risks of environmental lead exposure.

Question 19. What is the status of your water filter distribution? Have all 23,000 homes believed to have lead service lines received them? If not, when will all 23,000 homes receive them?

Response. WASA completed distribution of filters to residences identified as likely having a lead service line pipe, and will provide replacement cartridges through next spring. WASA is also providing water filters and replacement cartridges to residences that participate in the sampling program whose test results exceed 15 ppb, regardless of the service line pipe material composition or the presence of a member of the target population.

Question 20. How many of the homes tested that are not part of the 23,000 homes with known lead service lines have tested above 15 parts per billion for lead?

Response. For calendar 2004 through April 24, 10,526 property owners participated in the sample program. Of these, 7,266 of the addresses were identified as having a material other than lead. As you may know, the second draw sample is intended to capture the sample from water that has rested in the service line pipe for over 6 hours. Of these non-lead second draw samples:

- 6,238 tested 0–15 ppb;
- 642 samples tested >15–50 ppb;
- 264 samples tested >50–100 ppb;
- 85 samples tested >100–150 ppb;
- 37 samples tested over 150 ppb.

Question 21. During lead service line replacement, I understand that WASA is offering to also replace the homeowner's section of the lead service line at the homeowner's cost. One of the major benefits of this approach is that the entire lead service line is replaced AT THE SAME TIME. If a homeowner chooses to pay to have their portion of the lead service line replaced by WASA during replacement of the rest of the lead service line, are both portions replaced simultaneously? If not, why not, and how long are homeowners being asked to wait? Have you evaluated any health effects of not replacing them simultaneously given that it is commonly believed that replacing portions of lead service lines can actually increase lead levels for at least a short period of time?

Response. If a homeowner chooses to pay for replacing the private side replacement (inclusive of responding to WASA's initial inquiry, acceptance of a contractor's estimate, entering into a contract within necessary timeframes) both portions are replaced simultaneously.

The practice of cutting a lead service line in order to reattach it to copper pipe during the replacement of a lead service line pipe in public space is understood to result in temporarily elevated lead concentrations in tap water. Those elevated concentrations are understood to fall very dramatically following high water usage in the context of a proscribed program of customer flushing that follows the physical replacement. WASA relies upon the guidance of the EPA and the District of Columbia Department of Health with respect to the determination of any health effects, and we have suspended the practice of cutting lead pipes pending a determination from DOH.

RESPONSE BY JERRY JOHNSON TO ADDITIONAL QUESTION FROM SENATOR CRAPO

Question. What training in risk communication have you implemented, and does it include the CD-based program available from the Centers for Disease Control called, *CDCynergy: Emergency Risk Communication?*

You testified to the need for better teamwork if the Lead and Copper Rule is to be implemented effectively. Now that your agency and the other authorized agencies are working very closely in what has become a closely scrutinized effort, what improvements in teamwork have you learned? Also, how will you change routine procedures for working together to implement the Rule after the DC drinking water system returns below the Action Level?

Response. WASA has not used the CD-based program, CDCynergy: Emergency Risk Communication.

WASA has employed for a number of years Beverly Silverberg Communications, Inc. which has provided advice and training in crisis communications. WASA has also obtained the services of a team headed by Dr. Tee Guidotti, Professor and

Chair of the Department of Environmental and Occupational Health, School of Public Health and Health Services, and Director, Division of Occupational Medicine and Toxicology School of Medicine and Health Sciences, the George Washington University Medical Center. Dr. Guidotti is also the Co-Director of the Mid-Atlantic Center for Children's Health and the Environment, a pediatric environmental health specialty unit. The team of professionals working with WASA from the George Washington University also includes individuals with experience in risk communication.

The audit being undertaken by EPA and other inquiries that are currently underway will provide important information with respect to improving communication and coordination among relevant agencies. One conclusion that we believe we share with EPA even at this relatively early stage is that both the relatively routine communications on this issue that have been relatively frequent but informal are made more structured and formal.

STATEMENT OF DANIEL R. LUCEY, INTERIM CHIEF MEDICAL HEALTH OFFICER, DISTRICT OF COLUMBIA DEPARTMENT OF HEALTH

Good afternoon. My name is Daniel R. Lucey, MD, and I am the Interim Chief Medical Officer for the DC Department of Health. In the next 5 minutes, prior to responding to your questions, I would like to summarize my background and list several key points about the lead issues in Washington, DC.

I am a physician trained in adult medicine and infectious diseases with a Masters degree in Public Health. After serving in the military as a physician I joined the U.S. Public Health Service while working at the National Institutes of Health and the Food and Drug Administration. During 9/11 and the subsequent anthrax attacks I was the Chairman of the Infectious Disease Service at the Washington Hospital Center in DC. In 2002 I was involved with the smallpox vaccination program, in 2003 with SARS (traveling to Hong Kong and mainland China, and working in a hospital in Toronto), and in 2004 with avian influenza.

On February 10, 2004 I began work at the DC Department of Health (DOH) with a focus on biodefense. On February 13th I attended a Lead Task Force meeting. Every day since then I have worked on lead issues. Although not a lead expert, I have approached learning about the lead issues through an intensive process, much like learning about other previously unfamiliar diseases such as anthrax, SARS, and avian influenza.

On February 16th, I contacted the Director of the Centers for Disease Control and Prevention (CDC), Dr. Julie Gerberding, to request advice from lead experts at the CDC. Her response was immediate and outstanding. CDC assistance has been ongoing since that time.

On February 26th, the City Administrator, Mr. Robert Bobb, instructed me to direct the Department of Health response to lead issues. Later that day I completed and signed a Health Advisory letter from the Department of Health to the approximately 23,000 residences in DC with lead service lines. (Attachment 1) The advisory contained recommendations about drinking water and measuring blood lead levels in persons most at risk for lead poisoning in order to assess the health impact of increased lead in the water. To our knowledge, no such widespread health advisory on lead in drinking water has ever been issued in the United States. Our findings may be useful to other cities that find increased lead concentrations in their drinking water.

In order to provide blood lead level testing by the Department of Health, starting on February 28th at DC General Hospital, we mobilized many persons in the Department of Health. In addition, on March 1st, I contacted the U.S. Surgeon General, Dr. Carmona, to request personnel assistance. He responded immediately, and via Admiral Babb and the Commissioned Corps Readiness Force (CCRF), provided a team of Public Health Service officers over the next 4 weeks who worked long hours with us in clinics across DC. They also went to several hundred homes of persons at high risk of lead poisoning. On March 30th the DC DOH, CCRF and CDC published our preliminary results on blood lead levels in the CDC's Morbidity and Mortality Weekly Report (MMWR).

To summarize key points:

1. None of the 201 persons we tested who live in homes with the highest measured levels of lead in the drinking water (i.e. > 300 parts per billion (ppb)) had elevated blood lead levels. (Attachment 2 MMWR March 30, 2004).

2. From 2000–2003 the percentage of children less than 6 years of age with elevated blood lead levels (≥ 10 mcg/dl) continued to decline in DC both in homes with and without lead service lines. The percent of children with blood lead levels ≥ 5

mcg/dl did not decline in homes with lead service lines, although this percent did decline in homes without lead service lines. (Attachment 2 MMWR March 30, 2004).

3. Only 2 of the initial 280 children in home childcare facilities with lead service lines had elevated blood lead levels (Attachment 3).

4. Of the initial 4,106 persons who came to our clinics across DC for free blood lead level testing in our laboratory, 1,277 were young children <6 years old, of whom 16 had elevated blood levels. The initial 14 children have been found to live in homes with dust and/or soil lead levels exceeding EPA/HUD guidelines. The homes of the other 2 children are currently being evaluated. (Attachment 4).

5. According to the CDC, from 1976–1980, nearly 9 of 10 (88.2 percent) children 1–5 years old (adults now 24–28 years old) in the USA had blood lead levels that today are considered elevated, namely at least 10 micrograms/ deciliter ("≥ 10 µg/dl"). (Attachment 5).

6. The EPA "action level" for lead in drinking water of 15 parts per billion (or 0.015 mg/Liter) is *not* a health-based recommendation. According to the EPA:

> "This action level was not designed to measure health risks from water represented by individual samples. Rather, it is a statistical trigger that, if exceeded, requires more treatment, public education and possibly lead service line replacement"

(Attachment 6).

Thank you for your time and I will be pleased to respond to your questions.

GOVERNMENT OF THE DISTRICT OF COLUMBIA
Department of Health

ATTACHMENT 1:

FEBRUARY 26, 2004 LETTER FROM DOH INTERIM CHIEF MEDICAL OFFICER TO 23,000 DISTRICT RESIDENTS WITH LEAD SERVICE LINES

GOVERNMENT OF THE DISTRICT OF COLUMBIA

Department of Health

Office of the Director

26 Feb 2004

Dear Resident:

The District of Columbia Water and Sewer Authority (WASA) has identified your residence as one of approximately 23,000 probably having an underground lead service pipe that brings drinking water into your residence from the water mains under the streets. WASA has tested approximately 6,000 of these residences and found that about two-thirds of these residences are above the Environmental Protection Agency (EPA) Action Level for lead of 15 parts per billion (ppb), and one-third are below the Action Level. At this time, however, the Department of Health is not certain that a DC residence with a lead service line that tests below this EPA Action Level at any given time is always below this Action Level. Therefore, **out of an abundance of caution** the Department of Health is providing advice in this letter for all residences which have lead service pipes, and specific advice for all children under the age of six years and women who are pregnant living in these approximately 23,000 residences.

When your water is not used for a period of time, such as overnight, it may pick up increased amounts of lead from the lead service line or from your internal home plumbing. According to WASA and EPA, recommended efforts to decrease lead in this water include:

- *Use tap water for drinking or cooking only after other high water use activities, such as bathing, showering, flushing the toilet, or washing your clothes, so that a total of at least 10 minutes of running water from your faucets or pipes has occurred. This high water usage should let the water that was in contact with the underground lead service pipe pass through your home's plumbing.*

- *After this 10 minutes of use, let the water run from your kitchen faucet for 60 seconds, then collect drinking water in containers and store them in the refrigerator. About once a month, remove and clean the strainer/aerator device on your faucet to remove debris.*

- *Cold water should be used for drinking or cooking, as hot water will contain higher levels of lead. Cold water should be heated for making hot beverages or cooking. Do not use the water from the hot water faucet for drinking or cooking.*

Recommendations for children under six and women who are pregnant or breastfeeding

Your residence has been identified as one of 23,000 likely to have a lead service line, and your tap water may contain elevated levels of lead. Young children under six and unborn babies are especially vulnerable to the damaging effects of lead. Therefore, given our current state of information, we make the following additional recommendations. **Children under six years and women who are pregnant or breastfeeding should not drink unfiltered water, or use it to prepare infant formula or concentrated juice, in any of these 23,000 residences until the concerns regarding the lead levels in the water have been resolved.** We encourage continued breastfeeding, but at the same time recommend taking measures to avoid unfiltered water. We recommend that all children under six, and women who are pregnant, be screened for blood lead

825 N. Capitol Street, N.E. 4th Floor Washington D.C. 20002 (202) 442 –5999 *fax* (202) 442 - 4788

levels. To obtain free blood lead level testing through our Department of Health laboratory you can contact the Department of Health at (202) 671-0733. We ask that results of blood lead levels performed at other laboratories be provided to the Department of Health. The results of these blood tests will be analyzed by the Department of Health to see if there is a relationship between specific water lead levels and blood lead levels. Based on this information we may be able to reevaluate our recommendations about water precautions in these 23,000 residences.

Filters to remove lead from the water, and bottled water: 2 ways to decrease lead exposure

If you decide to purchase a home drinking water filter, please make sure that the option you choose is certified to remove lead. Although the DC Department of Health does not certify nor endorse specific home drinking water treatment filters, you should purchase a treatment filter certified to remove lead by an independent testing organization, such as the National Sanitation Foundation International (www.nsf.org/certified/DWTU) or the California Dept. of Health Services (www.dhs.cahwnet.gov/ps/ddwem/technical/certification/devices.html). Consider choosing a water treatment filter, called a "point-of-use" filter, that can be installed on your kitchen faucet or other site of drinking water. "Point-of-use" filters must be properly installed and operated according to manufacturers' instructions. Another option is a water pitcher specifically certified to remove lead.

Since not all bottled water is systematically tested for lead, if you decide to use bottled water you should use water that has been certified to be below the EPA Action Level for lead by organizations such as the International Bottled Water Association or the National Sanitation Foundation. You may wish to visit EPA's website www.epa.gov/region3/leaddc.htm for information on lead in drinking water, or contact the EPA Safe Drinking Water Hotline at 1-800-426-4791.

Lead paint can be another very important source of lead-exposure in the home

Lead paint was used prior to 1978 to paint the interior of homes, and exterior woodwork such as windows, doors, and porches. Lead paint is the most commonly identified cause of a child having an elevated blood lead level. Therefore, any of the residences in which a person is found to have an elevated blood lead level must also be tested for the presence of lead-based paint. Together we can work to get the lead out of DC homes. Peeling or flaking paint should be considered hazardous to a child until the paint is tested. Lead-based paint dust is hazardous if inhaled or ingested.

Remember: lead can cause harm to the developing brain of an unborn baby or young child, to the red blood cells that carry our oxygen, to the kidneys, and to other parts of the body. Current medications are effective only when the blood lead levels are very high, as occurs much more often from lead in paint than lead in drinking water.

We will continue to communicate with you as we obtain more information from ongoing tests of water and blood lead levels. We will keep you informed of any updated recommendations based on these ongoing tests. Our primary concern is your health and that of all the people of the District of Columbia.

Sincerely yours,

Daniel R. Lucey

Daniel R. Lucey, M.D.
Interim Chief Medical Officer

825 N. Capitol Street, N.E. 4th Floor Washington D.C. 20002 (202) 442 –5999 *fax* (202) 442 - 4788

GOVERNMENT OF THE DISTRICT OF COLUMBIA
Department of Health

ATTACHMENT 2:

BLOOD LEAD LEVELS IN RESIDENTS OF HOMES WITH ELEVATED LEAD IN TAP WATER —DISTRICT OF COLUMBIA, 2004

CENTERS FOR DISEASE CONTROL AND PREVENTION: MORBIDITY AND MORTALITY WEEKLY REPORT (MMWR)

MMWR Dispatch Vol. 53/March 30, 2004

MMWR

Morbidity and Mortality Weekly Report

MMWR Dispatch
Vol. 53 / March 30, 2004

Blood Lead Levels in Residents of Homes with Elevated Lead in Tap Water — District of Columbia, 2004

Lead exposure adversely affects intellectual development in young children and might increase the risk for hypertension in adults (*1*). In the District of Columbia (DC), of an estimated 130,000 residences, approximately 23,000 (18%) have lead service pipes (Daniel Lucey, MD, DC Department of Health [DCDOH], personal communication, March 24, 2004). The Environmental Protection Agency (EPA) requires water authorities to test tap water in 10–100 residences annually for lead. In March 2003, DC Water and Sewer Authority (WASA) expanded its lead-in-water testing program to homes with lead service pipes extending from the water main to the house. By late January 2004, results of the expanded water testing indicated that the majority of homes tested had water lead levels above EPA's action level of 15 parts per billion (ppb). On February 16, DCDOH requested CDC assistance to assess health effects of elevated lead levels in residential tap water. DCDOH also requested deployment of officers of the United States Public Health Service (USPHS) to assist in the investigations. This report summarizes the results of the preliminary investigations, which indicated that the elevated water lead levels might have contributed to a small increase in blood lead levels (BLLs). The investigation of elevated water lead levels is ongoing. In the interim, DCDOH has recommended that young children and pregnant and breast-feeding women refrain from drinking unfiltered tap water (*2*).

CDC's BLL of concern for children, 10 $\mu g/dL$, was adopted in 1991 in response to evidence associating BLLs ≥ 10 $\mu g/dL$ with adverse health effects (*3*). Adverse health effects have been reported recently at BLLs <10 $\mu g/dL$, particularly in vulnerable populations (e.g., infants and children) (*4,5*); no safe BLL has been identified (*6*). Longitudinal analysis was conducted to identify trends in BLLs in DC before and after changes in the water disinfection process by comparing homes with lead service pipes to homes without lead service pipes. Both the percentage of BLLs ≥ 10 $\mu g/dL$ and those ≥ 5 $\mu g/dL$ were examined over time. Cross-sectional analysis of BLLs of residents in homes with the highest water lead levels was conducted to determine if residents had BLLs ≥ 10 $\mu g/dL$.

Longitudinal Analysis of Childhood Blood Lead Screening Tests

WASA provided DCDOH and CDC with a list of homes (n = 26,141) with lead service pipes. During January 1998–December 2003, the DCDOH blood lead surveillance system recorded 84,929 BLLs. Of these, 43,314 (51%) tests were venous, and 6,794 (8%) were fingerstick; sample type was not listed on the remaining tests. All blood tests were used in this analysis. For each year of testing, these databases were linked by address. A total of 11,061 BLL laboratory requisition slips listed an address with a lead service pipe.

During 1998–2000, the percentage of BLLs ≥ 10 $\mu g/dL$ and ≥ 5 $\mu g/dL$ decreased substantially, regardless of the type of service pipe (Figure). During 2000–2003, the percentage of BLLs ≥ 10 $\mu g/dL$ in persons living in homes known to have lead service pipes decreased from 9.8% to 7.6% ($p = 0.008$). The percentage of BLLs ≥ 5 $\mu g/dL$ in persons living in houses without lead service pipes continued to decrease, from 22.7% to 15.6% (n = 14,152; $p<0.001$). However, the percentage of BLLs ≥ 5 $\mu g/dL$ in persons living in homes with lead service pipes did not decrease statistically significantly (from 696 [32.4%] to 405 [31.2%]; $p = 0.34$).

Cross-Sectional Study of Homes with >300 ppb Lead in Water

WASA provided the results of lead testing on water samples from 6,170 homes. Of these, 163 (3%) had lead levels >300 ppb in second-draw water collected after a change in water temperature, indicating that some of the lead in the water leached from water pipes outside the home. USPHS officers working in the DCDOH Incident Command structure contacted residents in the 140 (86%) homes that had telephones and arranged for visits to draw venous samples for BLLs. The DC Public Health Laboratory determined BLLs by using graphite furnace atomic absorption spectrophotometry for 184 persons in 86 households who consented to having blood drawn. Residents were provided with a water filter and infor-

FIGURE. Percentage of tests with elevated blood lead levels, by year and water-line type — District of Columbia, January 1998–September 2003

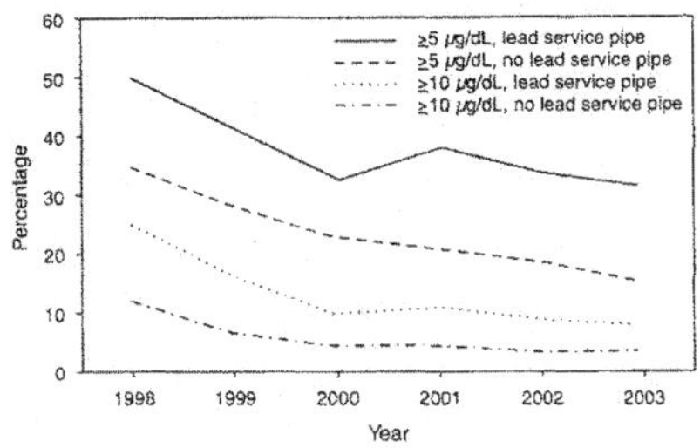

mation about reducing lead exposure. In addition, in 12 of the households contacted, 17 persons had a venous blood test drawn independently and reported to DCDOH since January 2004. These test results also were included in this analysis.

Of the 201 residents from 98 homes with water lead levels >300 ppb tested for BLLs, all had BLLs below CDC's levels of concern (10 µg/dL for children aged 6 months–15 years and 25 µg/dL for adults) (Table). Of the 201 residents, a total of 153 (76%) reported drinking tap water, and 52 households (53%) reported using a water filter. On February 26, 2004, DCDOH sent a letter to all DC homes with lead service pipes, recommending that young children and pregnant and breast-feeding women refrain from drinking unfiltered tap water (*2*).

Reported by: *L Stokes, PhD, NC Onwuche, P Thomas, PhD, JO Davies-Cole, PhD, T Calhoun, MD, AC Glymph, MPH, ME Knuckles, PhD, D Lucey, MD, District of Columbia Dept of Health. T Cote, MD, G Audain-Norwood, MA, M Britt, PhD, ML Lowe, MCRP, MA Malek, MD, A Szeto, MPH, RL Tan, DVM, C Yu, M Eberhart, MD, US Public Health Svc. MJ Brown, ScD, C Blanton, MS, GB Curtis, DM Homa, PhD, Div of Emergency and Environmental Health Svcs, National Center for Environmental Health, CDC.*

Editorial Note: The findings in this report indicate that although lead in tap water contributed to a small increase in BLLs in DC, no children were identified with BLLs ≥10µg/dL, even in homes with the highest water lead levels. In addition, the longitudinal surveillance data indicate a continued decline in the percentage of BLLs ≥10 µg/dL. The findings in this report suggest that levels exceeding the EPA action level of 15 ppb can result in an increase in the percentage of BLLs ≥5 µg/dL. Homes with lead service pipes are older, and persons living in these homes are more likely to be exposed to high-dose lead sources (e.g., paint and dust hazards). For this reason, in all years reported, the percentage of test results ≥10 µg/dL and the percentage of test results ≥5 µg/dL at addresses with lead service pipes were higher than at addresses without lead service pipes.

TABLE. Blood lead levels (BLLs) of residents in homes with >300 parts per billion in drinking water, by age group — District of Columbia, March 2004

	BLL (µg/dL)	
Age group (yrs)	Median	Range
1–5 (n = 17)	3	1–6
6–15 (n = 13)	2	1–4
16–40 (n = 56)	3	1–14
41–60 (n = 69)	4	1–20
≥61 (n = 46)	6	2–22
Total (n = 201)		

The findings in this report are subject to at least three limitations. First, the BLL surveillance data include multiple tests on the same person, and persons with lead poisoning are tested more frequently than those with low BLLs. Second, fingerstick tests are more subject than venous samples to contamination by ambient lead (*7*). Finally, neither the blood nor the water lead test results were collected from a randomized sample. Water was collected from homes with a high probability of having lead service pipes; the March 2004 BLL screening program was limited to families living in homes with the highest water lead levels, and the routine blood lead surveillance program focused on identifying children at highest risk for lead exposure. For these reasons, the percentages of BLLs ≥5 µg/dL or ≥10 µg/dL reported probably are higher than those found in the general population. However, none of these factors should affect the relative differences between percentage of tests ≥5 µg/dL by water line type, nor do they explain the change in trajectory of the percentage of tests ≥5 µg/dL by year after 2000.

The cause of the elevated water lead levels in DC is under review. Although the increase is associated temporally with the change in the disinfection process from chlorine to chloramines that occurred in November 2000, whether this change contributed to increased lead in the water is unknown.

Because no threshold for adverse health effects in young children has been demonstrated (*6*), public health interventions should focus on eliminating all lead exposures in children (*8*). Lead concentrations in drinking water should be below the EPA action level of 15 ppb. Officials in communities that are considering changes in water chemistry or that have implemented such changes recently should assess whether these changes might result in increased lead in residential tap water. EPA has asked all state health and environmental officials to monitor lead in drinking water at schools and day care centers. More information about lead poisoning is available from CDC at http://www.cdc.gov/nceh/lead/lead.htm.

Acknowledgments

This report is based in part on data collected by SB Adams, LC Cooper, PhD, KJ Elenberg, JM Gusto, MPH, JE Hardin, P Karikari-Martin, MPH, L Velazquez, PharmD, AA Walker, US Public Health Svc.

References

1. Agency for Toxic Substances and Disease Registry. Toxicological profile for lead. Alanta, Georgia: U.S. Department of Health and Human Services, Agency for Toxic Substances and Disease Registry, 1999. Available at http://www.atsdr.cdc.gov/toxprofiles/tp13.html.
2. District of Columbia Department of Health. Health advisory: lead in Washington, DC. February 26, 2004. Available at http://www.dchealth.dc.gov.
3. CDC. Preventing lead poisoning in young children: a statement by the Centers for Disease Control—October 1991. Atlanta, Georgia: U.S. Department of Health and Human Services, Public Health Service, CDC, 1991.
4. Canfield RL, Henderson CR Jr, Cory-Slechta DA, Cox C, Jusko TA, Lanphear BP. Intellectual impairment in children with blood lead concentrations below 10 μg per deciliter. N Engl J Med 2003;348:1517–26.
5. Bellinger DC, Needleman HL. Intellectual impairment and blood lead levels. N Engl J Med 2003;349:500–2.
6. Schwartz J. Low-level lead exposure and children's IQ: a meta-analysis and search for a threshold. Environ Res 1994;65:42–55.
7. Schlenker TL, Fritz CJ, Mark D, et al. Screening for pediatric lead poisoning: comparability of simultaneously drawn capillary and venous blood samples. JAMA 1994;271:1346–8.
8. Rogan WJ, Ware JH. Exposure to lead in children—how low is low enough? N Engl J Med 2003;348:1515–6.

GOVERNMENT OF THE DISTRICT OF COLUMBIA
Department of Health

ATTACHMENT 3:

BLOOD LEAD LEVELS TEST RESULTS FROM CHILDCARE FACILITIES IN THE DISTRICT OF COLUMBIA

GOVERNMENT OF THE DISTRICT OF COLUMBIA
DEPARTMENT OF HEALTH

BLOOD LEAD LEVEL TEST RESULTS FROM CHILDREN IN CHILDCARE FACILITIES WITH LEAD SERVICE LINES
FRIDAY, APRIL 2, 2004 - 5:00 P.M.

151 Childcare Facilities with lead service line in the District of Columbia as identified by WASA

- [illegible] Childcare facilities are closed or have no children currently enrolled
- [illegible] Childcare facilities have children that have been screened by private physician
- [illegible] Childcare facilities have been completed (416 Children Screened)
  - [illegible] below 10 mcg/dL
  - 2: above 10 mcg/dL
  - 136 results are pending lab results
- [illegible] will call back to make an appointment or give test results from private physician
- [illegible] Childcare facilities not tested
  - 6 Parents Refused
  - 2 facilities refused
  - 2: Unable to reach with visit or phone call

GOVERNMENT OF THE DISTRICT OF COLUMBIA
Department of Health

ATTACHMENT 4:

BLOOD LEAD LEVEL SCREENING RESULTS
FEB. 3 – APR. 1, 2004

Government of the District of Columbia
Department of Health

Blood Lead Level Screening Results

(Micrograms/deciliter ("mcg/dL")
As of February 3-April 1, 2004

Total number screened for blood lead levels	Total number of laboratory tests completed	Total number within DOH target population	Total number within DOH target population with lead service lines	Total number within DOH target population with high lead levels	Total number outside of DOH target population with high blood lead levels
4520	4106	1435 • 1277 (88.9%) children under the age of 6; • 85 (5.9%) women who are pregnant; and • 73 (5.1%) women who are nursing.	18.9% have lead service lines; • 17.9% of all children under the age of 6 • 20.0% of all women who are pregnant; and • 34.3% of all women who are nursing. 81.1% do not have lead service lines.	• Women who are nursing=2 & pregnant = 0 • 16 children under the age of 6 have elevated (10mcg/dL or higher) lead levels. o This is 1.3 % of all children under the age of 6; o 8 with lead service lines; and o 8 without lead service lines. Environmental home assessments for other sources of lead, such as lead paint, will be performed for all children with elevated blood lead levels. Of the initial 14 children whose homes have been tested, all 14 had dust and/or soil lead levels that exceed EPA and HUD guidelines.	• Of the 2636 residents outside of the DOH target population screened for blood lead levels only 4 have a blood lead level above 25 mcg/dL (the level of most concern for adults). Two residents have lead service lines. The blood lead level at which medication is offered is usually 45 mcg/dL.

Note: 35 ages being confirmed.

GOVERNMENT OF THE DISTRICT OF COLUMBIA
Department of Health

Blood Lead Levels for the Target Group (Pregnant, Nursing or <6 Years Old)
April 1, 2004
(N=1,435)

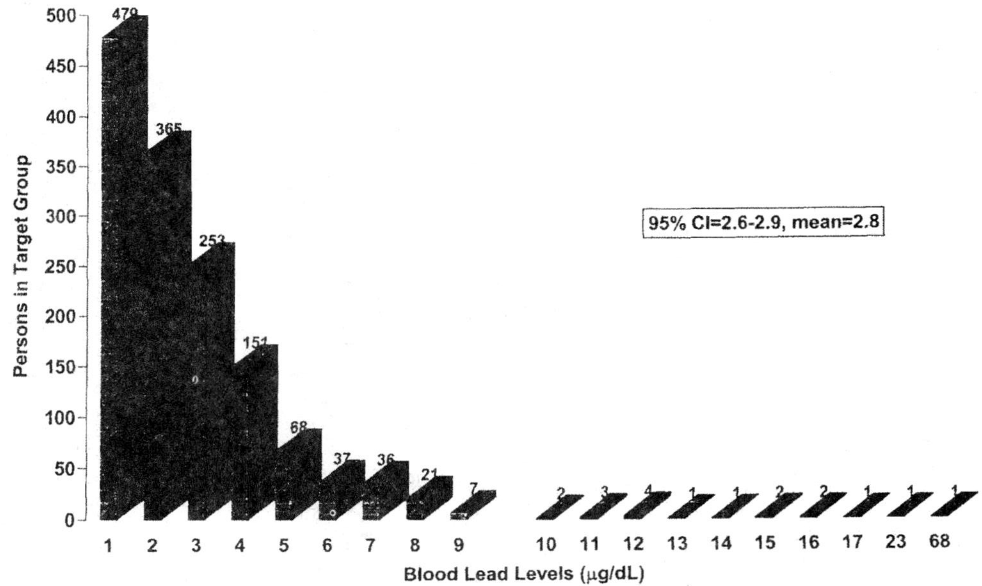

GOVERNMENT OF THE DISTRICT OF COLUMBIA
Department of Health

Blood Lead Level (μg/dL) Category by Age Category, as of April 1, 2004
(N=4,145)

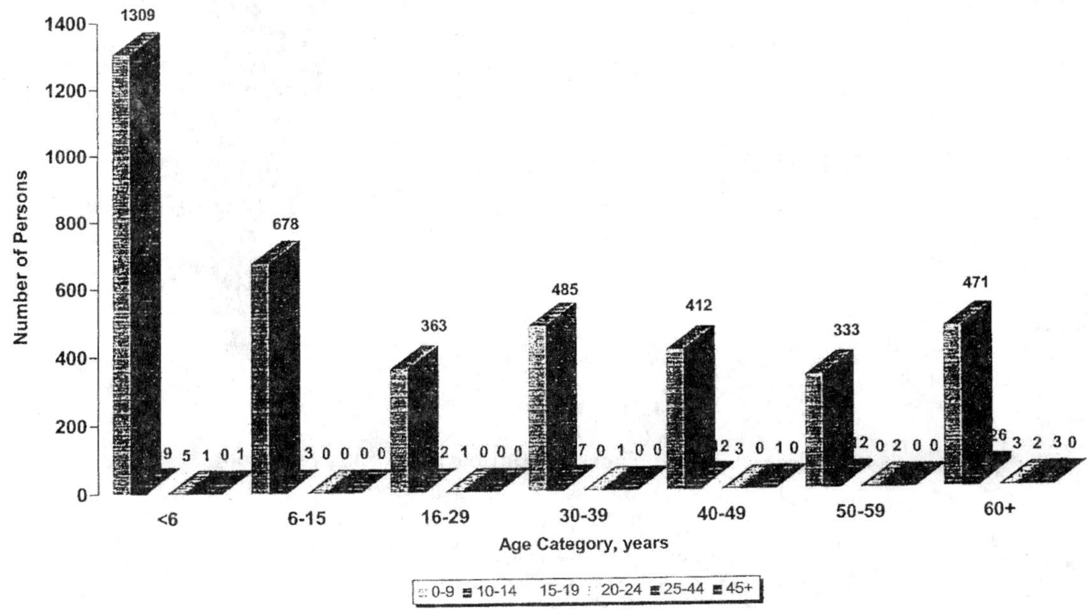

GOVERNMENT OF THE DISTRICT OF COLUMBIA
Department of Health

ATTACHMENT 5:

CENTERS FOR DISEASE CONTROL AND PREVENTION: NHANES PERCENTAGE OF U.S. CHILDREN 1-5 YEARS OF AGE WITH BLOOD LEAD LEVLES $\geq$ 10 mcg/dL: NHANES II AND III

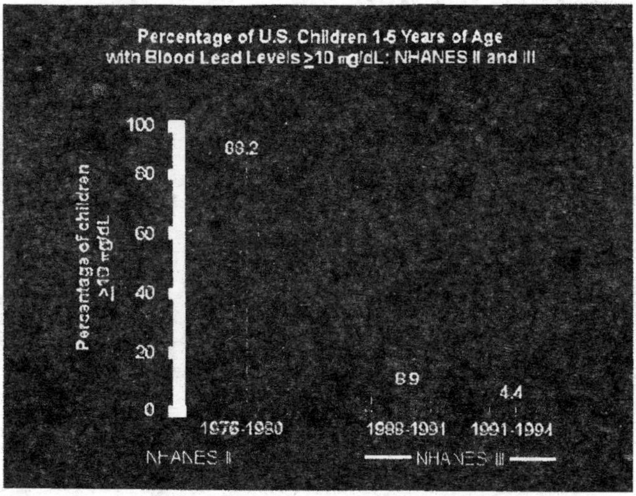
Percentage of U.S. Children 1-5 Years of Age
with Blood Lead Levels ≥10 mg/dL: NHANES II and III
Percentage of children
≥10 mg/dL
100
80
60
40
20
0
88.2
8.9
4.4
1976-1980
1988-1991
1991-1994
NHANES II
NHANES III

GOVERNMENT OF THE DISTRICT OF COLUMBIA
Department of Health

ATTACHMENT 6:

ENVIRONMENTAL PROTECTION AGENCY LEAD "ACTION LEVEL" DEFINITION

U.S. Environmental Protection Agency

Mid-Atlantic Region: Lead in Washington, DC Drinking Water

Serving Delaware, District of Columbia, Maryland, Pennsylvania, Virginia, and West Virginia

Contact Us | Print Version Search: GO

EPA Home > Region 3 > Region 3 Drinking Water > Region 3 Lead in DC Drinking Water > EPA Role and Structure

Lead in DC Drinking Water Home
Newsroom
Frequent Questions
Human Health
en Español
Laws/Regulations
Corrosion Control Research
Key Communications
Related Links

Drinking Water

EPA REGION 3'S OVERSIGHT ROLE AND REGULATORY STRUCTURE FOR ADDRESSING LEAD IN DRINKING WATER

LEAD AND COPPER ACTION LEVELS

The Action Level for lead is 0.015 milligrams per liter (mg/l) which is equivalent to 15 parts per billion (ppb). For copper, the Action Level is 1.3 mg/l or 1,300 ppb. This Action Level was not designed to measure health risks from water represented by individual samples. Rather, it is a statistical trigger value that, if exceeded, requires more treatment, public education and possibly lead service line replacement.

Extracted from http://www.epa.gov/dclead/oversight.htm. Last updated on Wednesday, March 31st, 2004

GOVERNMENT OF THE DISTRICT OF COLUMBIA
Department of Health

ATTACHMENT 7:

October 2002 Water and Sewer Authority - DC Department of Health Brochure "Living Lead Free in DC"

The District of Columbia
Water and Sewer Authority
and the
District of Columbia
Department of Health
acknowledge
National Lead Awareness Week
and its impacts on your health.
Living Lead-Free in D.C.
October 2002
DC WASA
of the District of Columbia
Williams, Mayor
DOH
DEPARTMENT OF HEALTH

WASA AND DOH ACKNOWLEDGE NATIONAL LEAD AWARENESS WEEK

What is Lead?

Lead is a common metal, used historically for a variety of purposes. It can be found throughout the environment in lead-based paint, air, soil, household dust, food, certain types of pottery, porcelain and pewter, water, and other sources. When water stands in lead pipes or copper pipes connected by lead-based solder for several hours, the potential exists for lead to dissolve into the water. This means the first water drawn from the tap in the morning, or later in the afternoon after returning from work or school, can contain fairly high levels of lead.

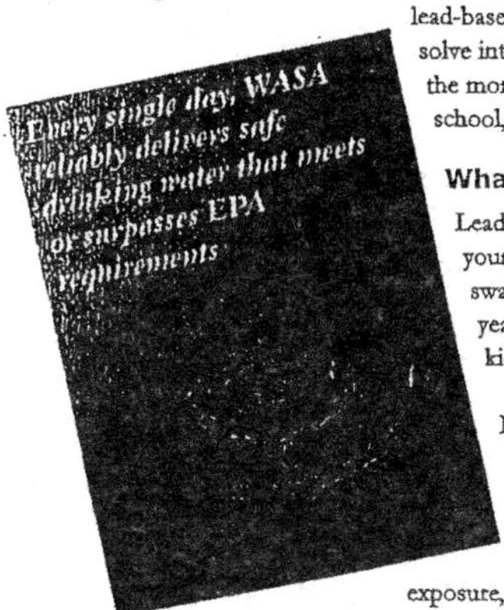

What Are the Health Effects of Lead?

Lead can pose a significant risk to health if too much of it enters your body. Even small amounts of lead can be harmful if swallowed or inhaled. If lead accumulates in the body over many years, it can cause damage to the brain, red blood cells, and kidneys.

Lead from chipping and flaking paint, if ingested, can cause significant health impacts especially for small children.

Lead in drinking water, although rarely the sole cause of lead poisoning, can significantly increase a person's total lead exposure, particularly the exposure of infants who drink baby formulas and concentrated juices that are mixed with water. The EPA estimates that drinking water can make up 20 percent or more of a person's total exposure to lead.

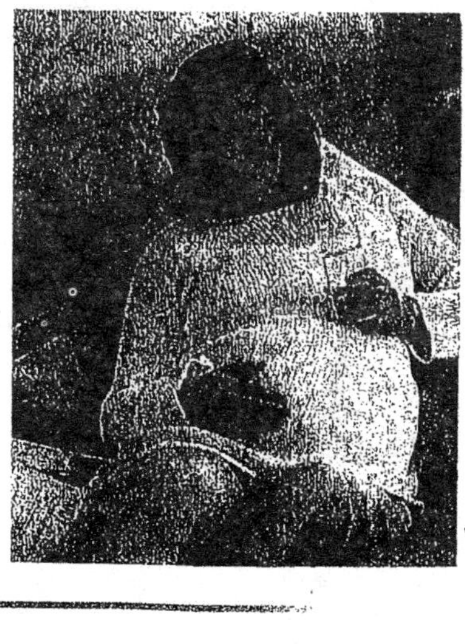

Does Lead Affect Everyone Equally?

The greatest risk is to young children and pregnant women. Amounts of lead that will not hurt adults can slow down normal mental and physical development of growing bodies. In addition, a child at play often comes into contact with sources of lead contamination like dirt and dust that rarely affect an adult. It is important to wash children's hands and toys often, and to try to make sure they only put food in their mouths.

Any child may be at risk, because lead is present in many sources. Children can be harmed by lead regardless of whether they live in a city, suburb or rural area, their family's economic background, or their ethnic background. They may be exposed to lead at home, school, day care, and on playgrounds.

The effects may not be obvious. But low levels of lead may damage the nervous system, including the brain, interfere with growth, harm hearing, lower IQ scores, make learning difficult. Exposure to low levels of lead may also affect a child's behavior, making him or her more excitable or less able to concentrate.

Lead in Drinking Water

The United States Environmental Protection Agency (EPA), the District of Columbia Water and Sewer Authority (WASA), and the DC Department of Health (DOH) want you to be fully informed about lead in your drinking water and lead from other sources. The most common cause of lead in drinking water is corrosion, a reaction between the water and lead pipes, fixtures containing lead such as brass and chrome-plated faucets or lead-based solder to connect copper pipes installed in interior household plumbing prior to 1987. In 1991, the EPA required public water utilities to test for lead. In areas where high lead levels were found, the utilities were required to take steps to reduce contamination.

Under federal law the water supplier is required to have a program in place to minimize lead in your drinking water. The Washington Aqueduct (WA) Division of the Army Corps of Engineers is the wholesale supplier of water to WASA. In compliance with federal requirements, WA has performed an optimal corrosion control study to minimize lead in the drinking water. The Optimal Corrosion Control Treatment (OCCT) was designated for WA by the EPA, and OCCT was implemented by WA in 1993. The OCCT implemented by the WA applies to the WASA water distribution system as WA is a wholesale supplier of drinking water and has no distribution system of its own. Under the OCCT requirements, a pH of 7.4 to 7.7 must be maintained at the entry points to the distribution system and a minimum pH of 7.0 is to be maintained in the distribution system. The purpose of the OCCT is to control the corrosivity of water and thus minimize leaching of lead or copper from lead service lines and customer plumbing into drinking water.

WASA regularly monitors for elevated lead and copper concentrations by collecting water samples at consumer taps. This monitoring also enables us to monitor the effectiveness of the OCCT program. The OCCT program in WASA's distribution system is effective in most cases and the lead concentrations found in drinking water have consistently been below EPA action level requirements since 1994. However, in the annual monitoring period ending June 30, 2002, the lead results indicate that although most homes have very low levels of lead in their drinking water, some homes in the community have lead levels above the EPA action level of 15 parts per billion (ppb).

In a partnership between WASA and DOH, we've prepared this Q&A (questions and answers) brochure to inform you of the effects of lead on your health, and to protect you and your loved ones by reducing your exposure to lead in drinking water as well as from other sources.

While lead isn't found in the District's water supplies or in treated water in the distribution system, some older homes built before 1950 may have elevated lead levels in their drinking water because of lead plumbing fixtures. The information on the following pages will help you determine if you have a problem, and if so, what to do about it.

WASA and DOH Acknowledge National Lead Awareness Week

Lead Paint in Homes Built Before 1978

Many houses and apartments built before 1978 have paint that contains high levels of lead (called lead-based paint). Lead from paint, chips, and dust can pose serious health hazards if not taken care of properly.

Federal law requires that individuals receive certain information before renting, buying, or renovating pre-1978 housing:

- Landlords have to disclose known information on lead-based paint and lead-based paint hazards before leases take effect. Leases must include a disclosure form about lead-based paint.
- Sellers have to disclose known information on lead-based paint and lead-based paint hazards before selling a house. Sales contracts must include a disclosure form about lead-based paint. Buyers have up to 10 days to check for lead.
- Renovators have to give you information before starting work.
- If you want more information on these requirements, call the National Lead Information Center at 1-800-424-LEAD (424-5323).

Identifying Lead Hazards

Lead-based paint is usually not a hazard if it is in good condition, and it is not on an impact or friction surface, like a window. It is defined by the federal government as paint with lead levels greater than or equal to 1.0 milligram per square centimeter, or more than 0.5% by weight.

Lead from paint chips, which you can see, and lead dust, which you can't always see, can both be serious hazards.

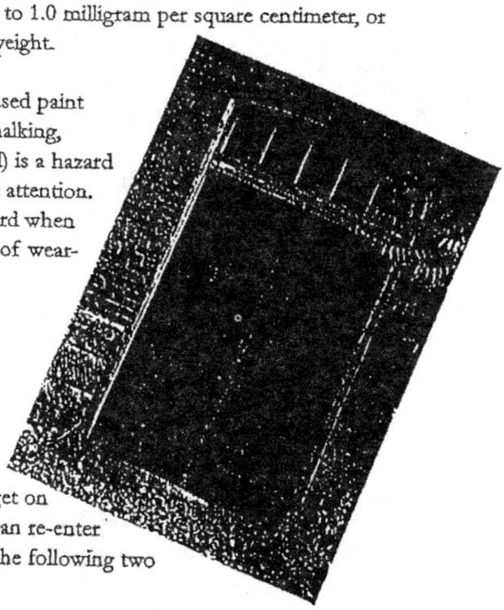

Deteriorating lead-based paint (peeling, chipping, chalking, crackling or damaged) is a hazard and needs immediate attention. It may also be a hazard when found on surfaces that children can chew or that get a lot of wear-and-tear, such as:

- Windows and window sills.
- Doors and door frames.
- Stairs, railings, banisters, and porches.

Lead dust can form when lead-based paint is dry scraped, dry sanded, or heated. Dust also forms when painted surfaces bump or rub together. Lead chips and dust can get on surfaces and objects that people touch. Settled lead dust can re-enter the air when people vacuum, sweep, or walk through it. The following two federal standards have been set for lead hazards in dust:

- 40 micrograms per square foot ($\mu g/ft^2$) and higher for floors, including carpeted floors.
- 250 $\mu g/ft^2$ and higher for interior window sills.

Lead in soil can be a hazard when children play in bare soil or when people bring soil into the house on their shoes. The following two federal standards have been set for lead hazards in residential soil.

- 400 parts per million (ppm) and higher in play areas of bare soil.
- 1,200 ppm (average) and higher in bare soil in the remainder of the yard.

The only way to find out if paint, dust, and soil lead hazards exist is to test for them. The most common methods used are described below.

Checking Your Home for Lead

You can get your home checked for lead in one of two ways, or both:

- A paint inspection tells you the lead content of every different type of painted surface in your home. It won't tell you whether the paint is a hazard or how you should deal with it.
- A risk assessment tells you if there are any sources of serious lead exposure (such as peeling paint and lead dust). It also tells you what actions to take to address these hazards.

Just knowing that a home has lead-based paint may not tell you if there is a hazard.

Hire a trained, certified professional who will use a range of reliable methods when checking your home, such as:

- Visual inspection of paint condition and location.
- A portable x-ray fluorescence (XRF) machine.
- Lab tests of paint, dust, and soil samples.

There are standards in place to ensure the work is done safely, reliably, and effectively. Contact your local lead poisoning prevention program for more information at the Department of Health (202-535-2690, or call 1-800-424-LEAD for a list of contacts in your area).

Home test kits for lead are available, but may not always be accurate. Consumers should not rely on these tests before doing renovations or to assure safety.

IMPORTANT!

Lead From Paint, Dust, and Soil Can Be Dangerous If Not Managed Properly

FACT: Lead exposure can harm young children and babies even before they are born.

FACT: Even children who seem healthy can have high levels of lead in their bodies.

FACT: People can get lead in their bodies by breathing or swallowing lead dust, or by eating soil or paint chips containing lead.

FACT: People have many options for reducing lead hazards. In most cases, lead-based paint that is in good condition is not a hazard.

FACT: Removing lead-based paint improperly can increase the danger to your family.

WASA AND DOH ACKNOWLEDGE NATIONAL LEAD AWARENESS WEEK

What You Can Do Now To Protect Your Family

If you suspect that your house has lead hazards, you can take some immediate steps to reduce your family's risk:

- If you rent, notify your landlord of peeling or chipping paint.
- Clean up paint chips immediately.
- Clean floors, window frames, window sills, and other surfaces weekly. Use a mop or sponge with warm water and a general all-purpose cleaner or a cleaner made specifically for lead. REMEMBER: NEVER MIX AMMONIA AND BLEACH PRODUCTS TOGETHER SINCE THEY CAN FORM A DANGEROUS GAS.
- Thoroughly rinse sponges and mop heads after cleaning dirty or dusty areas.
- Clean or remove shoes before entering your home to avoid tracking in lead from soil.

How Does Lead Enter Our Drinking Water?

Lead is unusual among drinking water contaminants in that it seldom occurs naturally in water supplies like rivers and lakes. Lead enters drinking water primarily as a result of the corrosion, or wearing away, of materials containing lead in the household plumbing and water service lines. These materials include lead-based solder used to join copper pipe, brass, and chrome-plated brass faucets, and in some cases, pipes made of lead that connect your house to the water main (service lines). In 1986, Congress banned the use of lead solder containing greater than 0.2% lead, and restricted the lead content of faucets, pipes and other plumbing materials to 8.0%.

Are There Screening Measures Available?

The level of lead in your child's blood can be measured, and early detection means early intervention. Measures include:

- A blood test can reveal if there's an elevated level of lead in your child's blood.
- A second blood test is usually done if a child's screening shows that lead may be present. X-rays and other tests may be necessary.
- Follow-up questions will be asked to learn about the child's behavior, health, and symptoms; anything the child has chewed on or swallowed; possible sources of lead; the child's diet; and/or family medical history.
- Other measures may include home inspection for lead sources, or counseling about how to protect children.

Should My Child Be Screened?

In general, all high-risk children need lead screening. For example, state or local health officials may consider a child at high risk if he or she:

- lives in an area that has a high number of older homes (built before 1950),
- lives in or regularly visits a home built before 1950,
- lives in or regularly visits a home built before 1975 that has recently been remodeled, or
- has had a brother or sister with lead problems.

What Else Can I Do to Protect My Child?

In your kitchen you can:

- Feed your child a well-balanced diet that's high in iron, calcium and vitamin C – it helps protect the body against lead.
- Don't store food in open cans.
- Don't use pottery for cooking or serving if you're unsure about its glaze.
- If you suspect lead, draw drinking and cooking water only from the cold tap after letting it run for a minute.
- Have your water tested.

In your home you can:

- Be alert for chipping and flaking paint.
- Use only safe interior paints on toys, walls, furniture, and other items.
- Replace plastic blinds made outside the U.S. with a type that is lead-free.

With your child:

- Don't allow your child to put things in his or her mouth that may be dirty or have paint on them.
- Keep children from chewing window sills or other painted surfaces.
- Don't allow your child to eat snow or icicles.
- Wash childrens hands often, especially before they eat and before nap time and bedtime.
- Keep play areas clean. Wash bottles, pacifiers, toys, and stuffed animals regularly.
- Make sure children eat nutritious, low-fat meals high in iron and calcium, such as spinach and dairy products. Children with good diets absorb less lead.

If you work with lead:

- Don't bring it home with you.
- Shower and change before coming home.
- Wash your clothes separately from your family's clothes.
- Follow all occupational safety guidelines for cleaning and storing work clothes and equipment.

What Actions Can I Take to Reduce Exposure to Lead in Drinking Water?

Despite our best efforts mentioned earlier to control water corrosivity and remove lead from the water supply, lead levels in some homes or buildings can be high. To find out whether you need to take action in your home, have your drinking water tested to determine if it contains excessive concentrations of lead. Testing the water is essential because you cannot see, taste, or smell lead in drinking water. Some local laboratories that can provide this service are listed at the end of this booklet.

WASA and DOH Acknowledge National Lead Awareness Week

If results of a water test indicate that drinking water drawn from a tap in your home contains lead above 15 ppb, then you should take the following precautions:

- Let the water run from the tap before using it for drinking or cooking any time the water in a faucet has gone unused for more than six hours. The longer water resides in your home's plumbing the more lead it may contain. Flushing the tap means running the cold water faucet until the waters gets noticeably colder, usually about 15–30 seconds. If your house has a lead service line to the water main, you may have to flush the water for a longer time, perhaps one minute, before drinking. Although toilet flushing or showering flushes water through a portion of your home's plumbing system, you still need to flush the water in each faucet before using it for drinking or cooking. Flushing tap water is a simple and inexpensive measure you can take to protect your family's health. It usually uses less than one or two gallons of water and costs less than $3.00 per month at half a penny per gallon. To conserve water, fill a couple of bottles for drinking water after flushing the tap, and whenever possible use the first flush water to wash the dishes or water the plants. The plumbing systems of high-rise buildings have more, and sometimes larger pipes than smaller buildings. (Most larger service lines and associated plumbing fixtures do not contribute significant amounts of lead.) If you live in a high-rise building, letting the water flow before using it may not work to lessen your risk from lead. Ask your landlord for help in locating any sources of lead and for advice on reducing the lead level if appropriate.
- Try not to cook with or drink water from the hot water tap. Hot water can dissolve more lead more quickly than cold water. If you need hot water, draw water from the cold tap and heat it on the stove.
- Remove loose lead solder and debris from the plumbing materials installed in newly constructed homes, or homes in which the plumbing has recently been replaced, by removing the faucet strainers from all taps and running the water from 3–5 minutes. Thereafter, periodically remove the strainers and flush out any debris that has accumulated over time.
- If your copper pipes are joined with lead solder that has been installed illegally since it was banned in 1986, notify the plumber who did the work and request that he or she replace the lead solder with lead-free solder. Lead solder looks dull gray, and when scratched with a key looks shiny. In addition, notify EPA Region's Safe Drinking Water Act Enforcement Branch at 215-814-5445 about the violation.

- Have an electrician check your wiring. If grounding wires from the electrical system are attached your pipes, corrosion may be greater. Check with a licensed electrician or your local electrical code to determine if your wiring can be grounded elsewhere. DO NOT attempt to change the wiring yourself because improper grounding can cause electrical shock and fire hazards.

How Do I Know if I Have a Lead Water Service Line?

The best way to determine if your service line is made of lead is by either hiring a licensed plumber to inspect the line or by contacting the plumbing contractor who installed the line. You can identify the plumbing contractor by checking the city's record of building permits which is maintained in the files of the Department of Consumer and Regulatory Affairs (DCRA) at 202-442-4642. WASA also maintains records of the materials located in the distribution system. Call 202-612-3440 for information.

How Can I Tell if the Plumbing in My House is Lead?

A licensed plumber can at the same time check to see if your home's plumbing contains lead solder, lead pipes, or pipe fittings that contain lead.

What if My Drinking Water Has Elevated Levels of Lead?

If a water test indicates that the drinking water coming from your tap contains lead concentrations in excess of 15 ppb after flushing, or after WASA has completed actions to minimize lead levels, then you may want to take the following additional measures.

- Purchase or lease a home treatment device. Home treatment devices are limited in that each unit treats only the water that flows from the faucet to which it is connected, and all of the devices require periodic maintenance and replacement. Devices such as reverse osmosis systems or distillers can effectively remove lead from your drinking water. Some activated carbon filters may reduce lead levels at the tap; however, all lead reduction claims should be investigated. Be sure to check the actual performance of a specific home treatment device before and after installing the unit.
- Purchase bottled water for drinking and cooking.

How Can I Get My Service Line Replaced?

If the service line that connects your dwelling to the water main contributes more than 15 ppb to drinking water, after WASA's comprehensive treatment program is in place, WASA is required to implement a multi-year program to replace the portion of the line we own. If the line is only partially owned by WASA, we are required to provide the owner of the privately owned portion of the line with information on how to replace the privately owned portion of the service line, and offer to eventually replace that portion of the line at the owner's expense. If we replace only the portion of the line that we own, we also are required to notify you in advance and provide you with information on the steps you can take to minimize exposure to any temporary increase in lead levels that may result from the partial replacement, to take a follow-up sample at our expense from the line within 72 hours after the partial replacement, and to mail or otherwise provide you with the results of that sample within three business days of receiving the results. An acceptable replacement is copper.

WASA AND DOH ACKNOWLEDGE NATIONAL LEAD AWARENESS WEEK

Has WASA Been Replacing Lead Service Lines in Public Space?

Under federal regulations, WASA is required to develop and implement a multi-year program to replace the portion of each lead service line that WASA owns if the line contributes lead concentrations of more than 15 ppb after the implementation of the comprehensive treatment program. The replacement program is underway and will continue until such time as the monitoring results show they no longer exceed 15 ppb. If you have questions about whether the service line serving your home contains lead or how we are carrying out the requirements of the lead regulations, please call WASA at 202-612-3440, between 8 a.m. and 4 p.m.

What Is WASA's Lead & Copper Program?

WASA has a number of drinking water quality monitoring programs, one of which is its Lead & Copper Program. One of the treatment objectives for DC's system is to control the corrosivity of water to minimize leaching of lead or copper from customer plumbing in the water. Therefore, WASA regularly monitors for elevated lead and copper concentrations by collecting water samples at consumer taps. Lead concentrations found in these sampling programs comply with EPA's requirements; however, tap water in some homes may contain higher levels of lead.

WASA's recent Lead & Copper Program hosted 53 volunteers who have single-family residences that are served by either lead services, internal lead plumbing or copper pipes with lead solder installed after 1982. During WASA's last sampling program in the summer of 2001 and June 2002, some of these homes tested above 15 ppb. In the District of Columbia, there are approximately 130,000 water service lines and 20,000 of these are lead services.

Who Can I Contact to Obtain More Information?

You can consult a variety of sources for additional information. Government agencies that can be contacted include:

- The WASA Water Quality Division (202-612-3440) can provide you with information about your community's water supply. A list of local laboratories that have been certified by EPA for testing water quality is below.
- The DC Department of Health (DOH) (202-535-2690) can provide you with information about the health effects of lead and how you can have your child's blood tested. Call for free blood lead screening, community outreach and education, medical follow-up services, lead-based paint inspection, and lead-based paint abatement. The DOH is holding free screenings for District children between the ages of 6 months and 6 years old, and expectant mothers. Remember that all children must be tested for daycare enrollment. Call 202-535-2690 for free in-home appointment testing, or walk-in testing is also available at 51 N Street, NE, Suite 3000.
- The DC Department of Consumer and Regulatory Affairs (202-442-4641) can provide you with information about building permit records that should contain the names of plumbing contractors that plumbed your home.
- Your family doctor or pediatrician can perform a blood test for lead and provide you with information about the health effects of lead.

The following is a list of some EPA-certified laboratories in your area that you can call to have your water tested for lead.

AMA Analytical Services, Inc.
4475 Forbes Boulevard
Lanham, MD 20706
301-459-2640

Anabell Environmental, Inc.
8648 Dakota Drive
Gaithersburg, MD 20877
301-548-9425

Envirometric Laboratories, Inc.
354 Hungerford Drive, Suite 100
Rockville, MD 20850
301-838-3091

GPL Laboratories, LLLP
202 Perry Parkway
Gaithersburg, MD 20877
301-926-6802

Metropolitan Environmental Testing Services, Inc.
179 Smallwood Village Center
Waldorf, MD 20602
301-870-1995

WSSC, LSG
12245 Tech Road
Silver Spring, MD 20904
301-206-7580

For more information, visit WASA's Website at **www.dcwasa.com.**

Para leer este folleto en Español, por favor visite nuestra página Web. **www.dcwasa.com.**

Si usted desea este material en Español, por favor llamamos al telefono gratis 202-612-3440.

DISTRICT OF COLUMBIA
Water and Sewer Authority
5000 OVERLOOK AVENUE, S.W.
WASHINGTON, D.C. 20032
DCWASA

OFFICIAL BUSINESS
PENALTY FOR MISUSE

First Class
U.S. Postage
PAID
Permit #4889
Suburban, MD

GOVERNMENT OF THE DISTRICT OF COLUMBIA
Department of Health

ATTACHMENT 8:

DC Department of Health Community Speaking Engagements:
Lead in the Water

DISTRICT OF COLUMBIA DEPARTMENT OF HEALTH

LEAD RELATED COMMUNITY SPEAKING ENGAGEMENTS

"LEAD IN THE WATER"

DATE/TIME	LOCATION	DEPARTMENT OF HEALTH SPEAKER
February 3, 2004 7:00 p.m.- 9:00 p.m.	Georgetown University MacDougal Auditorium	Dr. Walter Faggett
February 17, 2004 7:30 p.m.- 8:30 p.m.	St. Peter's Church 2nd and C Street, S.E.	Dr. Lynette Stokes
February 18, 2004 6:30 p.m.- 8:30 p.m.	Francis Gregory Library 3660 Alabama Avenue, S.E.	Dr. Lynette Stokes Dr. Daniel Lucey
February 26, 2004 6:30 p.m.- 8:30 p.m.	Highland Branch Library 115 Atlantic Street, S.W.	Dr. Daniel Lucey
March 2004	Executive Committee of the DC Chapter of the American Academy of Pediatrics at Howard University Medical Center	Dr. Daniel Lucey
March 2004	Medical-Chirurgical Society Board of Director's meeting at the National Medical Association to discuss lead strategies and community outreach efforts	Dr. Daniel Lucey Dr. Walter Faggett
March 2004	Meeting with the DC Medical Society of DC and the DC Hospital Association to discuss the recent lead incident in the District of Columbia. Both organizations sent out e-mail messages to their entire membership regarding the importance of reporting blood lead levels to the Department of Health and sent copies of the February 26, 2004 Health Advisory from the Department of Health to all 23, 000 residences in DC with lead service pipes.	Dr. Daniel Lucey
March 2, 2004 6:30 p.m.- 8:30 p.m.	Palisades Branch Library 4901 V Street, N.W.	Dr. Daniel Lucey
March 2, 2004 6:30 p.m.- 8:30 p.m.	St. Martin Catholic Church 1908 N. Capital Str., N.W.	Dr. Walter Faggett
March 8, 2004 7:00 p.m.- 9:00 p.m.	Congress Heights United Methodist Church 421 Alabama Avenue, S.E.	Dr. Daniel Lucey
March 10, 2004 2:00 p.m.- 3:00 p.m.	Office of Aging Senior Services Network 441 4th Street, Room 940	Dr. Thomas Calhoun
March 13, 2004 8:30 a.m.- 5:00 p.m.	Trinity College O'Connor Auditorium 125 Michigan Avenue, N.E.	Dr. Daniel Lucey
March 16, 2004 5:30 p.m.- 7:00 p.m.	Shiloh Baptist Church 1507 9th Street, N.W.	Dr. Daniel Lucey
March 22, 2004 6:30 p.m.- 8:30 p.m.	All Souls Church 1500 Harvard Street, N.W.	Dr. Daniel Lucey
March 23, 2004 6:30 p.m.- 8:30 p.m.	DC Department of Health Lead Advisory Committee meeting with local pediatricians and lead experts: Dr. Wolf (Children's National Medical Center), Dr. Collins (Howard University Medical Center), Dr. Paulsen (George Washington University Medical Center), and Dr. Nelson (Georgetown University Medical Center-Medstar Health).	Dr. Daniel Lucey Dr. Lynette Stokes Dr. Thomas Calhoun
March 23, 2004 6:30 p.m.- 8:30 p.m.	Hines Junior High School 335 8th Street, S.E.	Dr. Lynette Stokes
March 24, 2004 7:00 p.m.- 9:00 p.m.	G. Washington University Jack Morton Auditorium 805 21st Street, N.W.	Dr. Thomas Calhoun
March 25, 2005 11:00 a.m. - noon	Guest appearance on District Cable Channel 16 regarding the recent lead incident in the District of Columbia	Dr. Daniel Lucey Dr. Lynette Stokes
March 29, 2004 6:30 p.m.- 8:30 p.m.	Penn Baptist Church 3000 Penn Avenue, S.E.	Dr. Walter Faggett
March 31, 2004 6:30 p.m.- 8:30 p.m.	Ketchum Elementary School 1919 15th Street, S.E.	Dr. Daniel Lucey

Revised April 2, 2004.

GOVERNMENT OF THE DISTRICT OF COLUMBIA
Department of Health

ATTACHMENT 9:

DC Department of Health Lead Fact Sheet

Government of the District of Columbia

March 1, 2004

Lead *In the District of Columbia* Fact Sheet

What Are the Health Effects of Lead?

Lead can pose a significant risk to health if too much of it enters your body. Even small amounts of lead can be harmful if swallowed or inhaled. If lead accumulates in the body over many years, it can cause damage to the brain, red blood cells, and kidneys.

Lead from chipping and flaking paint, if ingested, can cause significant health impacts especially for small children.

Lead in drinking water, although rarely the sole cause of lead poisoning, can significantly increase a person's total lead exposure, particularly the exposure of infants who drink baby formulas and concentrated juices that are mixed with water. The EPA estimates that drinking water can make up 20 percent or more of a person's total exposure to Lead.

Should My Child Be Screened?

In general, all high-risk children need lead screening. Children, under the age of 6 may be at high risk if he or she:

- lives in an area that has a high number of older homes (built before 1950),
- lives in or regularly visits a home built before 1950,
- lives in or regularly visits a home built before 1975 that has recently been remodeled,
- has had a brother or sister with lead problems, or
- Resides in an area with reported elevated lead level in water.

What Else Can I Do to Protect My Child?

In your kitchen you can:

- Feed your child a well-balanced diet that's high in iron, calcium and vitamin C – it helps protect the body against lead.
- Don't store food in open cans.
- Don't use pottery for cooking or serving if you're unsure about its glaze.
- Have your water tested.

WASA and EPA Suggest:

- Draw water for drinking or cooking after other high water use activities, such as bathing, showering, flushing the toilet, or washing your clothes, so that a total of at least 10 minutes of flushing water from your lead service pipes has occurred.
- Flush your kitchen tap for 60 seconds, then collect drinking water in containers and store them in the refrigerator. About once a month, remove and clean the strainer/aerator device on your faucet to remove debris.
- Cold water should be used for drinking or cooking, as hot water will contain higher levels of lead. Cold water should be heated for making hot beverages or cooking, Do not use the water from the hot water faucet for drinking or cooking.

In your home you can:

- Be alert for chipping and flaking paint.
- Use only safe interior paints on toys, walls, furniture, and other items.
- Replace plastic blinds made outside the U.S. with a type that is lead-free.

With your child:

- Don't allow your child to put things in his or her mouth that may be dirty or have paint on them.
- Keep children from chewing window sills or other painted surfaces.
- Don't allow your child to eat snow or icicles.
- Wash children's hands often, especially before they eat and before nap time and bedtime.
- Keep play areas clean. Wash bottles, pacifiers, toys, and stuffed animals regularly.
- Make sure children eat nutritious, low-fat meals high in iron and calcium, such as spinach and dairy products. Children with good diets absorb less lead.

For More Information Call:
202.671.0733

Government of the District of Columbia
Anthony A. Williams, Mayor

GOVERNMENT OF THE DISTRICT OF COLUMBIA
DOH
DEPARTMENT OF HEALTH
James A. Buford, Director

Government of the District of Columbia

Lead *In the District of Columbia* Fact Sheet

If you work with lead:
- Don't bring it home with you.
- Shower and change before coming home.
- Wash your clothes separately from your family's clothes.
- Follow all occupational safety guidelines for cleaning and storing work clothes and equipment.

What You Can Do Now To Protect Your Family

If you suspect that your house has lead hazards, you can take some immediate steps to reduce your family's risk:

- If you rent, notify your landlord of peeling or chipping paint.
- Clean up paint chips immediately.
- Clean floors, window frames, window sills, and other surfaces weekly. Use a mop or sponge with warm water and a general all-purpose cleaner or a cleaner made specifically for lead. REMEMBER: NEVER MIX AMMONIA AND BLEACH PRODUCTS TOGETHER SINCE THEY CAN FORM A DANGEROUS GAS.
- Thoroughly rinse sponges and mop heads after cleaning dirty or dusty areas.
- Clean or remove shoes before entering your home to avoid tracking in lead from soil.

How Does Lead Enter Our Drinking Water?

Lead is unusual among drinking water contaminants in that it seldom occurs naturally in water supplies like rivers and lakes. Lead enters drinking water primarily as a result of the corrosion, or wearing away, of materials containing lead in the household plumbing and water service lines. These materials include lead-based solder used to join copper pipe, brass, and chrome-plated brass faucets, and in some cases, pipes made of lead that connect your house to the water main (service lines). In 1986, Congress banned the use of lead solder containing greater than 0.2% lead, and restricted the lead content of faucets, pipes and other plumbing materials to 8.0%.

Are There Screening Measures Available?

The level of lead in your child's blood can be measured, and early detection means early intervention. Measures include:

- A blood test can reveal if there's an elevated level of lead in your child's blood.
- A second blood test is usually done if a child's screening shows that lead may be present. X-rays and other tests may be necessary.
- Follow-up questions will be asked to learn about the child's behavior, health, and symptoms; anything the child has chewed on or swallowed; possible sources of lead; the child's diet; and/or family medical history.
- Other measures may include home inspection for lead sources, or counseling about how to protect children.

What Actions Can I Take to Reduce Exposure to Lead in Drinking Water?

Despite our best efforts mentioned earlier to control water corrosivity and remove lead from the water supply, lead levels in some homes or buildings can be high. To find out whether you need to take action in your home, have your drinking water tested to determine if it contains excessive concentrations of lead. Testing the water is essential because you cannot see, taste, or smell lead in drinking water.

For More Information Call:
202.671.0733

Government of the District of Columbia
Anthony A. Williams, Mayor

Government of the District of Columbia
DOH
DEPARTMENT OF HEALTH
James A. Buford, Director

GOVERNMENT OF THE DISTRICT OF COLUMBIA
Department of Health

ATTACHMENT 10:

DC Department of Health Lead Website

WELCOME TO WASHINGTON
District of Columbia

Department of Health

Lead in the District of Columbia

The DC Department of Health (DOH) is working with the DC Water and Sewer Authority (WASA) and city officials to respond to residents' concerns about lead in District of Columbia drinking water. Select the links below for up-to-date information on lead, lead testing, the health effects of lead exposure, and ways to avoid becoming exposed.

- District of Columbia Fact Sheet
- Blood Lead Levels Found in DOH Testing
- Telephone Numbers
  - Lead Screening (202) 671-0733
  - Lead in Water Testing (202) 787-2732
- DOH Interim Chief Medical Officer's Letter to Residents:

English* Chinese*
Vietnamese* Korean*
Amharic Spanish*

- DOH Fact Sheet on Lead

English* Chinese*
Vietnamese* Korean*
Spanish*

- Free Blood Screening Schedule
- Simple Steps to Protect Your Family from Lead Hazards
- Informative Links on Lead
  - http://www.cdc.gov/nceh/lead/CaseManagement/caseManage_chap3.htm
  - http://www.cdc.gov/health/lead.htm
  - http://www.epa.gov/air/urbanair/lead/index.html

* This document is presented in Portable Document Format (PDF) and a PDF reader is required for viewing. Download a PDF reader or learn more about PDFs.

GOVERNMENT OF THE DISTRICT OF COLUMBIA
Department of Health

ATTACHMENT 11:

DC Department of Health Lead Clinic Schedule

DISTRICT OF COLUMBIA DEPARTMENT OF HEALTH - LEAD INCIDENT BLOOD SCREENING SCHEDULE

Mon	Tue	Wed	Thu	Fri	Sat	Sun
1	2	3	4	5	6	7
8 **WARD 4 – Barnard Elem. School** 430 Decatur St., NW *(5:00 p.m. - 8:00 p.m.)* **Mobile Unit;** 12th and G Street, NE *(10:00 a.m. - 4:00 p.m.)* **DOH Lead Clinic 51 N St., NE** (Call for appointment: 202-671-0733 or Walk-in- 9:00 a.m. - 4:00 p.m.) **DC General Health Campus** 1900 Mass. Ave. SE *(5:00 p.m. - 9:00 p.m.)*	9 **St. Mary's Center** **2333 Ontario Road, NW** **(5:00 p.m. - 9:00 p.m.)** **Mt. Bethel Baptist Church** 75 Rhode Island Ave., NW Mobile Unit *(9:00 a.m. - 4:00 p.m.)* **Bridges Academy Daycare** 6117 Georgia Ave., NW (10:00 a.m. - 6:00 p.m.) **DOH Lead Clinic** 51 N St., NE (Call for appointment: 202-671-0733 or Walk-in- 9:00 a.m. - 4:00 p.m.)	10 **The Perry School** 128 M St., NW (10:00 a.m.-6:00 p.m.) **DOH Mobile Units** Various Daycare Centers (9:00 a.m. - 3:00 p.m.) **DOH Lead Clinic 51 N St., NE** (Call for appointment: 202-671-0733 or Walk-in- 9:00 a.m. - 4:00 p.m.) **DC General Health Campus** 1900 Mass. Ave. SE *(5:00 p.m. - 9:00 p.m.)*	11 **The Perry School** 128 M St., NW (10:00 a.m.-6:00 p.m.) **DOH Mobile Units** Various Daycare Centers (9:00 a.m. - 3:00 p.m.) **DOH Lead Clinic** 51 N St., NE (Call for appointment: 202-671-0733 or Walk-in- 9:00 a.m. - 4:00 p.m.)	12 **DOH Mobile Units** Various Daycare Centers (9:00 a.m. - 3:00 p.m.) **DOH Lead Clinic** 51 N St., NE (Call for appointment: 202-671-0733 or Walk-in- 9:00 a.m. - 4:00 p.m.)	13 **Trinity College** O'Connor Auditorium, 125 Michigan Ave. NE (8:30 a.m. - 5:00 p.m.) **WARD 5** **La Clinica del Pueblo** 2831 15th St. NW (9:00 a.m. - 1:00 p.m.)	14 **Sacred Heart Church** 1621 Park Rd. NW (11:00 a.m. - 3:00 p.m.)
15 **DOH Mobile Units** **Various Daycare Centers (9:00 a.m. - 3:00 p.m.)** **Mobile Unit;** 12th & G St., NE (10:00 a.m. - 4:00 p.m.) **DOH Lead Clinic** 51 N St., NE (Call for appointment: 202-671-0733 or Walk-in- 9:00 a.m. - 4:00 p.m.) **DC General Health Campus** 1900 Mass. Ave. SE *(5:00 p.m. - 8:00 p.m.)*	16 **DOH Mobile Units** **Various Daycare Centers (9:00 a.m. - 3:00 p.m.)** **St. Mary's Center** **2333 Ontario Road, NW** (6:00 p.m. - 9:00 p.m.) **Shiloh Baptist Church** 1507 9th St., NW (11:00 a.m. - 4:00 p.m.) DOH Lead Clinic 51 N St., NE (Call for appointment: 202-671-0733 or Walk-in- 9:00 a.m. - 4:00 p.m.)	17 **DOH Mobile Units** **Various Daycare Centers (9:00 a.m. - 3:00 p.m.)** **Mobile Unit;** 1000 5th St., SE (10:00 a.m. - 4:00 p.m.) **DOH Lead Clinic** 51 N St., NE (Call for appointment: 202-671-0733 or Walk-in- 9:00 a.m. - 4:00 p.m.) **DC General Health Campus** 1900 Mass. Ave. SE *(5:00 p.m. - 8:00 p.m.)*	18 **DOH Mobile Units** **Various Daycare Centers (9:00 a.m. - 3:00 p.m.)** **Children Hut Daycare** 510 Kennedy St., NW (10:00 a.m. - 2:00 p.m.) **DOH Lead Clinic** 51 N St., NE (Call for appointment: 202-671-0733 or Walk-in- 9:00 a.m. - 4:00 p.m.)	19 **DOH Mobile Units Various Daycare Centers (9:00 a.m. - 3:00 p.m.)** **Mobile Unit;** Canal St. & N St., SW (10:00 a.m. - 4:00 p.m.) **DOH Lead Clinic** 51 N St., NE (Call for appointment: 202-671-0733 or Walk-in- 9:00 a.m. - 4:00 p.m.)	20 **Anacostia Health Center** 1328 W St., SE (10:00 a.m. - 2:00 p.m.) **Hardy Recreational Center 45th & Q St., NW (9:00 a.m.-1:00 p.m.)** **Mobile Unit** -Hardy Recreational Center 45th & Q St., NW (9:00 a.m. - 1:00 p.m.) Hospital for Sick Children 1731 Bunker Hill Rd., NE (10:00 a.m. - 2:00 p.m.)	21
22 **Emory Recreational Center; 1337 14th St., NW (4:00p.m.-8:00p.m.)** **DOH Mobile Units Various Daycare Centers (9:00 a.m. - 3:00 p.m.)** **DOH Lead Clinic** 51 N St., NE (Call for appointment: 202-671-0733 or Walk-in- 9:00 a.m. - 4:00 p.m.) **DC General Health Campus** 1900 Mass. Ave. SE *(5:00 p.m. - 8:00 p.m.)*	23 **DOH Mobile Units** **Various Daycare Centers (9:00 a.m. - 3:00 p.m.)** **DOH Lead Clinic** 51 N St., NE (Call for appointment: 202-671-0733 or Walk-in- 9:00 a.m. - 4:00 p.m.)	24 **DOH Mobile Units** Various Daycare Centers (9:00 a.m. - 3:00 p.m.) **DOH Lead Clinic** 51 N St., NE (Call for appointment: 202-671-0733 or Walk-in- 9:00 a.m. - 4:00 p.m.) **DC General Health Campus** 1900 Mass. Ave. SE *(5:00 p.m. -8:00 p.m.)*	25 **DOH Mobile Units** Various Daycare Centers (9:00 a.m. - 3:00 p.m.) **DOH Lead Clinic** 51 N St., NE (Call for appointment: 202-671-0733 or Walk-in- 9:00 a.m. - 4:00 p.m.)	26 **DOH Mobile Units Various Daycare Centers (9:00 a.m. - 3:00 p.m.)** **DOH Lead Clinic** 51 N St., NE (Call for appointment: 202-671-0733 or Walk-in- 9:00 a.m. - 4:00 p.m.)	27 All Souls Church 1500 Harvard St. NW (10:00 a.m. - 4:00 p.m.)	28
29 **Mobile Unit;** 12th & G St., SE (10:00 a.m.-4:00 p.m.) **DOH Lead Clinic** 51 N St., NE **(Call for appointment: 202-671-0733 or Walk-in- 9:00 a.m. - 4:00 p.m.)** **DC General Health Campus** 1900 Mass. Ave. SE *(5:00 p.m. - 8:00 p.m.)*	30 YMCA Capital View 2118 Ridgecrest Court, SE (6:00 p.m. - 9:00 p.m.) **DOH Lead Clinic** 51 N St., NE (Call for appointment: 202-671-0733 or Walk-in- 9:00 a.m. - 4:00 p.m.)	31 **Hmpty Dumpty Daycare 32 Grant Circle NW (10:00 a.m. - 2:00 p.m.)** **Bethesda Baptist Church** 1808 Capital Street, N.E. 11:00 a.m. - 3:00 p.m. **DOH Lead Clinic** 51 N St., NE **(Call for appointment: 202-671-0733 or Walk-in- 9:00 a.m. - 4:00 p.m.)** **DC General Health Campus** 1900 Mass. Ave. SE *(5:00 p.m. - 8:00 p.m.)*	1 **LAST UPDATED ON 03.12.04**	2	3	4

DISTRICT OF COLUMBIA DEPARTMENT OF HEALTH - LEAD INCIDENT BLOOD SCREENING SCHEDULE - APRIL 2004

Mon	Tue	Wed	Thu	Fri	Sat	Sun
29 DOH Mobile Unit; 12th & C St., SE (10:00 a.m.-4:00 p.m.) Greater Southeast Community Hospital; 1310 Southern Ave., SE - Primary Care Clinic 1st Floor. Appointments only (202-574-7222) DOH Lead Clinic 51 N St., NE (Call for appointment: 202-671-0733 or Walk-in- 9:00 a.m. - 4:00 p.m.) DC General Health Campus 1900 Mass. Ave. SE (5:00 p.m. - 8:00 p.m.)	30 YMCA Capital View 2118 Ridgecrest Court, SE (6:00 p.m. - 9:00 p.m.) Greater Southeast Community Hospital; 1310 Southern Ave., SE - Primary Care Clinic 1st Floor; Appointments only (202-574-7222) DOH Mobile Units Various Daycare Centers (9:00 a.m. - 3:00 p.m.) DOH Lead Clinic 51 N St., NE (Call for appointment. 202-671-0733 or Walk-in- 9:00 a.m. - 4:00 p.m.)	31 Humpty Dumpty Daycare 32 Grant Circle NW (10:00 a.m. - 2:00 p.m.) DOH Lead Van Greater Southeast Community Hospital; 1310 Southern Ave., SE - Primary Care Clinic 1st Floor; Appointments only (202-574-7222) Bethesda Baptist Church 1808 Capital Street, N.E. 11:00 a.m. - 3:00 p.m. DOH Lead Clinic 51 N St., NE (Call for appointment: 202-671-0733 or Walk-in- 9:00 a.m. - 4:00 p.m.) DC General Health Campus 1900 Mass. Ave. SE (5:00 p.m. - 8:00 p.m.)	1 Greater Southeast Community Hospital; 1310 Southern Ave., SE – Primary Care Clinic 1st Floor; Appointments only (202-574-7222) DOH Lead Clinic 51 N St., NE (Call for appointment: 202-671-0733 or Walk-in- 9:00 a.m. - 4:00 p.m.)	2 Sherwood Recreational Center 10th & C Street, N.E. (6:00 p.m. - 9:00 p.m.) DOH Mobile Unit; 12th & C St., SE (10:00 a.m.-4:00 p.m.) Greater Southeast Community Hospital; 1310 Southern Ave., SE – Primary Care Clinic 1st Floor; Appointments only (202-574-7222) DOH Lead Clinic 51 N St., NE (Call for appointment. 202-671-0733 or Walk-in- 9:00 a.m. - 4:00 p.m.) DC General Health Campus 1900 Mass. Ave. SE (5:00 p.m. - 8:00 p.m.)	3	4
5 Greater Southeast Community Hospital; 1310 Southern Ave., SE - Primary Care Clinic 1st Floor; Appointments only (202-574-7222) DOH Lead Clinic 51 N St., NE (Call for appointment: 202-671-0733 or Walk-in- 9:00 a.m. - 4:00 p.m.) DC General Urgent Care Center; 1900 Massachusetts Avenue, SE (11:00 a.m. - 11:00 p.m.)	6 Greater Southeast Community Hospital; 1310 Southern Ave., SE - Primary Care Clinic 1st Floor; Appointments only (202-574-7222) DOH Lead Clinic 51 N St., NE (Call for appointment: 202-671-0733 or Walk-in- 9:00 a.m. - 4:00 p.m.) DC General Urgent Care Center; 1900 Massachusetts Avenue, SE (11:00 a.m. - 11:00 p.m.)	7 YMCA Calomiris 1906 Allison Street, NE (5:00 p.m. - 8:00 p.m.) Greater Southeast Community Hospital; 1310 Southern Ave., SE - Primary Care Clinic 1st Floor; Appointments only (202-574-7222) DOH Lead Clinic 51 N St., NE (Call for appointment: 202-671-0733 or Walk-in- 9:00 a.m. - 4:00 p.m.) DC General Urgent Care Center; 1900 Massachusetts Avenue, SE (11:00 a.m. - 11:00 p.m.)	8 Greater Southeast Community Hospital; 1310 Southern Ave., SE - Primary Care Clinic 1st Floor; Appointments only (202-574-7222) DOH Lead Clinic 51 N St., NE (Call for appointment: 202-671-0733 or Walk-in- 9:00 a.m. - 4:00 p.m.) DC General Urgent Care Center; 1900 Massachusetts Avenue, SE (11:00 a.m. - 11:00 p.m.)	9 YMCA National Capital 1711 Rhode Island Ave., NW (5:00 p.m. - 8:00 p.m.) Greater Southeast Community Hospital; 1310 Southern Ave., SE - Primary Care Clinic 1st Floor; Appointments only (202-574-7222) DOH Lead Clinic 51 N St., NE (Call for appointment: 202-671-0733 or Walk-in- 9:00 a.m. - 4:00 p.m.) DC General Urgent Care Center; 1900 Massachusetts Avenue, SE (11:00 a.m. - 11:00 p.m.)	10 DC General Urgent Care Center; 1900 Massachusetts Avenue, SE (10:30 a.m.- 7:00 p.m.)	11 DC General Urgent Care Center; 1900 Massachusetts Avenue, SE (10:30 a.m.- 7:00 p.m.)
12 Greater Southeast Community Hospital; 1310 Southern Ave., SE - Primary Care Clinic 1st Floor; Appointments only (202-574-7222) DOH Lead Clinic 51 N St., NE (Call for appointment: 202-671-0733 or Walk-in- 9:00 a.m. - 4:00 p.m.) DC General Urgent Care Center; 1900 Massachusetts Avenue, SE (11:00 a.m. - 11:00 p.m.)	13 Greater Southeast Community Hospital; 1310 Southern Ave., SE - Primary Care Clinic 1st Floor; Appointments only (202-574-7222) DOH Lead Clinic 51 N St., NE (Call for appointment: 202-671-0733 or Walk-in- 9:00 a.m. - 4:00 p.m.) DC General Urgent Care Center; 1900 Massachusetts Avenue, SE (11:00 a.m. - 11:00 p.m. M-F; 10:30 a.m.- 7:00 p.m. Sat. & Sun)	14 Covenant Baptist Church 3845 S. Capitol St., SE (5:00 p.m. - 8:00 p.m.) Greater Southeast Community Hospital; 1310 Southern Ave., SE – Primary Care Clinic 1st Floor; Appointments only (202-574-7222) DOH Lead Clinic 51 N St., NE (Call for appointment: 202-671-0733 or Walk-in- 9:00 a.m. – 4:00 p.m.) DC General Urgent Care Center; 1900 Massachusetts Avenue, SE (11:00 a.m. - 11:00 p.m.)	15 Greater Southeast Community Hospital; 1310 Southern Ave., SE - Primary Care Clinic 1st Floor; Appointments only (202-574-7222) DOH Lead Clinic 51 N St., NE (Call for appointment: 202-671-0733 or Walk-in- 9:00 a.m. - 4:00 p.m.) DC General Urgent Care Center; 1900 Massachusetts Avenue, SE (11:00 a.m. - 11:00 p.m.)	16 National City Christian Church 5 Thomas Circle, NW ((5:00 p.m. - 8:00 p.m.) Greater Southeast Community Hospital; 1310 Southern Ave., SE - Primary Care Clinic 1st Floor; Appointments only (202-574-7222) DOH Lead Clinic 51 N St., NE (Call for appointment: 202-671-0733 or Walk-in- 9:00 a.m. - 4:00 p.m.) DC General Urgent Care Center; 1900 Massachusetts Avenue, SE (11:00 a.m. - 11:00 p.m.)	17 DC General Urgent Care Center; 1900 Massachusetts Avenue, SE (10:30 a.m.- 7:00 p.m.)	18 DC General Urgent Care Center, 1900 Massachusetts Avenue, SE (10:30 a.m.- 7:00 p.m.)
19 Tifereth Israel Congregation 7701 16th St, NW (5:00 p.m. - 8:00 p.m.) Greater Southeast Community Hospital; 1310 Southern Ave., SE - Primary Care Clinic 1st Floor; Appointments only (202-574-7222) DOH Lead Clinic 51 N St., NE (Call for appointment: 202-671-0733 or Walk-in- 9:00 a.m. - 4:00 p.m.) DC General Urgent Care Center; 1900 Massachusetts Avenue, SE (11:00 a.m. - 11:00 p.m.)	20 Greater Southeast Community Hospital; 1310 Southern Ave., SE - Primary Care Clinic 1st Floor; Appointments only (202-574-7222) DOH Lead Clinic 51 N St., NE (Call for appointment: 202-671-0733 or Walk-in- 9:00 a.m. - 4:00 p.m.) DC General Urgent Care Center; 1900 Massachusetts Avenue, SE (11:00 a.m. - 11:00 p.m.)	21 Chevy Chase Community Center; 5601 Connecticut Ave., NW (6:00 - 9:30 p.m.) Greater Southeast Community Hospital; 1310 Southern Ave., SE - Primary Care Clinic 1st Floor; Appointments only (202-574-7222) DOH Lead Clinic 51 N St., NE (Call for appointment: 202-671-0733 or Walk-in- 9:00 a.m. - 4:00 p.m.) DC General Urgent Care Center; 1900 Massachusetts Avenue, SE (11:00 a.m. - 11:00 p.m.)	22 Greater Southeast Community Hospital; 1310 Southern Ave., SE - Primary Care Clinic 1st Floor; Appointments only (202-574-7222) DOH Lead Clinic 51 N St., NE (Call for appointment: 202-671-0733 or Walk-in- 9:00 a.m. - 4:00 p.m.) DC General Urgent Care Center; 1900 Massachusetts Avenue, SE (11:00 a.m. - 11:00 p.m.)	23 Greater Southeast Community Hospital; 1310 Southern Ave., SE - Primary Care Clinic 1st Floor; Appointments only (202-574-7222) DOH Lead Clinic 51 N St., NE (Call for appointment: 202-671-0733 or Walk-in- 9:00 a.m. - 4:00 p.m.) DC General Urgent Care Center; 1900 Massachusetts Avenue, SE (11:00 a.m. - 11:00 p.m.)	24 All Souls Church 2831 15th St. NW (9:00 A.M. - 1:00 P.M.) DC General Urgent Care Center; 1900 Massachusetts Avenue, SE (10:30 a.m - 7:00 p.m.)	25 DC General Urgent Care Center, 1900 Massachusetts Avenue, SE (10:30 a.m. - 7:00 p.m.)
26 Greater Southeast Community Hospital; 1310 Southern Ave., SE - Primary Care Clinic 1st Floor; Appointments only (202-574-7222) DOH Lead Clinic 51 N St., NE (Call for appointment: 202-671-0733 or Walk-in- 9:00 a.m. - 4:00 p.m.) DC General Urgent Care Center; 1900 Massachusetts Avenue, SE (11:00 a.m. - 11:00 p.m.)	27 Greater Southeast Community Hospital; 1310 Southern Ave., SE - Primary Care Clinic 1st Floor; Appointments only (202-574-7222) DOH Lead Clinic 51 N St., NE (Call for appointment: 202-671-0733 or Walk-in- 9:00 a.m. - 4:00 p.m.)	28 Greater Southeast Community Hospital; 1310 Southern Ave., SE - Primary Care Clinic 1st Floor; Appointments only (202-574-7222) DOH Lead Clinic 51 N St., NE (Call for appointment: 202-671-0733 or Walk-in- 9:00 a.m. - 4:00 p.m.) DC General Urgent Care Center; 1900 Massachusetts Avenue, SE (11:00 a.m. - 11:00 p.m.)	29 Greater Southeast Community Hospital; 1310 Southern Ave., SE - Primary Care Clinic 1st Floor; Appointments only (202-574-7222) DOH Lead Clinic 51 N St., NE (Call for appointment: 202-671-0733 or Walk-in- 9:00 a.m. - 4:00 p.m.) DC General Urgent Care Center; 1900 Massachusetts Avenue, SE (11:00 a.m. - 11:00 p.m.)	30 Greater Southeast Community Hospital; 1310 Southern Ave., SE - Primary Care Clinic 1st Floor; Appointments only (202-574-7222) DOH Lead Clinic 51 N St., NE (Call for appointment: 202-671-0733 or Walk-in- 9:00 a.m. - 4:00 p.m.) DC General Urgent Care Center; 1900 Massachusetts Avenue, SE (11:00 a.m. - 11:00 p.m.)	Last UPDATED 04/15/04	

GOVERNMENT OF THE DISTRICT OF COLUMBIA
Department of Health

ATTACHMENT 12:

DC Department of Health Lead Reporting Requirements Health Advisory to Local Healthcare Professionals

GOVERNMENT OF THE DISTRICT OF COLUMBIA
DEPARTMENT OF HEALTH
Childhood Lead Poisoning Prevention Program

REMINDER TO ALL DISTRICT OF COLUMBIA HEALTHCARE PROFESSIONALS

Tuesday, March 02, 2004

Dear Healthcare Professional:

As mandated by the District of Columbia Childhood Lead Poisoning Screening and Reporting Act (*Title XX of the Fiscal Year 2003 Budget Support Act of 2002*), all normal and abnormal serum lead levels must be reported to the District of Columbia Department of Health's (DOH) Childhood Lead Poisoning Prevention Program. The results should be faxed to DOH at 202-535-1398. For additional information, please contact the Childhood Lead Poisoning Prevention Program at 202-535-1911.

Additionally, DOH is providing free screenings for blood lead levels of children **less than six years of age**, pregnant women, and women who are nursing. The DOH does not screen children less than the age of six months. Screening services are available on a walk-in basis at 51 N. St, NE, Suite 300 or by appointment—202-671-0733. Testing hours of operation are from 9:00 a.m. to 4:00 p.m. All inquiries related to testing of water or service lines should be referred to WASA at 202-787-2732.

Sincerely,

Daniel R. Lucey, M.D., M.P.H.
Interim Chief Health Officer
District of Columbia
Department of Health

825 North Capitol Street, NE, 4th Floor, Washington, D.C. 20002, (202) 442-5999, (202) 442-47[illegible]3 Fax

GOVERNMENT OF THE DISTRICT OF COLUMBIA
Department of Health

ATTACHMENT 13:

DC Department of Health Lead Reporting Act
Title XX of the Fiscal Year 2003 Budget Support Act of 2002

TITLE XX. CHILDHOOD LEAD POISONING SCREENING AND REPORTING

Sec. 2001. Short title.

This title may be cited as the "Childhood Lead Poisoning Screening and Reporting Act of 2002".

Sec. 2002. Definitions.

For the purposes of this title, the term:

(1) "Child" or "Children" means an individual or individuals under the age of 6 years.

(2) "Elevated blood lead level" means an excessive absorption of lead concentration in whole blood of 10 mg/dl (micrograms of lead per deciliter) or greater.

(3) "Health care facility" means any institution providing individual care or treatment of diseases or other medical, physiological or psychological conditions, including, but not limited to, hospitals, clinics, laboratories, nursing homes, or homes for the aged or chronically ill, but excluding private medical offices.

(4) "Health care provider" means a physician, clinic, hospital, or neighborhood health center, licensed by the District of Columbia, that is responsible for providing primary care and coordinating referrals, when necessary, to other health care providers.

(5) "Lead-poisoned child" means a child with a confirmed blood lead level equal to or greater than 15 micrograms of lead per deciliter of blood, or such other lower threshold as the United States Centers for Disease Control and Prevention may establish.

(6) "Person" means an individual, a corporation, a partnership, firm, conservator, receiver, trustee, executor, or legal representative.

Sec. 2003. Childhood lead screening and reporting requirements.

(a) Each health care provider or facility shall inform the parent or guardian of every child under the age of 6 years in the District of Columbia, served by the provider or facility, of the requirement for periodic blood tests for lead poisoning as provided in this title and rules implementing this title.

(b) A health care provider or facility shall, unless parental consent is withheld or an identical test has already been performed within the last 12 months, perform a blood test for lead poisoning on every child who resides in the District of Columbia as part of a well-child care visit, once between ages 6 months and 9 months, and a second time between ages 22 months and 26 months. If a child's age exceeds 26 months, and a blood lead screening has not been performed, the child shall be screened twice prior to the age of 6 years.

(c) The laboratory that performed the tests pursuant to subsection (b) of this section shall forward all test results to the health care provider or facility where the blood sample was taken, and to the Mayor.

(d) The health care provider or facility shall forward all elevated blood lead level results immediately to the child's parent or guardian. Upon request of the Child's parent or guardian, the health care provider for facility shall provide written evidence of testing for lead poisoning that includes the date of the test and the tests results.

(e) The Mayor, pursuant to section 2006, shall issue rules governing the conditions under which a health care provider or facility shall administer additional lead screening tests exceeding the requirements of subsection (b) of this section, and the process for reporting lead screening results.

(f) Any agreement or contract entered into by the Medical Assistance Administration to provide its services through a health insuring or a managed care organization shall include a requirement for the organization to provide screening and reporting services pursuant to the provisions of this section and the rules implementing this section, or to provide reimbursement for those services.

(1) The agreement or contract shall explicitly include the provision of medical case management and other follow-up treatment of a Medicaid-enrolled, lead-exposed child as may be required to protect the child's health, or reimbursement for that management and treatment.

(2) The Medical Assistance Administration shall provide for coverage and reimbursement of an environmental investigation and source control measures necessary to eliminate any lead-based paint hazard to which a Medicaid-enrolled, lead-poisoned child is exposed in the child's home environment, including, but not limited to, paint stabilization and cleanup of any dust-lead hazard.

(g) The Mayor shall issue an annual report to the Council summarizing and analyzing the lead screening results obtained pursuant to this title. The report shall include recommendations based on or pertaining at a minimum to:

(1) The extent of compliance with the requirements of this section; and

(2) The incident and prevalence rates of childhood lead poisoning in the District of Columbia.

Sec. 2004. Enforcement.

(a) If the Mayor has reason to believe that there has been a violation of this title or of the regulations issued pursuant to this title, the Mayor may:

(1) Give written notice of the alleged violation, which shall include the provision of the law or regulation alleged to be violated, the facts alleged to constitute a violation, and an order that necessary corrective action be taken within a specified time set forth in the notice; or

(2) Impose civil or criminal fines and penalties in accordance with section 2005.

(b) Any party adversely affected by an action taken pursuant to subsection (a) of this section is entitled to a hearing before the Mayor upon filing with the Mayor, within 15 days from the date of the action, a written request for a hearing. The hearing shall be held in accordance with the requirements of Title I of the District of Columbia Administrative Procedure Act, approved October 21, 1968 (82 Stat.1204; D.C. Official Code § 2-501 et seq.).

Sec. 2005. Penalties.

(a) Any person who fails to comply with any of the provisions of this title shall be subject to a fine not to exceed $5,000 for each violation. Each and every day of the violation shall constitute a separate violation and the penalties prescribed shall be applicable to each separate violation unless otherwise indicated.

(b) Civil fines, penalties, and fees may be imposed as alternative sanctions for any infraction of the provisions of this title or the rules issued under authority of this title pursuant to Titles I through III of the Department of Consumer and Regulatory Affairs Civil Infractions Act of 1985, effective October 5, 1985 (D.C. Law 6-42; D.C. Official Code § 2-1801.01 et seq.).

(c) Any person knowingly or willfully violates this title shall, in addition to or in lieu of any civil penalty which may be imposed for the violation, be subject, upon conviction, to a fine of not more than $5,000 for each day of violation, or to imprisonment of not more than one year, or both.

Sec. 2006. Rules.

The Mayor, pursuant to Title I of the District of Columbia Administrative Procedure Act, approved October 21, 1968 (82 Stat. 1204; D.C. Official Code § 2-501 et seq.), shall issue rules to implement the provisions of this title. The Mayor is authorized to adopt, in whole or in part, guidelines issued by the Centers for Disease Control and Prevention and may consider such other materials relating to lead poisoning prevention, testing, and treatment, as deemed appropriate.

Sec. 2007. Fiscal impact statement.

Funds are sufficient in the fiscal Year 2002 through Fiscal Year 2005 budget and financial plan to implement this title because no additional staff or resources will be required. This title will have no effect on General Fund revenue.

GOVERNMENT OF THE DISTRICT OF COLUMBIA
Department of Health

ATTACHMENT 14:

DC Department of Health Lead Incident Command and Control Chart

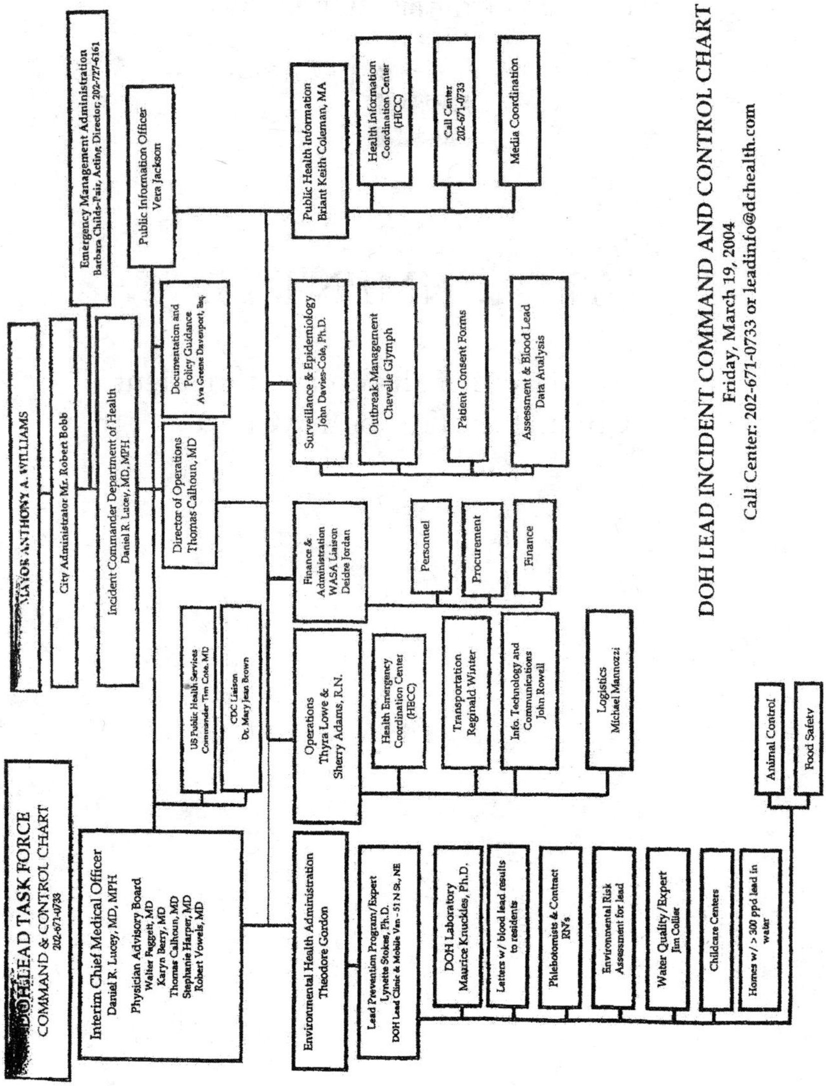

DOH LEAD TASK FORCE
COMMAND & CONTROL CHART
202-671-0733
MAYOR ANTHONY A. WILLIAMS
City Administrator Mr. Robert Bobb
Incident Commander Department of Health
Daniel R. Lucey, MD, MPH
Emergency Management Administration
Barbara Childs-Pair, Acting, Director, 202-727-6161
Interim Chief Medical Officer
Daniel R. Lucey, MD, MPH
Physician Advisory Board
Walter Faggett, MD
Karyn Berry, MD
Thomas Calhoun, MD
Stephanie Harper, MD
Robert Vowels, MD
US Public Health Services
Commander Tim Cote, MD
CDC Liaison
Dr. Mary Jean Brown
Director of Operations
Thomas Calhoun, MD
Documentation and
Policy Guidance
Ava Greene Davenport, Esq.
Public Information Officer
Vera Jackson
Environmental Health Administration
Theodore Gordon
Operations
Thyra Lowe &
Sherry Adams, R.N.
Finance &
Administration
WASA Liaison
Deidre Jordan
Surveillance & Epidemiology
John Davies-Cole, Ph.D.
Public Health Information
Briant Keith Coleman, MA
Lead Prevention Program/Expert
Lynette Stokes, Ph.D.
DOH Lead Clinic & Mobile Van - 51 N St., NE
DOH Laboratory
Maurice Knuckles, Ph.D.
Letters w/ blood lead results
to residents
Phlebotomists & Contract
RN's
Environmental Risk
Assessment for lead
Water Quality/Expert
Jim Collier
Childcare Centers
Homes w/ > 300 ppd lead in
water
Animal Control
Food Safety
Health Emergency
Coordination Center
(HECC)
Transportation
Reginald Winter
Info. Technology and
Communications
John Rowell
Logistics
Michael Mannozzi
Personnel
Procurement
Finance
Outbreak Management
Chevelle Glymph
Patient Consent Forms
Assessment & Blood Lead
Data Analysis
Health Information
Coordination Center
(HICC)
Call Center
202-671-0733
Media Coordination
DOH LEAD INCIDENT COMMAND AND CONTROL CHART
Friday, March 19, 2004
Call Center: 202-671-0733 or leadinfo@dchealth.com

RESPONSES BY DANIEL R. LUCEY TO ADDITIONAL QUESTIONS FROM SENATOR JEFFORDS

Question 1. I have been frequently told by parents—If I had only known, I could have taken precautions. I want to ask each of the witnesses at the table to tell me how you are responding to this questions—what explanation are you giving to parents whose children were totally unnecessarily exposed to lead in their drinking water and what steps are you taking to regain the trust of the citizens of Washington, DC?

Response. Yes, I have been asked the same questions at many of the DC-wide community meetings in which the Department of Health has participated. Since I only started working at the DC Department of Health on February 10, 2004, I cannot explain much of what happened in the Department of Health before that time, and I do not try to do so. Once I was designated on February 26th by the DC City Administrator, Deputy Mayor Robert Robb, to be the primary person at the DC Department of Health responsible for the lead-in-the-water issues then I emphasized what actions we are taking now, why we are taking these actions, and what the results are as we go forward together. I wrote and signed the February 26th letter to the approvimately 23,000 persons with lead-service lines and therein made specific Public Health recommendations that to my knowledge had never been made before on this scale. I spoke with both the Director of the Centers for Disease Control and Prevention and with the U.S. Surgeon General to request the respective expertise that their organizations were willing to provide immediately to the residents of the District of Columbia.

Efforts to regain trust included participating and answering questions in many community discussion groups across the city (please see attachment #8 in my April 7th testimony), speaking at many press conferences along with the Director of the DC Emergency Management Agency, DC Council members, the City Administrator, and with the Mayor. In addition we made multiple other educational efforts including the use of written pamphlets in multiple languages and significantly expanding lead-related information on our website. We worked with DC pediatric lead experts including those at Children's National Medical Center, Howard University Medical Center, Georgetown University Medical Center, and Georgetown University Medical Center. I also consulted with Obstetricians at Washington Hospital Center and with the American College of Obstetrics and Gynecology (ACOG) about lead issues.

Question 2. I know people who have homes with elevated lead levels, but not above 15 parts per billion. I know people who have homes with elevated lead levels well above 15 parts per billion who are not on your inventory of "known lead service lines." Given what we know about the harmful effects of lead, what is WASA and the DC Department of Health doing to help those residents take appropriate health precautions?

Response. I am not sure I understand the first part of the question because 15 parts per billion (ppb) is the definition of a home with a lead level above the EPA action limit. Should I understand this first part of your question to refer to the issue I have sometimes been asked at DC community meetings about homes with lead service lines, but measurements of lead in their drinking water that are not above 15 ppb? If so, then I emphasize that the original letter I wrote and signed on February 26, 2004 was sent to *all* approximately 23,000 District residents who had lead service lines regardless of whether their water lead concentrations were measured above or below 15 parts per billion. Free water filters to remove lead in the drinking water were given to everyone with a lead service line by WASA and free blood lead levels were offered by the DC Department of Health as specified in the February 26 letter. On the other hand, if you meant there are people "with elevated lead levels" in homes with water lead concentrations "not above 15 parts per billion" then I would reply that we provide the identical offer to everyone with an elevated blood lead level, namely a home environmental assessment to look for other sources of lead such as lead paint, or lead ceramics, or lead-containing stained glass, or certain medications, cosmetics, candies or other potential sources of lead exposure. The DC Department of Health made that commitment to look for other sources of lead, including lead paint, on page two of our February 26, 2004 letter.

Regarding homes with elevated lead concentrations over 15 ppb, even if they are not listed by WASA as having a lead service line, then WASA provides them with a water filter to remove the lead. The DC Department of Health continues to offer free blood lead level testing at five (5) standing clinics in DC, including one at the Department of Health on 51 N Street, NE. Another one of these five clinics is at the former DC General Hospital outpatient urgent-care clinic that is open 7 days a week, including evenings. A calendar for the month of May listing the locations and times of operation for these five clinics is attached as a document titled: District of Columbia Department of Health—Lead Blood Screening Schedule—May 2004.

Question 3. Your testimony provides several data points indicating that relatively small numbers of children had what you define as "elevated" blood lead levels. How do you define "elevated" and how did you select that number?

Response. The DC Department of Health uses the same definition of an elevated blood lead level, or level of concern, as that provided by the Centers for Disease Control and Prevention (CDC), namely 10 micrograms/deciliter (µg/dl) or higher.

Question 4. Based on the information in the April 17, 2003 New England Journal of Medicine study regarding blood lead levels in children, do you feel it is appropriate to ignore children and families with blood lead level test results above 0 but below 10?

Response. As a physician and Department of Health official who has worked and published for years in both laboratory and clinical research, I appreciate the value of studies such as the one cited. In particular, when medical papers have passed the peer-review process that is required for publication in one of the very best medical journals in the world, such as the New England Journal of Medicine, they warrant our attention. All scientific studies, however, require independent confirmation studies. The CDC is certainly aware of the paper you have cited. We have consulted regularly with lead experts at the CDC on this specific issue. We recognize that historically the value defining an elevated blood lead level has been set much higher by the CDC, such as between 1976–1980 when the CDC reported that 9 of 10 children (88.2 percent) between the ages of 1–5 years had blood lead levels of 10 µg/dl or higher.

Question 5. Can you describe in detail the testing program that the DC Department of Health has in place, any new testing programs that the Department has completed or has underway as a result of the lead in drinking water issue, and what your findings are?

Response. During the time I worked as the interim Chief Medical Officer at the DC Department of Health, from February 10–April 30 of this year, over 90 percent of my time was devoted to the lead-in-the-water issues, including development of the blood testing program. I am glad to summarize those testing programs. We established an Incident Command Structure for all lead-in-the-water issues (see attachment #14 in my April 7th testimony for the specific command structure details). In addition please find attached a table titled "Blood Lead Level Screening Results" with total results for 5,293 persons who have been tested at the DC Department of Health laboratory through our screening program as of May 6, 2004. The results include the fact that 37 children, two nursing mothers, and zero pregnant women have had an elevated blood lead levels. Of the 37 children, 13 live in homes with lead service lines and 24 do not. With one exception still under investigation, all homes of children and nursing mothers with elevated blood have shown lead dust and/or soil levels that exceed EPA and HUD guidelines. This environmental work on lead has been supervised by Lynette Stokes, Ph.D. at the DC Department of Health.

Also, please find attached a color graph that plots blood lead levels for 1,924 persons who are less than the age of 6, or women who are pregnant or nursing. The lowest measurable value in our laboratory assay is 1 µg/dl. The number of people at each value of blood lead level decreases at each blood lead level from 1 µg/dl to 9 µg/dl, with the mean (average) value being 3.0 µg/dl.

Since May 1st, Thomas Calhoun, M.D., is the physician responsible for lead-in-the-water issues at the DC Department of Health. I have consulted with him about the new testing programs since May 1st. There are now five (5) DC clinics where free blood lead level testing is provided. In addition, Dr. Calhoun has worked with a 6th site, at Children's National Medical Center, to offer free blood lead level testing to young persons at DC schools where elevated water lead concentrations were found. Dr. Calhoun is coordinating a program whereby the DC Department of Health will go visit these DC schools and offer free blood lead level testing to children under 6 years of age starting the week of May 17th. The April 29th letter that I wrote regarding water lead levels in DC schools, as well as the letter of February 26th, is posted with other lead-related information on our web site at *www.dchealth.dc.gov.*

Question 6. Your testimony implies that there is not a problem with elevated blood lead levels in children due to drinking water exposure. I believe you are basing that conclusion on the results of the voluntary testing program conducted in the city. I have read critiques indicating that it is impossible to judge the presence of a city-wide trend depending only on self-selected samples. Dr. Best also raised concerns with this approach during our discussions on the second panel.

During that panel, Senator Crapo and I both asked questions about your test results and the apparent conclusion that the people you have tested that were exposed to high lead levels do not show, in an overwhelming proportion, high blood lead levels. In our discussions with Dr. Best on this topic, you indicated that the majority of the people you tested were adults. Children, of course, are at the greatest risk of lead poisoning due to the higher rate at which they retain lead in their bloodstream when compared to adults.

Please explain how the Department of Health can draw conclusions about the severity of the issue at hand when they appear to be based only on self-selected, voluntary blood tests, or an evaluation of a primarily adult population which we know has a lower retention rate for lead in the body? Do you have any plans to conduct more widespread testing?

Response. In addition to the many clinics we set up all across the District for voluntary blood lead level testing since February 28th, we also went directly to the homes of two populations of people we considered potentially at high risk for lead toxicity. These two groups are persons living in homes with the highest levels of lead in their drinking water (> 300 parts per billion) and very young children in childcare facilities with lead-service lines. My April 7th testimony (attachment #2) provided data showing that none of 201 persons tested in homes with > 300 ppb of lead in their water had elevated blood lead levels as we published with the CDC in their March 30th Morbidity and Mortality Weekly Report (MMWR) Dispatch. In addition, in the same April 7th testimony (attachment #3) I included data showing that only 2 of 280 of the young children in these DC childcare facilities had elevated blood lead levels. These two specific outreach efforts are combined with the open blood lead testing program where adults are tested as well as children, an effort that now includes 1,752 children less than 6 years of age. Finally, we do plan to test more children under the age of 6 years, starting May 17–21, in DC schools where elevated levels of lead were reported on April 29th.

Question 7. In 1986, lead was discovered in drinking water in the Palisades section of Washington, DC. Residents were quoted as saying, "The runaround has been unbelievable . . . No one in the bureaucracy has even begun to take this seriously." The Director of water for the city stated that it was, ". . . premature to contact residents throughout the city" before the city developed a plan to handle and finance increased testing. I ask unanimous consent to insert several newspaper articles on this topic into the record. I find it unbelievable that no one at this witness table learned anything from this previous experience. Did anyone here refer to previous instances of lead contamination in the District when formulating a response plan? If so, please describe how you used this information, and if not, why not?

Response. When I started working this February 10th at the DC Department of Health I was not aware of the 1986 events at the Palisades. My immediate impression at the time was that as the Incident Commander for the Department of Health, to response to the crisis I needed to learn as much as possible about lead issues and take rapid action by offering free blood lead level testing. The then Director of the Department of Health, and his highest Deputy, were replaced by March 26th, 2004.

Responses by Daniel Lucey to Additional Questions from Senator Crapo

Question 1. You testified to the need for better teamwork if the Lead and Copper Rule is to be implemented effectively. Now that your agency and the other authorized agencies are working very closely in what has become a closely scrutinized effort, what improvements in teamwork have you learned? Also, how will you change routine procedures for working together to implement the Rule after the DC drinking water system returns below the Action Level?

Response. The exchange of information, and the critique of that information, has significantly improved both within the Department of Health and between other involved organizations as of February 2004. Medical aspects of the Department of Health have been better coordinated with the engineering aspects of the lead-in-the-water issues. Moreover, direct interaction between the DC Department of Health and the Environmental Protection Agency (EPA) has increased substantially, including on the Public Health and medical issues. This is an important change that needs to continue as part of routine procedures going forward.

Government of the District of Columbia
Department of Health

Blood Lead Level Screening Results
(Micrograms/deciliter ("mcg/dL"))
As of February 3-May 6, 2004

Total number screened for blood lead levels	Total number of laboratory tests completed	Total number within DOH target population	Percentage within DOH target population with lead service lines	Total number within DOH target population with high lead levels	Total number outside of DOH target population with high blood lead levels
5293	5293	1924 • 1752 (91.1%) children under the age of 6; • 95 (4.9%) women who are pregnant; and • 77 (4.0%) women who are nursing.	17.9% have lead service lines; • 17.0% of all children under the age of 6 • 19.0% of all women who are pregnant; and • 37.7% of all women who are nursing. 82.1% do not have lead service lines.	• Women who are nursing=2* & pregnant = 0 • 37** children under the age of 6 have elevated (10mcg/dL or higher) lead levels. o This is 2.2% of all children under the age of 6; o 13 with lead service lines; and o 24 without lead service lines. *One nursing woman retested and had second BLL that was not elevated. **One additional child from MD who attends a DC daycare had an elevated BLL. As of Friday, May 7, 2004, environmental assessments for other sources of lead, such as lead dust & paint have been performed on 44 residences. All children and nursing mothers with elevated blood lead levels (with the exception of 1 residence) have shown lead dust and/or soil levels that exceed EPA & HUD guidelines.	• Of the 3349 residents outside of the DOH target population screened for blood lead levels only 4 have a blood lead level above 25 mcg/dL (the level of most concern for adults). Two (2) of those four residents have lead service lines. 5 children ages 6-15 have elevated blood lead levels (10, 11, 12, 16, and 19 mcg/dL). None of these children live in homes with lead service lines. The blood lead level at which medication is offered is usually 45 mcg/dL.

Note: 20 ages being confirmed.

GOVERNMENT OF THE DISTRICT OF COLUMBIA
Department of Health

Blood lead levels for the Target Group (age < 6 yrs., pregnant, or nursing), May 6, 2004 (N=1,924)

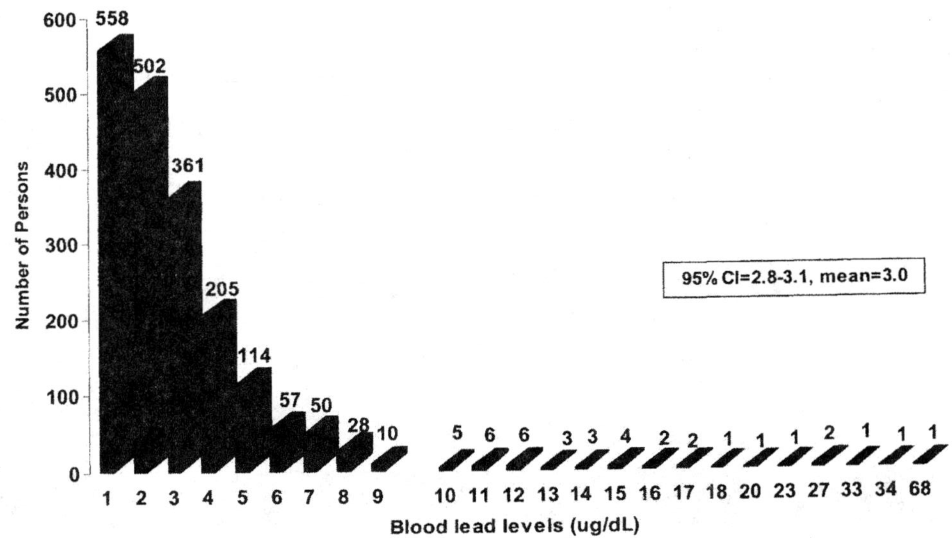

					1	2
3 Greater Southeast Community Hospital - Ward 8: 1310 Southern Ave., SE – Primary Care Clinic 1st Floor: Appointments only (202-574-7222) DOH Lead Clinic 51 N St., NE (Call for appointment: 202-671-0733 or Walk-in- 9:00 a.m. – 4:00 p.m.) Anacostia Health Center Ward 8,1328 W St., SE (8:00 a.m. – 4:00 p.m.) Congress Heights Clinic Ward 8, 3720 Martin Luther King, Jr. Avenue, SE (8:15 a.m. – 4:45 p.m.) DC General Urgent Care Center Ward 7; 1900 Massachusetts Avenue, SE (11:00 a.m. – 11:00 p.m.)	4 Greater Southeast Community Hospital - Ward 8; 1310 Southern Ave., SE – Primary Care Clinic 1st Floor; Appointments only (202-574-7222) DOH Lead Clinic 51 N St., NE (Call for appointment: 202-671-0733 or Walk-in- 9:00 a.m. – 4:00 p.m.) Anacostia Health Center Ward 8,1328 W St., SE (8:00 a.m. - 4:00 p.m.) Congress Heights Clinic Ward 8, 3720 Martin Luther King, Jr. Avenue, SE (8:15 a.m. – 4:45 p.m.) DC General Urgent Care Center Ward 7; 1900 Massachusetts Avenue, SE (11:00 a.m. – 11:00 p.m.)	5 Greater Southeast Community Hospital - Ward 8: 1310 Southern Ave., SE – Primary Care Clinic 1st Floor: Appointments only (202-574-7222) DOH Lead Clinic 51 N St., NE (Call for appointment: 202-671-0733 or Walk-in- 9:00 a.m. – 4:00 p.m.) Anacostia Health Center Ward 8,1328 W St., SE (8:00 a.m. – 4:00 p.m.) Congress Heights Clinic Ward 8, 3720 Martin Luther King, Jr. Avenue, SE (8:15 a.m. – 4:45 p.m.) DC General Urgent Care Center Ward 7; 1900 Massachusetts Avenue, SE (11:00 a.m. – 11:00 p.m.)	6 Greater Southeast Community Hospital - Ward 8; 1310 Southern Ave., SE – Primary Care Clinic 1st Floor; Appointments only (202-574-7222) DOH Lead Clinic 51 N St., NE (Call for appointment: 202-671-0733 or Walk-in- 9:00 a.m. – 4:00 p.m.) Anacostia Health Center Ward 8,1328 W St., SE (8:00 a.m. – 4:00 p.m.) Congress Heights Clinic Ward 8, 3720 Martin Luther King, Jr. Avenue, SE (8:15 a.m. – 4:45 p.m.) DC General Urgent Care Center Ward 7; 1900 Massachusetts Avenue, SE (11:00 a.m. – 11:00 p.m.)	7 Greater Southeast Community Hospital - Ward 8: 1310 Southern Ave., SE – Primary Care Clinic 1st Floor; Appointments only (202-574-7222) DOH Lead Clinic 51 N St., NE (Call for appointment: 202-671-0733 or Walk-in- 9:00 a.m. – 4:00 p.m.) Anacostia Health Center Ward 8,1328 W St., SE (8:00 a.m. – 4:00 p.m.) Congress Heights Clinic Ward 8, 3720 Martin Luther King, Jr. Avenue, SE (8:15 a.m. – 4:45 p.m.) DC General Urgent Care Center Ward 7; 1900 Massachusetts Avenue, SE (11:00 a.m. – 11:00 p.m.)	8 Congress Heights Clinic Ward 8, 3720 Martin Luther King, Jr. Avenue, SE (8:15 a.m. – 4:45 p.m.) DC General Urgent Care Center Ward 7; 1900 Massachusetts Avenue, SE (10:30 a.m.- 7:00 p.m.)	9 DC General Urgent Care Center Ward 7; 1900 Massachusetts Avenue, SE (10:30 a.m.- 7:00 p.m.)
10 Greater Southeast Community Hospital - Ward 8; 1310 Southern Ave., SE – Primary Care Clinic 1st Floor; Appointments only (202-574-7222) DOH Lead Clinic 51 N St., NE (Call for appointment: 202-671-0733 or Walk-in- 9:00 a.m. – 4:00 p.m.) Anacostia Health Center Ward 8,1328 W St., SE (8:00 a.m. – 4:00 p.m.) Congress Heights Clinic Ward 8, 3720 Martin Luther King, Jr. Avenue, SE (8:15 a.m. – 4:45 p.m.) DC General Urgent Care Center Ward 7, 1900 Massachusetts Avenue, SE (11:00 a.m. – 11:00 p.m.)	11 Greater Southeast Community Hospital - Ward 8: 1310 Southern Ave., SE – Primary Care Clinic 1st Floor; Appointments only (202-574-7222) DOH Lead Clinic 51 N St., NE (Call for appointment: 202-671-0733 or Walk-in- 9:00 a.m. – 4:00 p.m.) Anacostia Health Center Ward 8,1328 W St., SE (8:00 a.m. – 4:00 p.m.) Congress Heights Clinic Ward 8, 3720 Martin Luther King, Jr. Avenue, SE (8:15 a.m. – 4:45 p.m.) DC General Urgent Care Center Ward 7, 1900 Massachusetts Avenue, SE (11:00 a.m – 11:00 p.m.)	12 Greater Southeast Community Hospital - Ward 8; 1310 Southern Ave., SE – Primary Care Clinic 1st Floor; Appointments only (202-574-7222) DOH Lead Clinic 51 N St., NE (Call for appointment: 202-671-0733 or Walk-in- 9:00 a.m. – 4:00 p.m.) Anacostia Health Center Ward 8,1328 W St., SE (8:00 a.m. – 4:00 p.m.) Congress Heights Clinic Ward 8, 3720 Martin Luther King, Jr. Avenue, SE (8:15 a.m. – 4:45 p.m.) DC General Urgent Care Center Ward 7, 1900 Massachusetts Avenue, SE (11:00 a.m. – 11:00 p.m.)	13 Greater Southeast Community Hospital - Ward 8; 1310 Southern Ave., SE – Primary Care Clinic 1st Floor; Appointments only (202-574-7222) DOH Lead Clinic 51 N St., NE (Call for appointment: 202-671-0733 or Walk-in- 9:00 a.m. – 4:00 p.m.) Anacostia Health Center Ward 8,1328 W St., SE (8:00 a.m. – 4:00 p.m.) Congress Heights Clinic Ward 8, 3720 Martin Luther King, Jr. Avenue, SE (8:15 a.m. – 4:45 p.m.) DC General Urgent Care Center Ward 7, 1900 Massachusetts Avenue, SE (11:00 a.m. – 11:00 p.m.)	14 Greater Southeast Community Hospital - Ward 8; 1310 Southern Ave., SE – Primary Care Clinic 1st Floor; Appointments only (202-574-7222) DOH Lead Clinic 51 N St., NE (Call for appointment: 202-671-0733 or Walk-in- 9:00 a.m. – 4:00 p.m.) Anacostia Health Center Ward 8,1328 W St., SE (8:00 a.m. – 4:00 p.m.) Congress Heights Clinic Ward 8, 3720 Martin Luther King, Jr. Avenue, SE (8:15 a.m. – 4:45 p.m.) DC General Urgent Care Center Ward 7, 1900 Massachusetts Avenue, SE (11:00 a.m. – 11:00 p.m.)	15 Congress Heights Clinic Ward 8, 3720 Martin Luther King, Jr. Avenue, SE (8:15 a.m. – 4:45 p.m.) DC General Urgent Care Center Ward 7; 1900 Massachusetts Avenue, SE (10:30 a.m.- 7:00 p.m.)	16 DC General Urgent Care Center Ward 7; 1900 Massachusetts Avenue, SE (10:30 a.m.- 7:00 p.m.)
17 Greater Southeast Community Hospital - Ward 8; 1310 Southern Ave., SE – Primary Care Clinic 1st Floor; Appointments only (202-574-7222) DOH Lead Clinic 51 N St., NE (Call for appointment: 202-671-0733 or Walk-in- 9:00 a.m. – 4:00 p.m.) Anacostia Health Center Ward 8,1328 W St., SE (8:00 a.m. – 4:00 p.m.) Congress Heights Clinic Ward 8, 3720 Martin Luther King, Jr. Avenue, SE (8:15 a.m. – 4:45 p.m.) DC General Urgent Care Center Ward 7, 1900 Massachusetts Avenue, SE (11:00 a.m. – 11:00 p.m.)	18 Greater Southeast Community Hospital - Ward 8, 1310 Southern Ave., SE – Primary Care Clinic 1st Floor; Appointments only (202-574-7222) DOH Lead Clinic 51 N St., NE (Call for appointment: 202-671-0733 or Walk-in- 9:00 a.m. – 4:00 p.m.) Anacostia Health Center Ward 8,1328 W St., SE (8:00 a.m. – 4:00 p.m.) Congress Heights Clinic Ward 8, 3720 Martin Luther King, Jr. Avenue, SE (8:15 a.m. – 4:45 p.m.) DC General Urgent Care Center Ward 7, 1900 Massachusetts Avenue, SE (11:00 a.m – 11:00 p.m.)	19 Greater Southeast Community Hospital - Ward 8, 1310 Southern Ave., SE – Primary Care Clinic 1st Floor; Appointments only (202-574-7222) DOH Lead Clinic 51 N St., NE (Call for appointment: 202-671-0733 or Walk-in- 9:00 a.m. – 4:00 p.m.) Anacostia Health Center Ward 8,1328 W St., SE (8:00 a.m. – 4:00 p.m.) Congress Heights Clinic Ward 8, 3720 Martin Luther King, Jr. Avenue, SE (8:15 a.m. – 4:45 p.m.) DC General Urgent Care Center Ward 7; 1900 Massachusetts Avenue, SE (11:00 a.m. – 11:00 p.m.)	20 Greater Southeast Community Hospital - Ward 8; 1310 Southern Ave., SE – Primary Care Clinic 1st Floor; Appointments only (202-574-7222) DOH Lead Clinic 51 N St., NE (Call for appointment: 202-671-0733 or Walk-in- 9:00 a.m. – 4:00 p.m.) Anacostia Health Center Ward 8,1328 W St., SE (8:00 a.m. – 4:00 p.m.) Congress Heights Clinic Ward 8, 3720 Martin Luther King, Jr. Avenue, SE (8:15 a.m. – 4:45 p.m.) DC General Urgent Care Center Ward 7, 1900 Massachusetts Avenue, SE (11:00 a.m. – 11:00 p.m.)	21 Greater Southeast Community Hospital - Ward 8; 1310 Southern Ave., SE – Primary Care Clinic 1st Floor; Appointments only (202-574-7222) DOH Lead Clinic 51 N St., NE (Call for appointment: 202-671-0733 or Walk-in- 9:00 a.m. – 4:00 p.m.) Anacostia Health Center Ward 8,1328 W St., SE (8:00 a.m. – 4:00 p.m.) Congress Heights Clinic Ward 8, 3720 Martin Luther King, Jr. Avenue, SE (8:15 a.m. – 4:45 p.m.) DC General Urgent Care Center Ward 7, 1900 Massachusetts Avenue, SE (11:00 a.m. – 11:00 p.m.)	22 Congress Heights Clinic Ward 8, 3720 Martin Luther King, Jr. Avenue, SE (8:15 a.m. – 4:45 p.m.) DC General Urgent Care Center Ward 7: 1900 Massachusetts Avenue, SE (10:30 a.m.- 7:00 p.m.)	23 DC General Urgent Care Center Ward 7: 1900 Massachusetts Avenue, SE (10:30 a.m.- 7:00 p.m.)

STATEMENT OF THOMAS P. JACOBUS, GENERAL MANAGER, WASHINGTON AQUEDUCT, BALTIMORE DISTRICT, U.S. ARMY CORPS OF ENGINEERS

Good Morning, Chairman Crapo and Members of the Committee. I am Tom Jacobus, the general manager of Washington Aqueduct. Thank you for the opportunity to be here today.

Washington Aqueduct, which is a part of the Baltimore District of the U.S. Army Corps of Engineers, is a public water utility. We are regulated by the United States Environmental Protection Agency's Region 3 in Philadelphia.

Washington Aqueduct provides potable water not just to the District of Columbia Water and Sewer Authority, but also to Arlington County, Virginia and the city of Falls Church in Virginia as well. All funds for the operations and capital improvements at Washington Aqueduct come from its customers.

The provisions of the Safe Drinking Water Act and its associated regulations are the basis for all operations concerning the production, storage, and transmission of the drinking water produced and sold by Washington Aqueduct to its wholesale customers.

The Potomac River is the source of all water treated by Washington Aqueduct at its Dalecarlia and McMillan treatment plants. The treatment consists of chemically induced sedimentation using aluminum sulfate as the coagulant; filtration in dual media sand and anthracite coal filters; and disinfection using chlorine as the primary disinfectant and chloramines as a secondary disinfectant.

The primary objective of the treatment process is to produce and deliver water to the tap that is free of contaminants and pleasant to drink.

Three processes are simultaneously occurring to achieve that objective. First, organic and inorganic contaminants are removed during treatment. Then the water is disinfected so microorganisms that may have been in the raw water are killed. A disinfectant is carried along in the water (chloramine in this case) so that if the water encounters any bacteria in the distribution system or the building plumbing, the bacteria will be killed. Finally, the drinking water chemistry is adjusted as it leaves the treatment plants to make it less corrosive to the metals it will encounter in the distribution system and building plumbing.

The Washington Aqueduct's corrosion control has been accomplished by the use of lime to adjust the pH of the water. Tests done in conjunction with the promulgation of the Lead and Copper Rule and reported to EPA in 1994 demonstrated that pH control would be sufficient to achieve Optimal Corrosion Control Treatment for Washington Aqueduct's customers. In the years since it was first proposed, there has been a continuing involvement with EPA to refine the Washington Aqueduct's Optimal Corrosion Control Treatment and report on our ability to meet the pH targets.

However, the District of Columbia Water And Sewer Authority's sampling in accordance with the Lead and Copper Rule in the District of Columbia in 2001 and 2002 and the most recent intensive sampling in 2003 of water that has resided in lead service lines indicate that Washington Aqueduct must take immediate steps to adjust its optimal corrosion control treatment so that different results are achieved in the District of Columbia water distribution system.

To address this issue EPA Region 3, the District of Columbia Water and Sewer Authority and Washington Aqueduct's other wholesale customers in Virginia, the District of Columbia Department of Health, and Washington Aqueduct have formed a Technical Expert Working Group and developed an action plan to address the lead issue. I have attached a copy of the action plan dated March 10, 2004, as an enclosure to this testimony. A peer review panel appointed by U.S. EPA is reviewing this plan and their recommendations will be incorporated in a subsequent update.

In addition, teams have been formed to address production operations, distribution system operations, and risk communication to the public. Representatives of different agencies lead each of these teams. EPA leads the risk communications team; the District of Columbia Water and Sewer Authority leads the distribution system operations team; and Washington Aqueduct leads the production operations team. Representatives of all of the agencies participate on all of the teams. These teams will make a recommendation to be adopted by the technical expert working group that will result in a treatment change. While the exact type and dosage will be determined in the next couple of weeks as a result of the ongoing scientific analysis, it appears that the use of a phosphate-based corrosion inhibitor will be adopted.

Current plans are to begin a partial system application of a revised optimal corrosion control treatment this June followed by a full system application by September. There is optimism that this change will be effective in reducing the lead leaching, but it will take several months to measure the effects. Laboratory studies will be ongoing for many months in parallel with the revised treatment that may be used to further refine the change in treatment.

I would like to address the role of the Washington Aqueduct Wholesale Customer Board as it pertains to water treatment decisions. The Wholesale Customer Board governs the Washington Aqueduct's financial and strategic planning. The Board is comprised of the general manager of the District of Columbia Water and Sewer Authority, the Arlington County manager and Falls Church city manager. As part of the Board's oversight, there are technical committees that meet to evaluate engineering and financial operations throughout the year.

The board and the committees have worked very effectively to address difficult issues such as a new solids management strategy to meet a new discharge permit and to develop a disinfectant strategy to meet changes to the Disinfection Byproducts Rule. Based on what we have learned in the last 8 weeks, we will add corrosion control management as a permanent agenda item, just like we do now with filtered water turbidity, Total Coliform Rule compliance and Disinfection Byproduct Rule compliance. By doing this we will have a procedure in place for the Washington Aqueduct customers to share lead and copper data.

As the general manager of Washington Aqueduct, I intend to take two other actions. First I will make an adjustment to the structure of the Washington Aqueduct organization to integrate an existing water quality office and the capability of our plant operations branch, including our water quality laboratory. Second, I will ask our customers to participate in more frequent and more structured meetings that will improve an ongoing information loop involving them and our water quality office.

In the other two important treatment areas I mentioned earlier (i.e., disinfection and contaminant removal) we have worked very effectively with our customers to know at all times what the conditions are at the treatment plants and in their distribution systems and to implement systems responses on their behalf. The experience of the last several weeks concerning the application of the lead and copper rule indicates we need to give the third element (i.e., corrosion control) the same visibility.

I am confident Washington Aqueduct working with its customers and EPA can accomplish that.

Thank you again for the opportunity to be here today. I will be happy answer your questions.

Action Plan

To

Reduce the Occurrence of Lead

Leaching from Service Lines, Solder, or Fixtures

Into Tap Water

In the District of Columbia

And Arlington County and Falls Church, Virginia

Prepared by:

United States Environmental Protection Agency, Region 3
Washington Aqueduct, US Army Corps of Engineers
District of Columbia Water and Sewer Authority

With the cooperation of:

District of Columbia Department of Health
Arlington County Department of Public Works
Falls Church Department of Environmental Services

Assisted by:

CH2M Hill
Baker-Killam Joint Venture
HDR Engineering
US EPA Office of Research and Development

Introduction

In July 2002, the District of Columbia Water and Sewer Authority's (WASA) routine tap samples for lead exceeded the action level of 15 parts per billion (ppb). The 90th percentile level of lead went from 8 parts ppb the monitoring period before, to 75 ppb, with more than half the samples exceeding the action level. This milestone required that: 1) WASA collect more routine tap samples, more often; 2) WASA implement a public education program; and 3) WASA begin replacing lead service lines at a minimum rate of 7% of their inventory each year. The following monitoring period, ending June 30, 2003, WASA again exceeded the action level (40 ppb), requiring continuation of the public education program and lead service line replacement program as well as continue with the full routine tap monitoring.). With two monitoring periods of exceeding the action level, it became apparent that the elevated levels of lead were not a transient event and that the Washington Aqueduct and WASA would need to perform another corrosion control treatment study to determine a revised optimal corrosion control treatment for reducing lead levels in customer's tap samples.

WASA and EPA recognized the need to research why tap water lead levels increased suddenly. In the spring of 2003, EPA began the process to acquire an independent corrosion expert to help answer this question. The services of Dr. Marc Edwards of Virginia Tech were acquired in May of 2003. Dr. Edwards performed his investigations during the spring, summer and fall of 2003. He presented his written report to EPA Region III in October 2003 and presented his findings to EPA, WASA and the Washington Aqueduct in early November 2003.

WASA then developed a research strategy based on the recommendations of Dr. Edwards. This strategy was presented to the Washington Aqueduct, their Virginia wholesale customers and EPA in January 2004. WASA had begun implementing this strategy prior to the January 2004 presentation.

Again, WASA's routine tap sampling results for the six-month period ending December 2003 exceeded the lead action level (63 ppb). This necessitated the continuation of the public education and lead service line replacement programs. Both of these programs must continue until the 90th percentile lead levels are 15 ppb or below for two consecutive monitoring periods (12 months).

This Action Plan lays out the activities and targeted dates for their implementation for reducing lead levels in tap water of the District of Columbia and Arlington County and Falls Church, Virginia. This Plan may be revised or adjusted based on input from the Independent Peer Review Team being assembled by EPA's Office of Ground Water and Drinking Water. Revisions to this plan may also be made based on the feedback from studies carried out as outlined below.

Overview of the Technical Expert Working Group

The Technical Expert Working Group (TEWG) was formed to facilitate and expedite on-going research conducted by both WASA and the Washington Aqueduct (WA) of the Army Corps of Engineers as well as research already planned but not yet carried out. The U.S. Environmental Protection Agency Region III (EPA), WASA, WA and WA and the D.C. Department of Health (DC DOH) recognized that an overarching planning and coordination effort was needed to help ensure that all the necessary research was conducted, to ensure no redundant efforts took place and to move the work along as quickly as possible.

EPA, WASA , WA and DC DOH first gathered on a teleconference on February 5, 2004 to discuss research plans in place in both WASA and the Aqueduct. Another important purpose of the call was to develop a listing of any further work that needed to take place to find a treatment solution for re-optimizing the Aqueduct's optimal corrosion control treatment (OCCT). It was decided on that teleconference that the group should meet face to face to outline an overall strategy to this research effort. Staff and contractors for WASA and the Aqueduct, DC DOH, Arlington County and Falls Church Virginia, as well as EPA staff members from the Mid-Atlantic Regional Office, EPA's Office of Research and Development in Cincinnati, OH, and EPA's Office of Ground Water and Drinking Water gathered on the following Monday, February 9, 2004. The Group agreed to the strategy outline that formed the basis for this document as well as the Group's structure.

Organization of the Technical Expert Working Group

The TEWG consists of an overall Advisory team and three Technical Research Project Teams. Members of the Teams have representatives from all three entities. New members will be added to the Group as the need to add additional expertise is determined. The Technical Research Project Teams report to the Technical Advisory Team through a team leader. The Technical Advisory Team will provide for overall coordination of the Working Group and will be responsible for major decision points in the research process.

TEWG Members are:

EPA Region III

Rick Rogers, Chief, Drinking Water Branch

George Rizzo, Program Manger, District of Columbia Direct Implementation Program

Outside sub-contractor: Steven Reiber, Ph.D., and National Director for Water Research, HDR, INC., corrosion specialist

Outside sub-contractor: Gregory V. Korshin, Ph.D., Associate Professor, Environmental Eng. & Science, University of Washington; corrosion expert

Outside sub-contractor: Anne Camper, Montana State University; biofilm/disinfection/microbial corrosion expert (added early March)

EPA Office of Ground Water and Drinking Water
Jeffrey Kempic, Treatment Technology and Cost Team Leader, Standards and Risk Management Division

EPA Office of Research and Development
Michael Schock, National Risk Management Research Laboratory, Water Supply and Water Resources Division, Treatment Technology Evaluation Branch; corrosion expert

Centers for Disease Control and Prevention
Barry Brooks, Lead Poisoning Prevention Program, D.C. and Virginia Project Officer

District of Columbia Department of Health
James Collier, Chief, Bureau of Environmental Quality
Gregory Hope, Water Quality Control Branch
Jerusalem Bekele, Program Manager, Water Quality Division

Washington Aqueduct
Tom Jacobus, P.E., General Manager
Lloyd Stowe, P.E., Chief, Plant Operations Branch
Patty Gamby, Chief, Waterworks, Environmental, and Electrical Engineering Section
Elizabeth Turner, Chief, Washington Aqueduct Laboratory
Miranda Brown, Special Assistant for Water Quality
Outside Contractors, CH2M Hill
Outside sub-contractor: Vern Snoeyink, Ph.D., and Professor of Environmental Engineering, University of Illinois; corrosion expert

D.C. WASA
Michael Marcotte, Deputy General Manager and Chief Engineer
Kofi Boateng, Director, Department of Water Services
Rich Giani, Water Quality Manager
Outside Contractors: Baker-Killam Joint Venture

Arlington County, Virginia
David Hundelt, Department of Public Works, Water, Sewer, and Streets Division

Falls Church, Virginia
Robert Etris, Department of Environmental Services, Public Utilities Division

Technical Advisory Team: consists of Thomas Jacobus, P.E., General Manager, Washington Aqueduct; Michael Marcotte, P.E., Deputy General Manager, DCWASA, Rick Rogers, Chief, Drinking Water Branch, U.S. EPA Region III, James Collier, Chief, Bureau of Environmental Quality, DC DOH

Technical Research Project Teams:

Production Treatment Operations Team led by Washington Aqueduct

Distribution System Operations Team led by DCWASA
Communications Team led by U.S. EPA Region III

Mission and Priorities

The Group's mission is to develop a plan to reduce the corrosivity of treated drinking water in the District of Columbia to reduce consumers' overall lead exposure, through tap water to allowable levels. The mission of the TEWG is to determine, through bench-top, pilot scale and literature research, a revised optimal corrosion control treatment process for the Aqueduct. This re-optimized treatment will result in drastically reduced lead levels in the customer's tap water and WASA meeting EPA's lead action level. Seven main priorities of the Group are detailed below.

1. Communicate actions and progress to the community on a regular basis.

2. Choose a revised optimal corrosion control treatment (OCCT) based on desktop analysis.

3. Verify selected OCCT through partial system application and DC WASA and WA pipe loop studies.

4. Leave open the possibility of an immediate full system implementation.

5. EPA interim and final approval of selected re-optimization of corrosion control treatment.

6. Execute full system operations.

7. Use on-going pipe loop studies to refine chemistry and to determine the cause of elevated lead level.

The Research Process and Reaching a Treatment Solution Decision

The research strategy is designed to reach a treatment decision that will reduce the corrosivity of the water through a three-part plan. The three parts are:

1) Conducting a desktop analysis of all potential treatment options and make a recommendation for a revised corrosion control treatment process to the Technical Expert Working Group;
2) Conducting bench scale and pipe loops studies that will verify chemical dosage for the treatment options available and for the option selected during the desktop analysis; and
3) Communicate the strategy, the progress, the expected interim water quality effects and final outcomes to the public, the press, and government and community leaders.

The Aqueduct and WASA are responsible for developing a recommendation for a revised OCCT. It is EPA's responsibility to review and approve the revised OCCT based on input from experts from all parties. This will include feedback from an independent review panel selected by EPA's Office of Ground Water and Drinking Water in Washington, D.C. and from the DC DOH. EPA's and its contractor's involvement in the planning and implementation of this corrosion control treatment study is limited to coordination and advisory roles. Being involved from the very beginning, however, will enable a very rapid review and approval/disapproval decision on the treatment proposed.

The Technical Expert Working Group does not expect that any treatment selected and implemented will immediately reduce lead levels in the tap water. The Working Group expects lead levels to decrease over the course of implementing the revised treatment for at least several months. The Working Group also points out that, depending on the type of treatment selected, temporary changes in the aesthetic quality of the water may occur. These may be: a temporary release of rust from the water mains (red water) in some areas; an increase in calcium (lime) deposition in the water mains and in customers plumbing (if pH is adjusted higher with lime dosage); and an increase in indicator bacteria organisms, called total coliform bacteria. Some bacteria in the total coliform group are commonly among the population of harmless organisms that inhabit the rusty areas of water mains. When the rust is released, their presence may be detected in more samples than usual. The Washington Aqueduct and the Water and Sewer Authority will be taking every step possible to minimize the potential for these changes. One step will be performing an aggressive water main flushing program to help remove rust from the lines prior to applying the revised corrosion control treatment. They will monitor the water very closely during the partial treatment application and during the application of the revised treatment across the entire system distribution system.

What Consumers Should Do Until Lead Levels Decrease

Until the revised treatment scheme is in place and lead levels in tap water are reduced, the Technical Expert Working Group highly recommends that all consumers follow the tap flushing recommendations provided by EPA, WASA and the District of Columbia Department of Health (DCDOH) and DCDOH's special advisory for children, nursing mothers and pregnant women who live in homes with lead service lines. These are repeated on the following page for clarity:

Consumer advisory

Steps you should take today to reduce potential exposure to lead in drinking water:

All consumers:

- *Cold water should be used for drinking or cooking, as hot water will contain higher levels of lead. Cold water should be heated on the stove for making hot beverages or cooking. Boiling your water will not remove lead!*

- *Flush water lines that have not been used for more than six hours by running the cold water ("flush") for 60 seconds prior to using the water from a faucet for drinking or cooking.*

- *Periodically, remove and clean the strainer/aerator device on your faucet to remove debris.*

In addition, if you believe that you have a lead service line:

- ***The District of Columbia Department of Health (DC DOH) recommends that children under 6 years old and women who are pregnant or breast feeding should not drink unfiltered tap water until the concerns regarding the lead levels in the water have been resolved. DC DOH advises that unfiltered tap water should not be used for preparing infant formula or concentrated juices. In addition, DC DOH recommends that all pregnant women and children under 6 years old have their blood lead level tested. For more information on blood lead level screening, contact DC DOH at (202) 671-0733.***

- ***Draw water for drinking or cooking after another high water use activity such as bathing or washing your clothes so that a total of at least 10 minutes of flushing has occurred. (The large amount of water used will flush significant amounts of water from your home's pipes.)***
and
- ***Flush your kitchen tap for 60 seconds and then collect drinking water in clean containers and store in the refrigerator.***

If you still have concerns:

- ***Have your water tested by a certified laboratory or contact DC WASA***

Special recommendations for homes with children, pregnant women, and nursing mothers

Children and pregnant women are most at risk of adverse health effects from lead in drinking water. In homes with lead service lines, the District of Columbia Department of Health (DC DOH) recommends that children under 6 years old and women who are pregnant or breast feeding should not drink unfiltered tap water, nor should unfiltered water be used for preparation of infant formula or concentrated juices, until the concerns regarding the lead levels in the water have been resolved. DC DOH also recommends that all pregnant women and children under 6 years old have their blood lead level tested.

Three Part Research Strategy

Production Treatment Operations Team

The Production Treatment Operations Team is lead by the Washington Aqueduct and its consultants. They will carry out the desktop studies that will lead to a recommended treatment approach as well as pipe loop studies to verify that approach and implement a partial, then full treatment process to reduce the corrosivity of the water.

This section presents Washington Aqueduct's Corrosion Reduction Plan that has been designed to evaluate previous work/studies performed by Washington Aqueduct, treatment process alternatives, system implementation of the alternatives and, the benefit of full-scale implementation of the optimized treatment in reducing the current potential corrosion issues in WA customer distribution systems. The approach of this Plan consists of four sections:

- Desktop Analysis;
- Partial System Application;
- Full-Scale implementation; and
- Monitoring process optimization via pipe-loop studies.

Desktop Analysis

To meet the requirements of the Lead and Copper Rule (LCR, {141.82(c), 56 FR 26550}) the Washington Aqueduct conducted a LCR mandated "Lead and Copper Rule Corrosion Control Study" in June 1994. In addition, Washington Aqueduct conducted a "Caustic Soda Feasibility Study" in October 1997 and a "Corrosion Inhibitor Study for Dalecarlia and McMillan Treatment Plants" in May 1998.

Washington Aqueduct will conduct a review of these documents in light of the LCR and recent water quality changes in customers distribution systems. Specifically, the information will be reviewed and incorporated into documentation recommending an OCCT that will be submitted to EPA for approval. The review will include:

1. A recommendation of optimum carbonate passivation (elevated pH with sufficient alkalinity) while determining a feasible calcium carbonate precipitation potential (CCPP) that can be achieved for the Dalecarlia and McMillan water treatment plants (WTPs). The recommendation will evaluate a review of study findings to incorporate pH analysis and precipitation potential using either quicklime (current practice) and/or sodium hydroxide (caustic soda) and investigate the benefits and feasibility of moving the pH adjustment process upstream of the filters.
2. A determination of steps to be taken to optimize distribution system water chemistry prior to implementation of corrosion inhibitor feed facilities for a partial system and full-scale system application. The documentation review will evaluate optimized pH and chloramine residual levels that should be maintained to achieve maximum benefit from corrosion inhibitor feed.

3. Review existing corrosion inhibitor studies, literature and practices at other water treatment plants to determine comparable applications. Review findings with internal team (WA, CH2M Hill and Dr. Vern Snoeyink) and develop recommended corrosion inhibitor feed plan. Present the plan to the Technical Expert Working Group (TEWG) consisting of members from EPA, DC Department of Health (DC DOH), DC WASA, Washington Aqueduct, Falls Church, Arlington County and, consultants representing DC WASA and Washington Aqueduct.
4. The chemical regime that provides the greatest lead reduction while minimizing adverse impacts will be selected for partial system application.
5. Prepare OCCT document and submit to EPA for approval.

Partial System Application

Using the results obtained from the desktop analyses and input from electrochemistry and recirculating pipe loops from the Distribution System Operations Team, DC WASA and Washington Aqueduct will conduct a partial system application of what emerges as the preferred alternative in the Fourth (4th) High pressure zone. The 4th High pressure zone has elements consistent with other service zones in the WASA system including lead service lines (LSLs) as well as unlined cast iron pipe. Temporary chemical feed facilities will be set-up in the Fort Reno pumping station. The temporary chemical feed facilities will be designed and installed for delivering the selected corrosion control treatment. The temporary facilities will be automated and will target the pH necessary for the treatment scheme previously determined during the desktop study. During this period, samples will be collected utilizing existing LCR sampling sites in addition to others that will be established prior to partial system applications start-up. This monitoring will significantly increase the number of water samples collected compared to normal system monitoring. The collected samples will be evaluated for:

- pH (field and laboratory);
- Alkalinity (as calcium carbonate {$CaCO_3$});
- Total and dissolved calcium;
- Turbidity;
- Total and dissolved lead;
- Temperature (field);
- Chlorine species;
- Dissolved oxygen;
- Oxidation/reduction potential
- Silica;
- Aluminum;
- Iron
- Phosphate;
- Nitrogen species;
- Total coliform/E. coli; and
- Heterotrophic plate count bacteria

It is anticipated that the partial system application will be conducted over a period of three (3) months beginning June 1, 2004 and concluding at the end of August 2004. Following completion of the partial system application all temporary testing equipment (chemical feed pumps, monitoring equipment, chemical storage units, etc.) will be removed from the Fort Reno facility and the pump station will be returned to pre-existing condition,

At the conclusion of the first three (3) months of the partial system application, a recommendation will be submitted to the TEWG for conducting full system implementation.

Full System Implementation

The full-scale implementation of the selected corrosion reduction plan will begin 3-months after initiation of the partial system application assuming that there are no lingering adverse effects associated with the partial system application. Following implementation of the agreed upon full-scale solution (by the TEWG) samples will be collected utilizing existing LCR sampling sites in addition to others that will be established prior to full-scale system application start-up. The collected samples will be evaluated for:

- pH (field and laboratory);
- Alkalinity (as calcium carbonate {$CaCO_3$});
- Total and dissolved calcium;
- Turbidity;
- Total and dissolved lead;
- Temperature (field);
- Chlorine species;
- Dissolved oxygen;
- Oxidation/reduction potential
- Silica;
- Aluminum;
- Phosphate;
- Nitrogen species;
- Iron
- Total coliform/E. coli; and
- Heterotrophic plate count bacteria.

Pipe Loop Studies

Pipe loop studies will be conducted beginning in mid April 2004. The pipe loop studies will be utilized as a technical tool to determine corrosion inhibitor dose, pH, inhibitor type and, system control. The pipe loops will be constructed at the Dalecarlia water treatment plant with the testing material consisting of DC WASA LSLs. LSL segments (approximately 6-inches in length) will be inserted into the assembly and one will be removed on a tri-monthly basis and sent to EPA Research and Development (R&D) Cincinnati, OH for inner scale analysis. The pipe loop assemblies will include unlined cast iron (if possible) segments as well as brass faucets.

The present pipe loop testing scenario will include the following five (5) assemblies:

1. A control loop with three (3) parallel segments of LSLs and coupons that utilizes Dalecarlia finished water;
2. One loop assembly (with three (3) parallel segments) run at pH 7.7 (+/- 0.1 pH units) with phosphate addition and full-scale chloramine addition (currently approximately 3.5 mg/L combined chlorine);
3. One loop assembly (with three (3) parallel segments) run at pH 7.7 (+/- 0.1 pH units) with phosphate addition and reduced chloramine addition (approximately 2.5 mg/L combined chlorine);
4. One loop assembly (with three (3) parallel segments) run at pH 7.7 (+/- 0.1 pH units) with phosphate addition and reduced chloramine addition (approximately 2.0 mg/L combined chlorine); and
5. One loop assembly (with three (3) parallel segments) run at high pH and reduced chloramine addition (approximately 2.5-3.0 mg/L combined chlorine);

The pipe loops will be run at a flow rate commensurate with full-scale system service line rates for a period of sixteen (16) hours and subsequently shutdown for a period of eight (8) hours to replicate typical household usage. During the 8-hour shut down six (6) hours will be quiescent with a trailing two (2) hour period for sample collection. The pipe loops will be controlled by a program logic control (PLC) methodology and therefore, self-controlled.

Samples will be collected during the two (2) hour window and evaluated for:

- pH (field and laboratory);
- Alkalinity (as calcium carbonate {$CaCO_3$});
- Total and dissolved calcium;
- Turbidity;
- Total and dissolved lead;
- Temperature (field);
- Chlorine species;
- Dissolved oxygen;
- Silica;
- Aluminum;
- Phosphate;
- Nitrogen species; and
- Iron.

Should distribution system conditions change following full-scale implementation, the pipe loop strategy may be changed accordingly.

Distribution System Operations Team

The Distribution System Operations Team's plan is to develop an extensive distribution system monitoring program. The monitoring plan will be used to assess conditions before and after a revised treatment change. Either as a part of that plan, or as a separate monitoring plan, WASA will design a sampling plan for purposes of assessing extent of the lead levels in homes without lead service lines.

Distribution System Baseline Monitoring

To provide a more extensive water quality database with regards to current water quality conditions, distribution system managers will supplement monitoring by collecting and analyzing additional samples. Parameters collected will relate to corrosion control as well as microbiological activity. One treatment option being considered is to further raise the pH. Depending on conditions, calcium carbonate can precipitate in the distribution system at elevated pH levels. Monitoring for precipitated calcium carbonate in the distribution system will be conducted. Analysis of the precipitate will be conducted to determine if the calcium carbonate is the result of precipitation or from undissolved lime. Gathering this baseline information will allow WASA to monitor these levels if the ph adjustment is selected as the revised optimal treatment. Also, water quality parameters from the existing Corrators (remote testing stations) will be obtained. Corrators can provide extensive data on metal leaching due to water corrosion in the distribution system.

Water quality monitoring will continue when the Aqueduct makes its annual switch from chloramination to free chlorine in spring of 2004. A comparison of the data with chloraminated water and free chlorine water will be made to assess differences in water quality and corrosion chemistry between the two disinfectant schemes. Distribution system monitoring will gather the information necessary to assess aluminum levels for later analysis on impacts to the make-up of pipe wall scale.

Oxidation/reduction potential (redox potential) will also be monitored. The purpose of this is to detect any changes in redox potential that may occur with any changes made to treatment. This will be done during the annual shift to free chlorine in the spring of 2004 and after the switch back to chloramines. Monitoring redox potential during partial and full system deployment of a revised corrosion control treatment will also indicate if the revised treatment will control redox potential shifts.

The Distribution System Operations Team will develop a sampling protocol for monitoring these parameters, as well as lead and copper levels throughout the partial system application and full system application of the revised corrosion control treatment. This will include a plan for analyzing the acquired data.

WASA will also be developing a plan for sampling and analysis of homes and other buildings - including schools and day care centers - in the District that are not known or suspected to be served by lead service lines to determine whether homes and other buildings not served by lead service lines may also be experiencing excessive concentrations of lead in the water. This sampling will follow EPA protocol. The Production Treatment Operations Team will use this data as input into the effectiveness of the revised corrosion control treatment.

Monitoring Effects of Revised Corrosion Control Treatment

Monitoring Effects of Revised Corrosion Control Treatment

The Distribution System Operations Team will continue to monitor for parameters described in the baseline monitoring program, as well as incorporate a detailed first flush tap water sampling protocol to monitor lead concentrations in customers homes. This plan will be a part of or may serve as their routine tap sampling program performed as required by the Lead and Copper Rule. This plan will be ready for monitoring the partial treatment system application area by the time the partial system application is scheduled to begin (targeted for June 1, 2004). The full monitoring plan, covering the entire distribution system, including the Virginia wholesale customers' distribution systems, will be completed by the expected start-up date for full system deployment of the revised corrosion control treatment (targeted for September 1,2004).

Analysis of Differences Between the WASA, Arlington County and Falls Church Distribution Systems

The Distribution System Operations Team will conduct a comparative analysis of the three distribution systems served by Washington Aqueduct. Experience to date indicates that the optimal corrosion control treatment employed by the Aqueduct has had different results in the various distribution systems. To ensure that the revised OCCT is effective throughout the service area, the team will look at existing data and perform additional sampling as necessary to make a recommendation to the Production Operations Team.

Flushing

For proper maintenance, distribution systems must be occasionally flushed to remove particulate matter and biofilm that will develop over time. This is especially important in light of an adjustment to the system chemistry. Each of the jurisdiction's distribution systems will undergo flushing at the direction of the distribution system manager to prepare for the revised optimal corrosion control treatment.

Study of Flushing Protocols

One recommendation was that WASA investigate the effects of flushing instructions for sample collection and their effects on sample results. Higher flushing rates may cause more pipe scouring and contribute more lead to the water sample, especially in particulate form. WASA conducted lead profiling studies at several residences in the District in December 2003, and January, February and March 2004. During these studies, several consecutive water samples are collected from the residence plumbing over a period of several hours. These samples are then analyzed to determine how lead levels vary over time and the fraction of particulate lead of the total. By tracking the variability, better flushing procedures for sampling were developed. The revised flushing procedures were incorporated into new instructions for residents.

Metal Loadings to Sewage Treatment Plant

A recommendation was to examine loadings to the Blue Plains Sewage Treatment Plant to determine if lead, zinc and copper levels increased suddenly after the Aqueduct's conversion to chloramination in November 2000. WASA conducted this analysis and found no indication of an increase in lead, copper or zinc levels detected in wastewater influent.

Galvanic Corrosion Related to Water Meter Replacement

A recommendation was made that galvanic corrosion of lead from pipes and fixtures could have been made worse by the installation of new water meters if dielectrics were used to electrically isolate customers' plumbing systems from the water mains in the street. WASA is conducting an investigation on this possibility. Early indications are that dielectrics were not used in the meter installations. More meter inspections are being conducted to fully rule out this possibility. Additionally, recommendations have been made that galvanic corrosion between lead and copper plumbing may be accelerated in the presence of chloramines. The existing copper meter setters that are directly connected to the lead service lines were reportedly installed in the 1980s. Confirmation that the meter setters were not recently installed is being confirmed.

Lead Profiling

In order to determine possible treatment solutions, it is important to obtain profiles of lead concentrations over time at the tap. The primary goal for profiling is to determine if the lead is in a particulate or dissolved form, as well as to determine if it is coming from the lead service line or from within the home. Particulate forms may indicate lead is breaking off of the pipe wall, while dissolved lead may indicate other possible causes such as dissolution through biological activity (i.e., biofilms) or chemical reactions. WASA will also collect data on other mineral particulates such as aluminum, iron and zinc to determine if these minerals are playing a role in increasing lead release from pipes and fixtures. The objective of this task is to obtain lead profile data from 10 residential homes containing lead service lines that have exceeded lead action levels (15 ppb).

The Distribution System Operations Team will also implement lead profiling in homes without lead service lines to more accurately gauge the characteristics of lead leaching from faucets, valves and fixtures and will conduct the scheduled profiles in homes with lead service lines partially replaced.

Follow-up lead profiling will be undertaken in April after WA switches to free chlorine, to determine if changes in lead leeching occur during the free chlorine period. This may help determine if there is a difference between lead leaching rates when free chlorine is used and when chloramines are used.

Pipe Scale Analyses

Pipe Scale Analyses

EPA's Office of Research and Development in Cincinnati, Ohio will be conducting detailed analyses on the mineral build-up, or scale, that currently exists in District of Columbia lead service line pipe taken from service. This investigation will include x-ray diffraction analysis to determine the mineral make-up of the scale. Confirmatory analysis will be done using Xanes spectroscopy performed at Argonne National Laboratories. A batch of lead service line pipe-wall scale will be sent to U.S. Geological Survey laboratories for complete elemental analysis. That will assist with the identification of minerals present at very low levels. Mineral composition of the pipe-wall scale may provide an important clue as to the cause of elevated lead levels in tap water.

Lead Leaching Rates

A study will be completed by researchers from the University of Washington to determine the rate that lead leaches from bronze alloy fixtures and lead service lines into water sitting in the lead pipe or fixtures. The purpose of this study is to determine how quickly, under current corrosivity conditions, the lead concentrations build up. This information will help in adjusting advice to consumers on how often they should flush their plumbing if they do not follow the advice to flush once per day and collect water in containers for a day to several days worth of drinking and cooking needs.

Electrochemical Pipe Loop Study

This section provides an overview of the short-term recirculation loop testing programs that DCWASA will undertake to provide additional data to the Production Treatment Operations Team. The purpose of these tests is to expediently screen a select group of treatment strategies to determine and verify chemical dosages. This pipe loop system will also be used to try to gain an understanding of the observed increases in lead levels in the customer's tap water since the establishment of OCCT.

Two separate recirculation loop-testing configurations will be performed

1) Electrochemical recirculation loop configuration
2) Water quality/stagnation flow loop configuration

The testing will include several treatment options (discussed below) and data will be obtained regarding water quality and operating conditions of each chemical treatment system.

Materials to be evaluated

The results of the lead profile studies indicate that lead service lines are the major contributor to high lead concentrations in drinking water. Therefore, the focus of the short-term bench testing program will be to study corrosion of lead service lines. Copper will also be evaluated as part of the electrochemical testing program because of the ease and relative low expense of adding copper coupons to the recirculation loops. WASA will follow with additional testing of copper, brass, and lead tin/solders as part of the water quality/stagnation flow loop tests at a later time once the lead service line evaluations are complete.

Chemical Regimes

(for both electrochemical and water quality/stagnation flow loop configurations)

The following chemical amendment strategies will be evaluated:

Strategy 1 – Control condition consisting of chloraminated finished water without any additional chemical treatment. (*This regimen will serve as a baseline comparison for the other test strategies.*)

Strategy 2 – Carbonate passivation adding lime to raise the pH to 8.5 - 9.0 (*Based on carbonate solubility model predictions, this chemical amendment will initiate the development of protective calcite scales on the lead service lines.*)

Strategy 3 - Corrosion inhibitor in the form of phosphoric acid at a dose of 10 mg/L as phosphate with the pH at 7.5. (*Phosphates are a documented inhibitor on many lead bearing surfaces. Given the short duration of the testing a high phosphate dose is necessary in order to rapidly passivate the lead surfaces and assess whether phosphate has a likely value as a corrosion inhibitor. Demonstration testing (if performed) would be performed at a lower dosage more typical of distribution system usage.*)

Strategy 4 - Corrosion inhibitor in the form of a 50/50 blended ortho/polyphosphate at a dose where the total phosphate is 2 mg/L with pH adjustment between 7.8 -8.0. (*Polyphosphates may have value in limiting copper pitting problems, as well as minimizing other distribution system problems (i.e. colored water). However, they may also enhance the mobilization of lead via sequestration. This testing regimen will determine whether polyphosphates can be used without potential adverse impact to lead release rates.*)

Strategy 5 – Establish chloramines residual at 1.0 mg/L and use phosphoric acid addition per Strategy 3. (*This strategy will determine the effectiveness of phosphate inhibitors in the presence of a lower concentration of chloramines.*)

Every effort will be made to insure that the inorganic chemistry of the recirculated water does not vary during the course of the testing regimens. This will include frequent water change-outs, and daily monitoring of basic water quality parameters such as carbonate distribution, pH , chloride and sulfate levels.

Electrochemical Testing

Electrochemical testing will be performed in accordance with the methodology for "Pipe Section Flow Cells" contained in "Internal Corrosion of Water Distribution Systems", second edition published by AWWARF.

Seven recirculation loops will be constructed, one for each chemical strategy and two for testing alternate scenarios to help determine the cause of the elevated lead levels.

Water Quality/Stagnation Flow Loop Configuration

The lead service lines are generally ¾ to 1-inch diameter. Approximately 3 linear feet is required to obtain a 250 ml sample for ¾-inch piping. A piping rack with 9 ft of lead piping will be needed in order to provide a sufficient volume for sampling. A pump will be provided to recirculate water at a maximum flow rate of 2 gpm corresponding to a velocity of 1.2 ft/sec.

Five loops will be constructed, one for each chemical strategy. Each loop will consist of a reservoir (35 gallon container), a transfer pump, appropriate valving, a lead piping specimen approximately 9 ft long and sampling cocks. A 3 ft section of the lead piping specimen will be equipped with an electrode that will continuously measure the electro potential. The electrical potential will be continuously monitored from each pipe segment.

Communications Team

The Communications Team will be lead jointly by EPA Region III and the District of Columbia Department of Health. They will carry out their responsibilities with input from the entire working group, agency public affairs specialists, community involvement staff, and any private contractor who may be hired to provide specialized communication of technical information, particularly health issues, to the general public on a large scale. Public involvement in the development and implementation of this plan will be encouraged and will play a key role in the plan's success.

The main objectives of the communications team will be:

- Highlight short-term solutions for reducing risk to lead exposure: Whatever the treatment solution selected, it may still be until the end of calendar year 2004 before lead levels begin to decline. The communications team will need to continue reminding consumers to follow the recommendation to flush their tap prior to collecting water for drinking. The Team will also highlight information regarding the distribution of point of use filters and how WASA will handle replacement filter cartridges until lead levels decrease below the action level.
- Regular progress updates will be communicated to news media and posted on the Internet.
- Communicate to the public advance notice of potential temporary shifts in aesthetic qualities of the water resulting from anticipated treatment adjustments during this research. This may include red (rusty) water, increases in total coliform bacteria (harmless indicator organisms), or increased deposition of calcium carbonate in customers' plumbing.
- Describe research strategies of public interest, such as identifying sampling locations for a partial system deployment and posting results, as appropriate.
- Communicate information and implementation plans of particular interest to environmental justice groups, neighborhood advisory councils and other community organizations.
- Continue to describe the goals of the Lead and Copper Rule so that people learn to accurately differentiate the lead action level from a health-based standard.
- Coordinate with public health agencies to help them communicate actual health-effects concentrations from lead in drinking water if a consensus number is

reached through the use of EPA's integrated exposure uptake biokinetic model for lead exposure

Schedule

The recommended target dates of the approach milestones are presented below.

Initiation of electrochemical pipe loop studies	March 15, 2004
Distribution of desktop analysis preliminary findings and recommendations to Working Group	March 24, 2004
Complete desktop options analysis and decision Point for treatment selection	April 1, 2004
Begin unidirectional flushing in partial distribution system application area	April 1, 2004
Treatment selection presented to full TEWG	April 15, 2004
Initiation of flow-through pipe loop studies	April 15, 2004
EPA tentative approval of revised OCCT	May 1, 2004
Implement comprehensive public communications plan	May 1, 2004
Hold Public Information Sessions on treatment option	May 2-14, 2004
Initiate partial distribution system application	June 1, 2004
Initiate full distribution system unidirectional flushing program	June 1, 2004
Begin system-wide deployment of revised OCCT	Sept. 1, 2004
Final EPA approval of revised OCCT	Dec. 31, 2004

Independent Peer Review Conducted (March 24, 2004 – Dec. 31, 2004)

RESPONSES BY THOMAS P. JACOBUS TO ADDITIONAL QUESTIONS FROM SENATOR INHOFE

Question 1. How much would it cost to add zinc orthophosphate to the water produced at the Washington Aqueduct on a per household basis?

Response. Since our testimony on April 7, we have continued to develop the design and refine our cost estimates. We now expect that the engineering and construction costs to implement the addition of zinc orthophosphate and to conduct studies involving pipe loops made from lead service lines taken from the District of Columbia distribution system will be approximately $3,000,000. Since this corrosion inhibitor is a new chemical not previously used, its cost will have to be added to our chemical supply budget. We expect to pay in the range of an additional $1,100,000 per year to add zinc orthophosphate to the treatment process. Washington Aqueduct is a wholesaler that serves approximately 1,000,000 people in our service area of the District of Columbia, Arlington County and Falls Church. We do not bill on a household basis and are unable to provide costs on that basis.

Question 2. With respect to the addition of other forms of phosphate, does zinc add a secondary anti-corrosive or protective property that would be beneficial for a system such as Washington's?

Response. Initially the Technical Expert Working Group made up of representatives of the District of Colombia Water and Sewer Authority, Washington Aqueduct, Virginia customers of Washington Aqueduct, the United States Environmental Protection Agency, and the District of Columbia Department of Health recommended phosphoric acid as the corrosion inhibitor. EPA organized an Independent Peer Review panel consisting of nationally renowned experts on corrosion control who had direct experience with similar situations. Members of the peer review panel suggested that the zinc orthophosphate might work more quickly to passivate the lead service lines and solder joints and brass and bronze fixtures. Both phosphoric acid (commonly known as orthophosphate) and zinc orthophosphate operate similarly and we would expect both to work. Since, however, the reduction of lead concentration in tap water is of utmost concern, we chose to take the advice of this panel and selected the zinc orthophosphate as the corrosion inhibitor to be used.

Question 3. How quickly could the Corps implement the addition of zinc-orthophosphate?

Response. We plan to do a partial system application in a small portion of the District of Columbia distribution system starting June 1, 2004 followed by a full system application on or about July 15, 2004.

Question 4. Do you know of any water systems that have had a problem with elevated lead levels after they began adding zinc orthophosphate?

Response. Because the chemistry of the water inside the pipes of the distribution system is complex and it is influenced by the nature of the source water, it is not possible to make a firm prediction of how soon the passivation will occur and lead levels will decrease. Consultants working with Washington Aqueduct have reported that in some cases there might be a slight increase in lead levels before the eventual decrease begins. In most cases, after 6 to 18 months, utilities saw a decrease in lead concentrations after the addition of zinc orthophosphate. This reduction was eventually sufficient for those systems to be in compliance with the Lead and Copper Rule's action level at the 90th percentile.

RESPONSES BY THOMAS P. JACOBUS TO ADDITIONAL QUESTIONS FROM SENATOR JEFFORDS

Question 1. I have been frequently told by parents—If I had only known, I could have taken precautions. I want to ask each of the witnesses at the table to tell me how you are responding to this question—what explanation are you giving to parents whose children who were totally unnecessarily exposed to lead in their drinking water and what steps are you taking to regain the trust of the citizens of Washington, DC?

Response. Washington Aqueduct is a wholesaler of water and does not deal directly with the citizens in its service areas in the District of Columbia or Virginia unless it does so in conjunction with the appropriate water utility officials. Certainly in this current situation that has been widely reported and discussed since February, Washington Aqueduct has accompanied the District of Columbia Water and Sewer Authority along with the District's Department of Health to many public meetings to explain exactly the nature of the problem and the interim and longer-term actions being taken to protect public health. We believe that the public's trust will be gained on an individual-by-individual basis once they understand what caused the elevated levels to occur and the steps we are taking to change the water chemistry and the time that we expect to take. Through a variety of means (e.g., pubic meetings, City Council hearings, information sent directly to households as well as doing the science and engineering to make the chemistry changes at the treatment plants) we are fully involved and committed to rebuilding that trust.

Question 2. In 1986, lead was discovered in drinking water in the Palisades section of Washington, DC. Residents were quoted as saying, "The runaround has been unbelievable. . . . No one in the bureaucracy has even begun to take this seriously." The Director of water for the city stated it was, ". . . premature to contact residents throughout the city" before the city developed a plan to handle and finance increased testing. I ask unanimous consent to insert several newspaper articles on this topic into the record. I find it unbelievable that no one at this witness table learned anything from that previous experience. Did anyone here refer to previous instances of lead contamination in the District when formulating a response plan? If so, please describe how you used this information, and if not, why not?

Response. The Lead and Copper Rule, which went into effect in 1991, was designed to overcome the situation you described in the Palisades in 1986. In meeting the requirement of that rule, Washington Aqueduct includes corrosion control treatment as an integral part of its overall treatment process. That treatment, referred to as Optimal Corrosion Control Treatment (OCCT) was approved by the United States Environmental Protection Agency Region 3, with the expectation, based on scientific analysis conducted by Washington Aqueduct and reported to EPA, that lead concentrations in tap water would be below the action level in accordance with the rule. In the current situation with elevated levels beyond the permissible action level the Washington Aqueduct's response plan on the treatment side has been to reevaluate OCCT. In so doing Washington Aqueduct is working with EPA and its wholesale customers and consultants. The response has been rapid and targeted on the problem at hand. We expect that the partial system application of revised chemistry will begin in a portion of the District of Columbia's distribution system on June 1, 2004 with the full system application to begin on or about July 15, 2004.

RESPONSES BY THOMAS P. JACOBUS TO ADDITIONAL QUESTIONS FROM SENATOR CRAPO

Question 1. You testified to the need for better teamwork if the Lead and Copper Rule is to be implemented effectively. Now that your agency and the other authorized agencies are working very closely on what has become a closely scrutinized effort, what improvements in teamwork have you learned? Also, how will you change routine procedures for working together to implement the Rule after the DC drinking water system returns below the Action Level?

Response. We have had a very strong working relationship internal to the Washington Aqueduct organization as well as with technical and management officials representing our Wholesale Customers. It is designed to quickly react to situations in the treatment process or within the distribution system that could cause a violation of a regulatory threshold. But compliance with the Lead and Copper Rule is not measured by a single event or exceedance as are other rules promulgated under the Safe Drinking Water Act. Therefore we now realize that we need to schedule regular periodic meetings to evaluate specifically the effectiveness of our corrosion control treatment and ask the customers to share their distribution sampling data with us at the same time they send it to their regulatory agencies. This will allow all four entities, Washington Aqueduct, the District of Columbia Water and Sewer Authority, Arlington County and Falls Church, to understand trends in each other's jurisdictions. Since there is no chemical difference in the water produced by Washington Aqueduct as it goes to all customers, we need to be prepared to make adjustments that are appropriate and effective in each of the systems.

STATEMENT OF GLORIA BORLAND, DUPONT CIRCLE PARENTS

I am here because I am the Mother of a 22 month-old girl, who has lead in her blood at twice the national average. I have been a DC resident for 22 years and a homeowner in Dupont Circle for 15 years. Most of the people you see here in the audience today are DC parents worried about lead from the water harming their young children.

Your letter asked me to tell you what we parents believe "would be the most effective way for government to *communicate* and respond to the sort of information that is now slowly coming to light." Also, what do we parents "suggest the DC government must do to *reinstate the trust* of the citizens in their water supply?"

First of all, communication and trust—those two must go hand in hand.

I'm sure if the Washington Post's David Nakamura had not exposed this scandal, our young children today on April 7 would still be drinking leaded water. And WASA would still be hiding this crisis from us.

Communicating is not rocket science, it is the easiest and simplest thing to do when there is leadership in an organization willing to do it. The problem is not in "the process" of communication, the problem lies in deception. WASA leaders wanted to operate under the radar screen.

WASA has a public relations department right below the chain of command of their General Manager. WASA managers, lawyers, and board of directors made the decision to not communicate truthfully, to cover up and manipulate for years. They deceived us. They tried to hide extraordinarily high levels of lead poison in our water supply, thus putting our young children at risk. The EPA and the Army Corps of Engineers went along with this deception, in violation of their Federal oversight responsibilities.

How was this crisis communicated?

For most DC parents of young children, our day of infamy was Saturday, January 31, 2004, when we read the headline of the *Washington Post* that morning and were hit with the shocking bomb that our infants, toddlers and young children have been secretly poisoned by lead in the drinking water in our homes.

We discovered that lead in the water can stunt fetus' and young children's growth and mental development, and cause learning disabilities. Formula-fed infants my get as much as 40 to 60 percent of their lead exposure from water. Lead in young children lowers their IQ. Lead has a negative effect on children's ability to learn—lowers average IQ 5–15 percent depending on severity and length of exposure. When lead enters the brain of a child, it causes long-term learning and behavioral problems. Once the baby's brain has been damaged by lead, it is irreversible. Lead can remain in the child's body for decades.

I was pregnant in 2001 and 2002 when high lead levels first became noticed by WASA. The right and legal thing for WASA to do was to issue an emergency warning to the public, and to obstetricians and pediatricians to warn their patients not to drink tap water. It does not matter if WASA hadn't identified the source; that kind of research could take months, even years. In the interest of public safety, you issue the warning to the public to take the precaution, and then you take the time and spend the resources to figure out the cause. When firemen see a house burning, their first priority is to save peoples lives, put the fire out, then they begin their investigation into the cause of the blaze. WASA got it backward. They wanted to figure out the cause of the lead crisis first, before trying to save our babies' lives.

All WASA had to do was warn me and other mothers, don't drink your water without a filter or buy bottled water. I had a PUR water filter back when I was pregnant. But I wasn't diligent about changing the cartridge all the time. Like most Moms, we drank plenty of water, we were so very careful about everything. I nursed my baby, but I also gave her formula with tap water. When I learned about lead in the water, I wanted to cry. I had been so careful, I even gave up coffee for Gods sake, and now I hear about the lead!

My daughter Imiloa Borland

Pre-school students at Early Childhood Development Center – Dupont Circle

My daughter attends a very good pre-school and day care center in Dupont Circle. The Early Childhood Development Center has an enrollment of 65 students from infants to 4 year-olds and is run by the First Baptist Church at 16th and O Streets N.W. When the building was built in 1989, as part of receiving its national accreditation, they had an outside firm test the water. There was no harmful levels of lead in their water and the center passed with flying colors.

Last year, WASA tore up the street next to the center and changed or repaired some pipes. WASA did their construction work without warning and without cooperation with the center. The staff came to work the next day and found the gymnasium floors had been flooded; the mats had to be replaced costing the day care center $2,000. WASA's work caused a classroom toilet to back up feces, and that room had to be steam cleaned at the center's own expense.

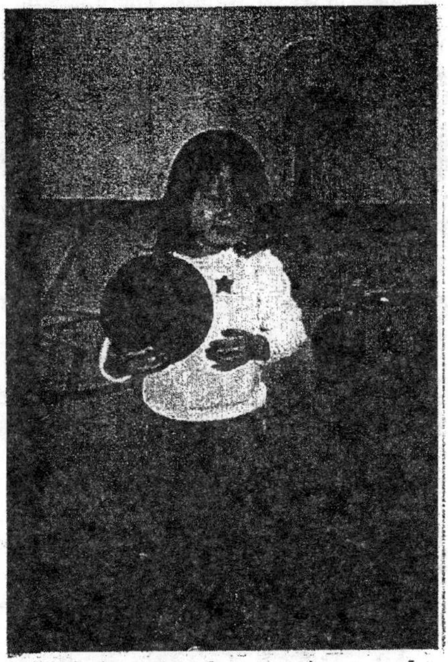

A child playing in the center's gym, where WASA caused $2,000 in water damage.

My daughter using the bathroom sink at the center

Center Director Mrs. Henderson instructing toddlers to drink from the new bottled water dispenser

When this current lead crisis in the water broke, the center's executive director Helen Henderson did the responsible thing and called back the same private firm to re-test the water in the day care center. Some of the sinks and water fountains had high levels of lead at 3,100 parts per billion and 5,900 parts per billion. When I saw the letter Mrs. Henderson sent to all the parents with the water test results, I thought there must have been a typo. This was not 15 parts per billion, the legal threshold for high lead levels, but in the thousands!!! I couldn't believe the lead in their water almost reached 6,000 ppb!

The center immediately went to bottled water. Brita water filters remove lead up to 150 ppb—filters don't work when lead is in the thousands.

As of Monday April 5, no one from WASA has contacted Mrs. Henderson, nor has anyone from the District government sent her a letter or even made a phone call. Since this crisis broke, no one from WASA or the District has offered to help her. All this talk of the city reaching out to day care centers is baloney.

I know there is some controversy on what are safe numbers: 15 ppb, is it too low, is it too high? I saw the Fox News Channel story last week that the push by environmentalists to crack down on chlorine caused DC lead problems. While scientists can still debate what are proper lead levels in water, there can be no doubt that 5,900 ppb is an emergency! Especially at a Day Care center with 65 young children.

HOW COULD THIS HAVE HAPPENED TO OUR YOUNG CHILDREN?

Most parents have been perplexed as to why an entity entrusted with public safety would lie, and then cover up their lies; what is their motive? When I spoke over the weekend to my very wise friend Joe Louis Ruffm, III father of a 3-year-old boy, living in Chevy Chase, DC. Joe said WASA wanted to protect their bond rating. Communicating the truth would have brought their assets down.

So this is like Enron! Enron was only about money. This is about money, bond ratings. But the consequences here, the innocent victims are the lives of our next generation.

I want you Senators to see the victims—our young children, when bond ratings get in the way of public safety. Look into the faces of our babies (see the photos on display): they paid the price for managers who decided bond ratings were more important than our babies brain development, their nervous system and their IQ.

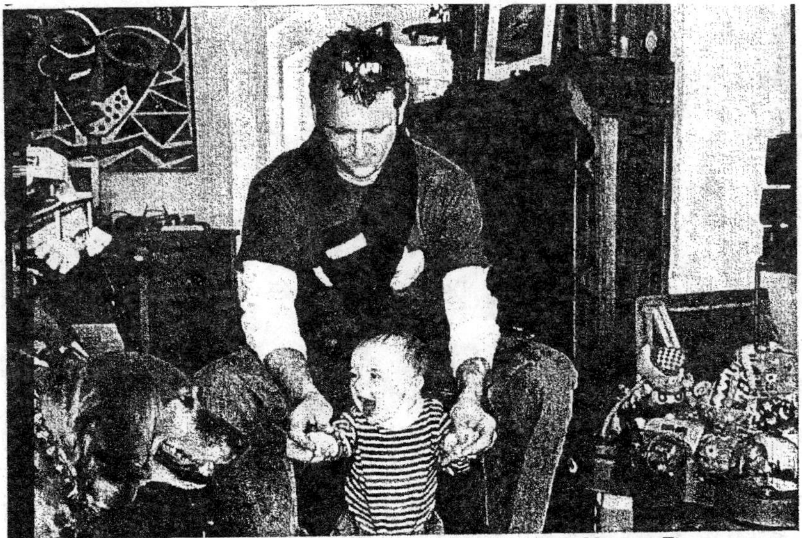

Paul McKay, co-founder of www.purewaterdc.com and his son Finn

STORIES FROM PARENTS

Here are some stores from parents all over the city. This lead water crisis has a dramatic impact on every parent of young children. To prove that this is wide spread, there is a petition with over 1,000 signatures from DC parents being presented to the Senate today.

When the lead crisis erupted, parents quickly responded by signing a petition on *www.purewaterdc.com*, that demanded the city take immediate action to restore safety in our water system. The petition also called for the firing of WASA managers and the reinstatement of the whistle-blower, Seema Bhat. The petition was a way for parents to fight back and to let our officials know we were outraged. Over 1,000 signatures were added in just 2 weeks.

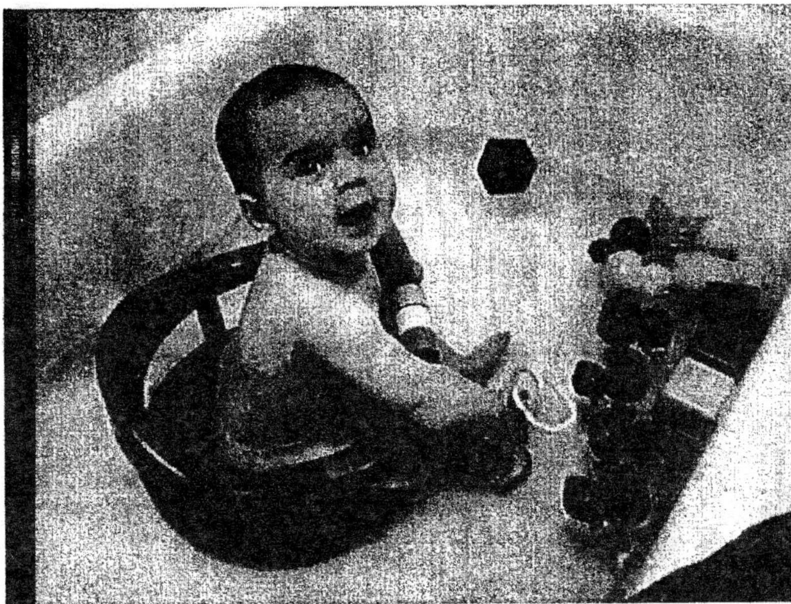

1 year-old, London Brown in her bath...surrounded by the contamination

Theresa Brown lives in LeDroit Park, right near Howard University. She says,

"I am the mom of London, she just turned one last week. I am outraged, and would like to know if any of the WASA officials had wives, daughters, daughters-in-law, sisters who were pregnant or had young children during this heinous cover-up and how they feel about allowing their families to be exposed. And if they don't have relatives or loved ones at risk, then perhaps that explains their ineptitude.

I feel completely and utterly betrayed. They have a responsibility to the citizens of this city, especially to those young babies and children who are completely defenseless. You cannot put a price on brain damage . . . how about if we deliberately caused brain impairment to their kids or grandkids and see how they like it?"

Hinkle Family

I am writing to express my anger that my family, particularly my two children aged 7 and 3, has been drinking and cooking using contaminated water for an unknown length of time, and that WASA officials knew of this problem but did not inform us. It is unconscionable that these officials allowed families to expose their babies and young children to lead. Over the past 4 years, my son has been exposed to lead contamination in utero, through breastfeeding, through his formula, and now through even the drinking fountains at his daycare. Can anyone at WASA assure me that my son will not suffer harmful effects from this exposure?

Desa Sealy Ruffin, son Miles, and Joe Louis Ruffin, III

Desa Sealy Ruffin (wife of Joe, who is mentioned earlier) lives in Chevy Chase, DC. She says,

"I have been so mad that I haven't really been able to speak all that much. I can only say that I think the District, WASA and EPA have all broken a really fundamental covenant with the citizens in the District of Columbia and I now no longer trust them to do anything. I think they are evil or as a Jamaican friend put it so aptly, wicked. I know a friend was horrified because she was pregnant during this whole thing. The fact that the head of WASA didn't want to spend the money to replace the lead service lines burns me up beyond belief. I think that the 3 agencies conspired to keep us in the dark. They should be facing criminal charges."

Charlie Horne and his Dad

Valerie Jablow, mother of Charlie Horn, lives on Capitol Hill and she says,

"I live in a city that is full of taps dispensing water with high lead levels. My son drinks water from his daycare every day of the week, and they don't know what the lead levels there are. Similarly, we like to go to the libraries, the public pool, stores, and restaurants in our neighborhood all the time—and who knows if the water they dispense is full of lead and thus unsafe to drink? In a few years my son will go to the public school a few blocks from our house—in addition to wondering about the quality of his education, I now will have to worry about the quality of the water he drinks while there."

Parker Heath-Imirowicz, 16 months old

Terrance Heath is father to 16-month old Parker, living in Logan Circle. Terrance says,

"I am writing to express my outrage at the discovery that my family, and particularly our 16 month old son, have been drinking and cooking with water potentially contaminated with lead: I am dismayed because after 10 years as a DC resident it never occurred to me that my family actually lives in a city where we cannot or should not drink the water; I am outraged because I found out about this potential contamination not from WASA, but from the Washington Post. Our son is adopted, and thus was bottle fed (water mixed with formula) as an infant, and as a toddler. Early this year we learned that his first lead screening result indicated elevated lead level.

As a DC resident and parent, I have no faith in WASA officials to make decisions for the benefit of my family's health. I believe that oversight is clearly needed, since WASA officials cannot be trusted to do their jobs. Beyond that, I believe that those responsible at WASA should be held accountable for the harm that has resulted or may result from their decisions. There can be no excusing the decision to put our children at risk."

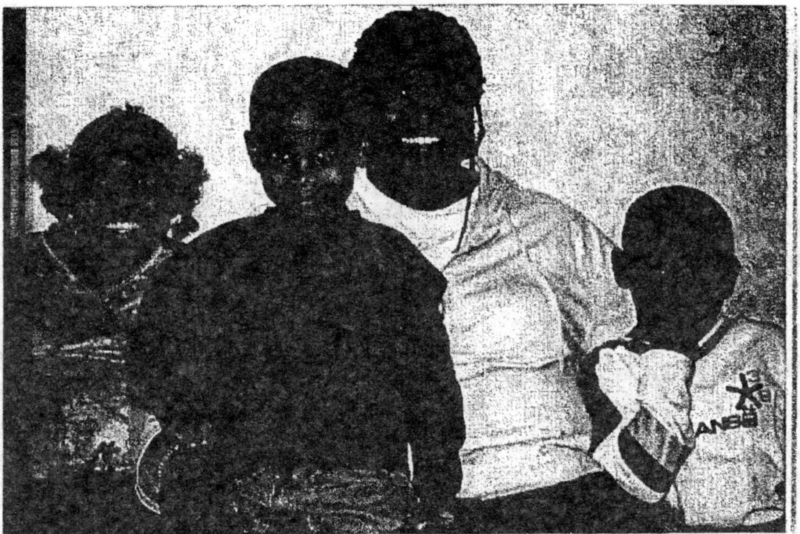

Ronnette Bristol and her family

Ronnette Bristol, lives in Northeast DC in an apartment building. She has 4 children ranging in age from 3 to 9, and just found out that her building has not been tested for lead. She says, "we are buying lots of bottled water, until we can get someone to come out and test our apartment building."

Ronnette is very worried about the quality of water in her apartment and she wants her children to have the cleanest and safest water possible.

Lyubov Gurjeva and her toddler son

Lyubov Gurjeva is originally from Russia. She says,

"I never believed DC water was safe to drink. When I arrived in DC 2 years ago friends told me that there had been numerous problems earlier. But the fact that the situation has been bad for a long time does not make it more acceptable. I am glad this matter has attracted so much public attention this time. We need safer water for DC."

Xin Chen and her two children

Dupont resident Xin Chen and her husband Brett Sylvester were informed by WASA last August that they had a lead service line and told to test their water. They submitted two sets of water samples for testing on August 15, 2003. They never received their test results. When,

"the lead story broke out, every time we talked to them on their hotline, we got different answers. Sometimes they told us they lost our samples. Sometimes they said our results were OK but they couldn't give them to us on the phone, sometimes they simply said we should re-test. We've just had the most unpleasant experience with WASA. I don't trust them and I don't trust their test results"

(No picture provided)

Maria DePaul and her husband Ethan Premysler live on Capitol Hill. They have a 3-year-old son and they are expecting the birth of their second child any day now. Their son has elevated lead levels in his blood. They have an exposed basement and they clearly see a lead water main. Maria and Ethan examined their pipes and believe it is an old lead line. When they called WASA, they were told their house was classified with a copper line and not a lead line. Maria and Ethan called numerous times to argue that they did indeed have a lead line and that WASA's records were wrong. WASA would not budge. Maria and Ethan requested that WASA send someone over to take a look, and see that they indeed had lead pipes, but WASA refused. Ethan says he is completely frustrated, "you are completely on your own, we called so many times, so many times, so many times. WASA will find legal loop holes so they do not have to help you out."

Many parents complained about the cost of buying expensive filters and bottled water. People don't mind buying bottled water in an emergency for a couple days or a couple weeks. But when a couple weeks turns into a couple months, with still no end in sight, people are beginning to feel the financial burden of buying bottled drinking water for every day use. When parents heard that WASA managers now want to raise rates 5 percent, they were aghast at the absurdity of the request. This move clearly shows WASA managers still don't get it.

HOW TO RESTORE TRUST

My feeling and that of other parents since that day of infamy is outrage! We expected our elected leaders in the District to quickly step in, fire the managers at WASA and respond with swift action to fix our water crisis. That. would have been a rightful first step on the long journey of restoring trust with the citizenry. Instead our District government leaders made the political decision to protect WASA managers, and engage in a finger-pointing game directed at everyone else. EPA and Army Corps of Engineers were supposed to be looking out for our interest by being the watch dog over WASA. Instead it appears they looked the other way and our children today have to suffer the consequences. I hope EPA and Army Corps leaders remember the faces of our babies when they carry out their daily oversight duties.

Senators, the Federal regulations you already have in place for communicating to the public, I think are fine. The problem here was not a failure of process, but a problem of cover up, fire anyone on staff who become a whistle-blower, perhaps to protect their bond ratings.

When I hear Jerry Johnson say "In hindsight, maybe we should've done more to inform people," my blood boils. The pat answer prepared by lawyers and PR people are not going; to fool the thousand families who signed that petition. We are not stupid. We know your bond ratings were more important to you than having our babies get lead in their nervous system.

Senators, we are not soccer Moms, or NASCAR Dads, we are a new demographic.

We are urban by choice. We're smart, we're parents with young children and because of our children we are involved in our neighborhood communities. We understand the need for companies to make a profit. Many of us own stocks and bonds for our children's 529 college plans. But when you endanger the lives of our babies, we are going to fight back with incredible force.

WASA, Don't think by hiring an outside health consultant, we will go away satisfied. We know a PR stunt when we see one. This just reinforces the premise that current managers don't possess the mandatory basic knowledge of "what is safe water to drink," and thus they have no business working in the water industry.

The seed of the problem here is a management culture instilled by Jerry Johnson, Mike Marcotte and board chairman Glenn Gersten. To restore trust, do what is done in the private sector, remove managers for extremely poor judgment, failed performance, and put in new management. These are the necessary first step toward restoring trust. New managers are now in place at Enron, Adelphia and Worldcom. Gersten is a Wall Street attorney, so he understands why old board members at the NYSE had to be replaced in order to restore public confidence in the institution. Martha Stewart maybe going to jail for covering up her lie. This management cover up has caused more damage to the public than Martha: they put the lives of our young babies at terrible risk.

We know our mayor is asking the Federal Government for more money to solve this problem. I don't think it makes sense to put good money in the hands of bad managers.

Drastic action needs to take place to restore safety in our water supply. I am an entrepreneur and believe in the free enterprise system. So I don't say this lightly. Most of the parents that signed the petition presented to you today also believe in the private markets. We don't normally seek government help for every little problem. But the problems and dysfunction at WASA are so huge, they require Federal intervention.

We parents encourage the U.S. Senate to institute its powers to begin the process of putting WASA under Federal control, Federal receivership. A new management team and new board of directors needs to be put in place to work on solving this lead crisis. And we need to make sure the Army Corps and EPA are listening to our demands for better communication, a two-way dialog with the public they are entrusted to serve.

Yes, we need Federal help and Federal dollars to solve this crisis, but that goes hand-in-hand with new management and Federal control of the system, until our water is deemed drinkable again.

Thank you.

(In addition to raising her daughter, Gloria Borland is also a media entrepreneur. She is currently developing a new national television series profiling women entrepreneurs called "She's the CEO" *www.sheceo.com*. Gloria was born in Kodiak, Alaska, raised in Honolulu, Hawaii and chooses to live in the Nation's Capitol.)

RESPONSE BY GLORIA BORLAND TO ADDITIONAL QUESTION FROM SENATOR JEFFORDS

Question. If there were three or four actions you could have the government agencies immediately take that were present here today (at the hearing), what would they be?

Response. (1) Immediately fire the top leadership of WASA; (2) Publicly develop a lead removal process that's transparent; and (3) Have DC water certified by an independent authority every year—very important.

RESPONSE BY CHRISTOPHER MCKEON TO ADDITIONAL QUESTION FROM SENATOR JEFFORDS

Question. If there were three or four actions you could have the government agencies immediately take that were present here today (at the hearing), what would they be?

Response. No one at WASA has been reprimanded, fined, demoted, or fired because of what happened with DC's water. Ditto for EPA. Ditto for Army Corps. They all had something to do directly with the lead crisis. But that level of unaccountability is unacceptable: what if this hadn't been lead but something more immediately harmful? So here is what needs to happen, in the order it needs to happen in:

(1) The EPA, working with the DC Department of Health, needs to put new people in charge at WASA. Now.

(2) Army Corps of Engineers needs to get folks from the community—not just elected officials, or WASA people, but regular folks—involved with their decision making in regard to the District's water. More specifically, citizens of DC need to be in on *every* decision that Army Corps makes in regard to our water, and they need to have a voice there equal to that of WASA and the army corps itself.

(3) The EPA Region 3 office is too far away to deal effectively with DC's water. EPA should form a new office here in the District to have oversight over District water. Given our role not only as the Nation's Capital but also as a major tourist destination for people from all over the world, Washington, DC deserves to have real oversight over its water, not a half-hearted rubber stamping of every WASA decision, as shown in the lead crisis.

Ok, that's my two cents' worth—thanks Gloria. Let me know what transpires.

RESPONSE BY VALÉRIE JABLOW TO ADDITIONAL QUESTION FROM SENATOR JEFFORDS

Question. If there were three or four actions you could have the government agencies immediately take that were present here today (at the hearing), what would they be?

Response. (1) EPA: More stringent notification requirements for water contamination, i.e., a one-page letter listing the actual contamination level; and the changes in the so-called acceptable contamination levels to better protect women and children. Public notification would be triggered at these lower levels; Changes in the testing guidelines which allowed WASA to replace higher lead water samples with lower lead samples. (2) DC Department of Health: better monitoring and enforcement of environmental hazards related to children and daycares. For example, we are required to provide lead testing data on our children's health forms but this data is not being analyzed or tracked to understand potential problems.

The DC Department of Health should be responsible for informing parents and offering testing services on a much wider basis than was actually done.

There should also be significant monitoring of (and penalties for non-compliance) public utilities providing services to daycare centers and schools—i.e. no mobile lead testing unit was sent to FBC, no one seems concerned about putting FBC on a priority list, and WASA should have tested (and be planning to replace) the piping and water systems at daycare centers and schools first.

Lastly, I just want to reiterate the issue that parents now have a lack of confidence in WASA and their management, and would like to see another agency or entity monitoring this situation to ensure that decisions and improvements are made moving forward.

STATEMENT OF JODY LANARD, M.D., RISK COMMUNICATION CONSULTANT, PRINCETON, NJ

Senator Crapo and Members of the Subcommittee:

Thank you so much for asking me to testify about needed improvements in public communications regarding the detection of lead in Washington DC water.

I am Jody Lanard, a psychiatrist from Princeton, New Jersey, specializing for almost 20 years in the rapidly evolving field of risk communication. My own contributions are mostly in the crisis communication branch of the field. I work independently, with my husband Peter Sandman, a former academic who is one of the early founders of risk communication. My biography and c.v. follow the testimony.

Here is a list of the crisis communication strategies which underlie my critique of WASA's communications with the public. Expanded descriptions of these strategies, references to articles from which they were derived, and references to case studies using (or failing to use) these strategies are appended at the end of this statement.

1. Don't over-reassure.
2. Put reassuring information in subordinate clauses.
3. Err on the alarming side.
4. Acknowledge uncertainty.
5. Share dilemmas.
6. Acknowledge opinion diversity.
7. Be willing to speculate.
8. Don't overdiagnose or overplan for panic.
9. Don't aim for zero fear.
10. Don't forget emotions other than fear.
11. Don't ridicule the public's emotions.
12. Legitimize people's fears.
13. Tolerate early over-reactions.
14. Establish your own humanity.
15. Tell people what to expect.
16. Offer people things to do.
17. Let people choose their own actions.
18. Ask more of people.
19. Acknowledge errors, deficiencies, and misbehaviors.
20. Apologize often for errors, deficiencies, and misbehaviors.
21. Be explicit about "anchoring frames."
22. Be explicit about changes in official opinion, prediction, or policy.
23. Don't lie, and don't tell half-truths.
24. Aim for total candor and transparency.
25. Be careful with risk comparisons.

(Adapted from Sandman and Lanard's crisis communication articles)

I'm going to tell you about some good and bad risk communication practices, and give you my critique of the DC Water and Sewage Authority's public communications. But first, here's my bottom line: for the most part, WASA did fairly typical, rather ordinary, well-intentioned public relations-oriented communication—as most agencies are inclined to do, despite the fact that it regularly backfires. My colleague Peter Sandman and I like to call this, only half-joking, a "knee-jerk under-reaction," as a way of turning the tables on officials who often feel the public responds to scary new situations with "knee-jerk over-reactions." Sometimes investigation reveals a genuine intentional cover-up, but much more often, the communication strategy is based on false beliefs:

- about how the public learns new information,
- about responsible early speculation,
- about how much anxiety the public can bear,
- about how reassuring to be in the face of uncertainty.

For instance, WASA's outrage-inducing delay in informing the public was rationalized by WASA officials in at least two conventional ways: the desire to have "complete" information before releasing it, and the desire to prevent public panic. WASA's communication practices are misguided approaches when used in uncertain, potentially scary situations. It is painful to watch agencies walk over the public relations cliff in a crisis communication situation, especially given what other agencies, such as the Centers for Disease Control, have worked so hard to learn and to distill into easily accessible nearly free training programs.

The U.S. EPA was one of the first agencies to take the new field of risk communication seriously, publishing articles about it when it barely had a name, starting in the mid 1980s. And the American Waterworks Association put out a risk communication training video in 1992, entitled "Public Involvement: a Better Response to

People's Concerns About Water Quality," subtitled: "an interview with Peter Sandman on how water providers can translate risk communication concepts into mechanisms to build effective relationships with the public." So risk communication is not a new concept in environmental and water management. But each new generation of managers discovers it anew—often as a result of botching a crisis.

What WASA did was, unfortunately, common every-day bad risk communication. I have had no access to internal agency documents or communications, so if there was deliberate self-serving intent to hide important information from the public, I do not know it. But officials' public comments so far suggest only that WASA was unskilled in the difficult, counter-intuitive strategies of crisis communication—and despite the ramping up of crisis communication planning and training since September 11, this is extremely common.

RISK COMMUNICATION

My analysis of WASA's communication rests partly on Sandman's re-definition of risk, drawing on the risk perception work of Slovic, Fischoff, and others, that states: Risk = Hazard + Outrage. This notion was first published in the EPA Journal in 1987. The rest of my analysis is informed by the crisis communication work I and others have done since September 11, 2001.

Health and safety workers usually define risk as "probability times magnitude"—How bad is the worst case? How likely is it to happen? The public—including the experts when they are at home—mostly see "risk" in terms of what Sandman calls "outrage factors," and others more recently call "fear" or "dread" factors: Is this hazard being imposed on me voluntarily? Do I have control over it? Is it industrial, or is it natural? Can I trust the people who are managing the hazard? Have they been open and honest with me? Do they show concern for my worries? Does the hazard effect particularly vulnerable populations, like children? Does it disproportionately fall on oppressed and powerless groups? Is it particularly dreaded, like cancer or AIDS?

On most risks, most of the time, people are apathetic or in denial. That was the starting position of most of the Washington DC population about lead, in October 2002 when WASA first widely released information about elevated lead levels in some homes' water. Normally, you can't easily scare people about such hazards as obesity, high blood pressure, not wearing seatbelts, not wearing condoms—or lead poisoning. So when people suddenly get upset about a risk they have long been ignoring, there are usually new "outrage factors" (or "fear factors") driving the reaction.

The main problems with WASA's communication about lead in the water were: initially, trying to attack apathy with information alone, and without scaring anyone; and later, neglecting public outrage—especially its outrage at them, the official sponsor of the outrage.

1. Trying to attack apathy with information alone

Trying to attack apathy—let me define this as a profound *lack* of outrage—with information alone—especially information that doesn't signal a change in a familiar situation—is a big reason for communication failure when people are not upset about a potentially serious risk. This is what WASA failed to factor in, in its October 2002 Lead Awareness Week brochure. WASA cites this brochure to demonstrate that it did not try to hide the lead problem from the public—and technically they didn't hide it. And they did send notices to the actual homes which tested high for lead in the water. But they did not signal that it was a potential problem for the public at large.

WASA's brochure was entitled, "The District of Columbia Water and Sewer Authority and the District of Columbia Department of Health Acknowledge National Lead Awareness Week and Its Impacts on Your Health. Living Lead-Free in DC." The explicit purpose of the brochure, based on its title, was to acknowledge National Lead Awareness Week.

This brochure has excellent educational content. But it only weakly signals to the public that there are new reasons to take this information seriously. The title makes it sound like the PR department decided to use "National Lead Awareness Week" as a news peg for sending out information about what to do about lead—major lead sources like paint and dust; less significant sources like lead in the water. Pretty picture of water on the cover. A smiling pregnant woman holding a glass of water on page 2. Low down on page 3 comes this sentence:

> "However, in the annual monitoring period ending June 30, 2002, the lead results indicate that although most homes have very low levels of lead in their

drinking water, some homes in the community have lead levels above the EPA action level of 15 parts per billion."

By the time a reader gets to this sentence, the context of the brochure suggests that "some homes" are very few, and "above the EPA action level" is only a little above. The cheerful, informative tone of the preceding pages, in context with the celebratory title of the brochure, does not signal, "DO something! This is a surprising change in our findings! Take this seriously!"

The next mention of the actual water problem is at the bottom of page 7: "Despite our best efforts mentioned earlier to control water corrosivity and remove lead from the water supply, lead levels in some homes or buildings can be high." CAN be high? Didn't they know? How many homes or buildings so far? HOW high? You cannot tell from the brochure. But on January 31, the *Washington Post* reported that "some homes" meant 4,075 homes, and "how high?"—2,287 homes were above 50 parts per billion, way over the EPA action level. A much scarier way for the public to find this out. Yes, public anxiety would have increased at any point that they heard about this, and public anger too; but WASA lost the opportunity to help the public cope with its anxiety, and WASA generated much more anger, by letting the story break unexpectedly. Feeling blindsided gets translated into a belief that the hazard is much more serious: this is a very robust finding in risk communication research.

I can understand WASA's reluctance to lay out this information explicitly early on. On March 2, *The Washington Times* wrote that WASA General Manager Jerry Johnson "wanted more comprehensive analysis of the test data before unnecessarily alarming the public." Well, they still don't know the full extent of the problem. They still do not know if many people, especially, children, have been affected. They wouldn't even get much of a hint until March 30, when the CDC published a small amount of mostly reassuring data. It's hard to tell people bad news, and then add, "Not only that, but we don't know how bad, and we don't know what it means, or what to do!"

But an official at WASA trained in crisis communication could have shared the anguish of this uncertainty with the public:

> "I'm so sorry to tell you that we're finding a lot of unexpected high lead levels in water coming out of the taps in lots of homes. We don't know yet why this is happening. We don't know yet whether any people, especially children, have increased blood lead levels because of this. We don't even know all the recommendations we want to make to you, because we feel you deserve to know this information quickly, so we'll just give you some preliminary precautionary recommendations. We'll be learning things over the next weeks that we'll wish we had known months ago. We may make mistakes, or retract things we've already said, or change our minds as new information and guidance come in. But we're committed to sharing this with you early. We know you'll be worried; we share that worry; and we will bear this together and get through it."

This would have expressed confidence in the public—a compliment the public might well have returned, *along with* its appropriate anger at you, and its anxiety. You can't skip the part where they are angry at you, you can only manage it better. But telling the public you don't think they can handle bad news—"we didn't want to panic people"—is insulting, patronizing, and it generates mistrust.

The third mention of an actual problem is on page 10.

> "WASA's recent Lead and Copper Program hosted 53 volunteers who have single-family residences that are served by either lead services, internal lead plumbing or copper pipes with lead solder installed after 1982. During WASA's last sampling program in the summer of 2001 and June 2002, some of these homes tested above 15 ppb. In the District of Columbia, there are approximately 130,000 water service lines and 20,000 of these are lead services."

As an imaginary recipient of this brochure, I react to this thinking:

> "Well, if they knew this in the summer of 2001 and in June of 2002, and they are only telling me now in the context of acknowledging National Lead Awareness Issue, and they are not mentioning how elevated the levels were, this must be not be very important new information."

You cannot tell if officials are even a little worried about a developing situation. I'd love to know data about how many people responded to this brochure by testing their water, or testing their houses for lead, or screening their children. It was a very good brochure in terms of what to do if you are worried about lead, which many people should be. It just didn't give readers a new reason to do it, if they were not worried about lead to start with.

In my Mandarin Chinese classes at Princeton University, we learned a little word that you put at the end of a sentence to signal, "New situation! Not business as usual!" The word is "le." There is no "le" in the brochure. Terrific information. Lots of action people can and should take. But no alarm bells, just business as usual.

2. Neglecting public outrage—especially outrage at "you", the official sponsor of the outrage

Neglecting or disparaging the public's outrage is one of the main problems in poor risk communication when people are already angry or upset.

By February 5, WASA and other officials were sounding defensive—the beginning of their own knee-jerk reaction to the public's outcry over the story. Spokesman Johnnie Hemphill insisted,

> "We certainly didn't do anything to hide this information. . . . we have done everything we were supposed to, from beginning to end . . . It's clearly a challenge for WASA and homeowners, but it isn't a crisis."

Not recognizing and acknowledging that this was indeed a crisis—a crisis of public confidence and fear—was insulting to the public. Not as bad as "There's no need to panic," but still disrespectful. Hemphill was leaning on the technical side of hazard versus outrage—the actual effort to assess and mitigate the lead hazard. If the October brochure suggested that WASA did not know how to send a signal to apathetic people that a new problem had arisen, WASA statements after January 31 revealed that they didn't know how to listen or talk to angry people either.

Since I have watched many good officials do this wrong and then learn, I am probably more sympathetic to how hard it is, how counter-intuitive it is, to engage in compassionate, responsive, human crisis communication when people are attacking you! You feel like a good person, engaged in thankless tasks with inadequate resources; you feel like you've been trying to get people to take lead seriously forever and suddenly they are accusing *you* of not taking it seriously! And you get defensive. (I just illustrated a risk communication strategy called "telling people stories about themselves," very useful when trying to get through to angry worried people.) Hemphill's reactions are as natural as the public's reactions. I hope I can help some of the people who are angry at WASA to understand this, just as I hope I can help WASA see that they genuinely did a lot of communication things wrong, and made public outrage and fear much worse than it could have been, and that there are learnable strategies for doing it better.

On February 13, in a letter to its customers, WASA General Manager Jerry Johnson sounded like he was minimizing the extent of the potential problem:

> "There are about 130,000 water service pipes in the District. . . . The vast majority of those are not lead service pipes. Our initial efforts are concentrating on the relatively small percentage of our customers served by lead service lines. . . . 23,000 homes . . ." That's about 18 percent of the homes. I'm sure that doesn't sound relatively small to WASA when they try to figure out a budget for mitigating all those lead service lines, and it doesn't sound relatively small to the public when they live in a neighborhood served by lots of these lines. So Johnson sounds defensive, like he's minimizing the problem. In a world where we don't want a single child to be damaged by lead, it sounds callous and uncaring to refer to 23,000 homes as "a relatively small percentage." I will bet that Jerry Johnson is not actually callous and uncaring, but in his defensive posture, he sounded that way.

In the same letter, Jerry Johnson says that the houses served by lead service lines "may have increased levels of lead in their tap water." He certainly must mean that all of those houses are at risk—which is the right message. But he still isn't saying how many houses he already knows have elevated lead in the water. And in the next paragraph he discusses how in spring and summer of 2002, "samples indicated that some households experienced increased lead levels above the [EPA] 'action level'". The indefinite words—"may have," "some households," "increased lead levels"—all sound evasive, and are likely to evoke both alarm about the extent of the exposure, and anger about an attempt to minimize it. Can you imagine the U.S. Postal Service saying, "Out of the billions and billions of letters mailed every day, we have found 'some letters' which contain anthrax spores"? Or the U.S. CDC saying, "We have found 'some patients' with SARS"? The crisis—not the hazard crisis, but the outrage crisis, the crisis in confidence—was in full swing by February 13, and WASA was still doing mostly public relations, trying to reassure.

We have come up with a concept called the Risk Communication Seesaw. If you—the official—sit on the over-reassuring, minimizing side of the seesaw, I—the public or your critics—will heavily sit on the alarming side. If you sit more toward the fulcrum, and share some of my fears, and validate my anger, and openly acknowledge the worrisome news while also giving me information that is hopeful or reassuring, I will put it in perspective better, I will feel less patronized, I will bear my worries better, and paradoxically I will blame you less—after I get through telling you how angry I am! You can't skip that step with the public.

Now I heard that WASA held a lot of public hearings where they let people yell at them. This is excellent crisis management. But most of the quotes I've read of officials responding at those meetings sound defensive, bureaucratic, and technocratic. The public gave you outrage and you gave them back hazard. I'm not saying to respond only to the outrage—you have to address the hazard, but you are in no danger of forgetting to do that. But bend over backward to acknowledge and validate people's feelings, show some of your own anguish, express your wishes that you had responded differently, express your regrets, express your hopes about managing the problem, ask people even more for their ideas and for their help, tell them stories about what other members of the public have told you (and I do not mean complimentary stories)—these are all ways officials can let the public feel they have been heard and even understood. Learn how to apologize—

> "I'm so sorry we didn't break this story months ago, so people would have been spared months of drinking so much leaded water. I'm so sorry we tried to deal with this ourselves instead of involving the public early, so people could take their own precautions sooner."

Using good risk communication, Johnson could have written to his customers:

> "I have been appalled for months that about 4,000 houses—out of about 6,000 tested—had elevated lead in their water. That's about 66 percent! We can only guess that about the same percentage of the rest of the houses served by lead service lines may have elevated levels too. Even though most of districts homes are not served by lead lines, there are 23,000 homes I am worried about until we find out if they have elevated lead too. I wish I could tell you not to worry while I work on this problem. But it's your drinking water, and of course you have a right to be worried."

Two other examples of neglecting outrage:

On February 19, WASA posted an alert on its website entitled, "Lead Service Line Flushing Clarification," outlining a change in previous guidelines for how long to let taps run before drinking the water. The recommendation increased in an alarming direction—from "one or 2 minutes" to "10 minutes to protect against high levels of lead in drinking water." Why the first recommendation was now seen to be inadequate is not clear; issues on this recommendation between WASA and EPA are not clear; but they are not my focus when reading this "clarification."

The word "clarification" is odd: the previous recommendation was clear, and the new recommendation is equally clear. The new recommendation isn't a clarification, it is a change. A revision. The old recommendation may or may not have been an error. The new recommendation may be based on evolving knowledge, or a re-thinking of old knowledge. But a clarification it isn't. Tell us what it is!

This alert notice let people know that the precautions they thought were adequate for quite a while had not been adequate, and therefore they had been exposing themselves to more hazard than they thought. This is upsetting! WASA's alert, while clearly for the purpose of telling people the new recommendation, could have added a couple lines of regret that the new recommendation hadn't been made sooner, and an acknowledgment that it is frustrating and upsetting for people to find that their precautions had been inadequate.

Dr. Vicki Freimuth, who was director of communications for the CDC during the anthrax attacks and the SARS outbreak, describes how during anthrax, evolving knowledge was perceived as mistakes—and that this was largely because of failure to acknowledge uncertainty all along. Several important risk communication strategies to reduce public alarm in response to changing information are:

- use anticipatory guidance: warn people that information and recommendations are likely to change as we learn more, or have more time to analyze what we already know, or consult with more experts; warn people (regretfully!) that some of what we know will turn out wrong.
- acknowledge uncertainty all along.
- show your own humanity: express the wish that you knew more, and that you didn't have to put the public through anxiety-provoking changes.

While some of these techniques can raise anxiety at first, they also let you share the public's worry, and help them bear it, rather than trying to squelch the public's worries and leaving them alone with their fears.

My last example of WASA ignoring public outrage and fear comes from a statement by Glenn S. Gerstell, Board Chairman of WASA, on about February 27. In this statement, Gerstell says he is "pleased" that nearly 99 percent of school water samples are below the EPA action level, and he is "pleased" that WASA has caught up with its backlog of voicemails. "Pleased" is a PR kind of word. I'd vote for "relieved." This is a minor quibble, but I use it to illustrate that PR and crisis communication are different. Gerstell also wrote that he and other top officials "have conducted numerous media interviews to communicate facts and findings as we get them." Separate from my comment that the "facts" very often did not include numbers of houses affected, or degree of lead elevation, I want to point out that this view of communication—communicating facts to the public—is probably less than half of good crisis communication. Listening to the public, acknowledging human feelings—your own and the public's—is a very large part of what makes crisis communication work when people are angry and afraid.

3. Some examples of spectacular risk communication from other crises

WASA General Manager Jerry Johnson has been quoted as saying he believes in using "facts to overcome fears" to educate the public. I hope I can convince him to use even the *scary*-sounding facts, and to go beyond the facts and help people bear their fears. It is part of the job, as Mayor Guiliani demonstrated so magnificently on September 11, and as superb risk communicators in public health do.

Here are some examples of very good risk communication, which illustrate validating public emotion, acknowledging uncertainty, using anticipatory guidance, showing your own humanity, and not prematurely over-reassuring people.

In June 2003, North Carolina had its only confirmed SARS patient. State Epidemiologist Jeff Engel responded with a series of news conferences. At one of them, a reporter asked if all the news coverage had the potential to cause more hysteria and fear. Dr. Engel replied:

> "We need to involve our community in all aspects of public health. Certainly a disease like SARS, so new, so frightening, should instill fear. Fear is an appropriate response for me as a public health physician, for everyone in the community. We need to transfer that fear into positive energy, and keep the facts out in front of hysteria. . . . I think [the media's] response is appropriate. This is a new disease, it spreads person to person, it can kill, it has a high case-fatality rate. That is newsworthy!"

Two months later, Dr. Engel made essentially the same empathic statement about Eastern Equine Encephalitis (EEE). Here he is in the August 24, 2003 *Fayetteville Observer*:

Dr. Jeff Engel, a State epidemiologist with Health and Human Services, said the State has documented "only 12 or 13 human infections since 1964." The most in one year was three in 1989. . . .

Though human infections are rare, Engel emphasized precautions.

"Fear is appropriate. I mean, my God, here you have a mosquito that can kill," Engel said. "What we are trying to do through you guys, the media, is use that fear in a positive way. We are trying to get information out there."

The local Wal-Mart sold out of insect repellant after the EEE news conference. Dr. Engel generated preparedness, not panic.

On March 14, 2003, 2 days after the World Health Organisation issued a rare global alert, WHO spokesman Dick Thompson said:

> "With relatively few SARS deaths, one might think we are overreacting but when you don't know the cause, when it strikes hospital staff, and moves at jet speed . . . until we can get a grip on it, I don't see how it will slow down . . . It's highly contagious. It's bad."

And one of the best risk communicators I know, CDC Director Dr. Julie Gerberding, often acknowledges uncertainty, and balances reassuring information with caution, by putting the reassurance in a subordinate clause. This shows what we mean by balancing on the fulcrum of the risk communication seesaw.

Early in the SARS outbreaks Dr. Gerberding was asked if SARS could be bioterrorism. She answered, "While we have lots of reasons to think that the SARS outbreaks are not due to terrorism, we're keeping an open mind and being vigilant." Other officials said only the first half: "There is no evidence of a terrorist attack." Dr. Gerberding's version is paradoxically more reassuring; we know she is still looking, just in case.

Later in the SARS outbreak, Dr. Gerberding reassured us and cautioned us at the same time, saying, "Although we haven't seen community transmission of SARS, we're not out of the woods yet."

Sometimes, when people hear my examples, or my re-writes of what I think officials should have said, they ask me, "Well, aren't some people just naturally inclined to do crisis communication well? Is it really something you can learn and practice?" I usually answer by telling them my favorite Julie Gerberding story:

One day during SARS, there had been a really weird newspaper article quoting an astrobiologist from Wales that SARS and other viruses might come from outer space, on meteor dust. (It had to be a very quiet day on the SARS front when newspapers had space for this strange notion.) At a CDC telebriefing, CNN's Miriam Falco said, "Dr. G., I just have to ask you about this outer space thing. What do you think?" Dr. G. answered, with a wicked twinkle in her eyes, "Although we have no evidence that SARS is from outer space, we're keeping an open mind." The reporters in the room roared with laughter—in recognition of her signature way of acknowledging uncertainty and not over-reassuring.

Crisis communication is hard, but learn-able. As a field, it is a moving target; we are learning and trying out new strategies all the time, and seeing what works—and what doesn't work. I'm not sure if WASA officials can learn it, but I am hopeful they can. Some of what I recommend may backfire on you too—and I will feel terrible when that happens. Some of it may turn out wrong. I wish I knew everything there was to know about crisis communication, and I wish it was easier to learn.

So even though WASA officials think they are doing good communication, I hope they will be keeping an open mind as they consider other ways. And even though many in the public think that WASA did egregious communication (the closest I come to agreeing with that is the delay in informing the general public), I think WASA mostly did ordinary conventional "bad" risk communication. And this applies to me too: Even though I think WASA's communication mistakes are pretty run-of-the-mill, I will be keeping an open mind as I learn more about how they actually managed the lead crisis.

Thank you.

STATEMENT OF DANA BEST, M.D., M.P.H., DIRECTOR, SMOKE FREE HOMES PROJECT, MEDICAL DIRECTOR, HEALTHY GENERATIONS PROGRAM, ASSISTANT PROFESSOR, GEORGE WASHINGTON UNIVERSITY SCHOOL OF MEDICINE AND HEALTH SCIENCES, CHILDREN'S NATIONAL MEDICAL CENTER

LEAD'S EFFECTS ON CHILDREN, PREGNANT WOMEN, AND NURSING MOTHERS

Thank you for the opportunity to present testimony to you today regarding the effects of lead on children, pregnant women and nursing mothers. I am a board-certified pediatrician and preventive medicine physician, with expertise in pediatric environmental health. I hope that I can provide the committee with some useful and important information about lead, lead poisoning, and current research on the topic.

Children's Hospital is a 279-bed pediatric inpatient facility located in the District of Columbia. For more than 130 years, Children's has served as the only provider dedicated exclusively to the care of infants, children, and adolescents in this region. It is our mission to be preeminent in providing health care services that enhance the well-being of children regionally, nationally, and internationally. The Children's system includes a network of nine primary care health centers located throughout the city, and a number of pediatric practices throughout the region, providing stable medical homes for thousands of children. We operate numerous regional outpatient specialty centers in Maryland and Virginia, providing access to high quality specialty care in the communities we serve. We are proud to be the region's only Level I pediatric trauma center. Children's Hospital serves as the Department of Pediatrics for George Washington University School of Medicine and Health Sciences, and runs a highly respected pediatric residency program, providing education and experience to the next generation of pediatricians, pediatric subspecialists, and pediatric researchers. We conduct significant research within the Children's Research Institute, with funding from the National Institutes of Health, the Health Resources Services Administration, the Department of Defense, the U.S. Environmental Protection Agency, and many other public and private funders.

INTRODUCTION AND A BRIEF HISTORY

Lead is a bluish-white metal of atomic number 82. Its isotopes are the end products of each of the three series of naturally occurring radioactive elements. It is soft, malleable, and resistant to corrosion, which makes it ideal for use in plumbing, pottery, tools, etc. Alloys of lead include pewter and lead solder.[1]

Use in ancient Rome. Lead pipes used as drains from the Roman baths and bearing the insignia of Roman emperors, are still in service.[1] Debate over the contribution of lead poisoning to the fall of the Roman Empire persists, but it is generally accepted that lead was widely used in plumbing, pottery, and cooking vessels. One potential source of lead poisoning in Roman times was the practice of boiling unfermented grape juice in lead pots. The resulting sugar and lead-laden syrup was added to wine to improve taste. The Romans recognized that lead was harmful, and identified the dangers of breathing fumes from lead furnaces and drinking water from the areas of lead mines; the connection of lead cooking vessels to lead poisoning is less well-described.[2]

Use in gasoline. Tetraethyl lead, the "antiknock" compound in leaded gasoline, was first described in 1854. In 1921, the emerging auto industry found it to be an effective, inexpensive gasoline additive that reduced engine "knock", a pernicious problem. Even in 1921 the poisonous effects of lead ingestion had been described and many public health authorities warned against this use of tetraethyl lead, particularly since other effective anti-knock gasoline additives were available. Nevertheless, due to cost reasons, tetraethyl lead was used. In 1922 the U.S. Public Health Service warned of the dangers of leaded fuel, and the scientific community added further concerns. In 1923, Thomas Midgley, the primary proponent of leaded fuel, suffered from acute lead poisoning and several workers at plants that made tetraethyl lead died.

In 1926, a committee appointed by the U.S. Surgeon General to review the harms of tetraethyl lead called for regulation of the product and for further studies funded by Congress. Those studies were never funded and never performed. Further evidence of the harms of lead continued to be published, but leaded gasoline was not phased out until 1986, and lead-containing motor fuel additives were not banned until 1996.[3] This belated public health success resulted in a significant drop in the blood lead levels of U.S. children: in 1976, when the standards were implemented, the average blood lead level in children was 15 mcg/dl; in 1991, those levels had dropped to 3.6 mcg/dl.[1]

Use in paint.[4] Lead has been used for centuries to make paint whiter, last longer, and cover better. The harm from lead in paint to children was first noted in the English literature in 1887.[5] In 1904, child lead poisoning was linked to lead-based paints,[6] and as a result, many countries began banning lead-based interior paints. Lead continued to be used in paints in the U.S., however, including paint used on cribs. In 1914 the death of a Baltimore boy due to lead poisoning from chewing on his crib railing was described, and other cases continued to be reported.[7,8] In 1992 the League of Nations banned lead-containing interior paint but the United States did not adopt the ban. In 1943 it was reported that eating lead-containing paint chips causes physical, neurological, behavior, learning and intelligence problems in children. Finally, in 1971, the Lead-Based Paint Poisoning Prevention Act was passed and finally implemented in 1977. As a result of these delays in banning leaded paint, many U.S. homes still contain lead paint. With the banning of leaded fuel, lead paint is now the primary source of childhood lead poisoning in the U.S.

Lead in water. Federal regulation of drinking water quality began in 1914, when standards for bacteriological levels were set; lead as a water contaminant was not regulated until much later, in 1962.[9] Most of the lead in water comes from industrial releases, urban runoff, and atmospheric deposits. While these sources of environmental lead are small, in comparison to other sources such as leaded gasoline, they can be significant, depending on water conditions. pH, grounding of household electrical systems to plumbing, and water additives can increase the leaching of lead from pipes and increase the solubility of the leached lead.[3] In most cities in the U.S., lead in tap water is due to the corrosion of lead-containing materials, such as lead pipes, in water distribution systems and household plumbing.[10] In terms of lead in water as a source of childhood lead poisoning, discussions of oral lead ingestion do not separate dust sources or paint chips from lead in the water supply, making it extremely difficult to discriminate between lead poisoning from household paint and lead poisoning from lead-contaminated water supplies. It is highly likely that lead-contaminated water can contribute to lead poisoning of children. However, no studies of lead in water as the sole source of environmental lead were found.

Lead in other sources. Other sources of lead include cosmetics (such as kohl), folk remedies, pottery, cans with lead-soldered seams, contaminated vitamins, and herbal remedies. In communities in which lead smelters or other industrial applications of lead exist, special attention should be paid to contaminated air, water, and workers' clothing. Anyone who works with lead should change clothing and shoes and shower before leaving work. Lead soldiers, hand-made munitions, and other hobbies can be a source of lead. Vinyl mini blinds were identified as a source of lead and removed from the market in 1996. (See Appendix 1)

LEAD POISONING, OR, "THERE ARE NO SAFE LEVELS OF LEAD"

Critical periods in human development.[11] The developing embryo, fetus, and child are growing and changing rapidly. If, during this rapid period of change, the fetus or child is exposed to a poison of some kind, development can be deranged. These "critical windows of exposure" are specific periods of development during which the embryo or fetus is undergoing some process, such as the development of arms and legs between days 22–36 of pregnancy, when thalidomide damages their development.[12, 13] There are many other examples of this effect, including tobacco smoke and behavioral effects, and alcohol and fetal alcohol syndrome. The critical period associated with harm from lead poisoning is brain and nervous system development, which begins in early pregnancy and continues until at least age 3 years. Since different parts of the nervous system are responsible for different functions, and since these different nervous system parts develop at different times, the timing of lead exposure can lead to different effects.[14]

Differences between children and adults. Children's behaviors expose them to more lead dust through hand-to-mouth exploration, greater exposure to potentially lead-laden soil, and closer contact with lead dust and paint chips on the floor. Children also absorb lead more efficiently than adults through their digestive systems: children absorb 40–50 percent of ingested lead while adults only absorb 10–15 percent.[15] In addition to greater absorption of lead from the digestive tract, the bones of infants and children are absorbing calcium at a high rate as they grow. Lead is chemically similar enough to calcium that it can be stored in bone, to be released gradually into the blood stream, providing an "internal source" of lead poisoning.[16] There is similar evidence that lead and iron can occupy the same molecular sites, contributing to anemia and providing another "internal source" of lead. Another significant difference between children and adults is in the rate of their metabolisms. Children have significantly faster metabolisms, which means that they breathe faster and ingest proportionately more food and water.[16] This difference means that in similar environments, children are exposed to a greater extent to contaminants. For example, the average infant drinks 5 oz of breast milk or formula per kilogram of body weight, an amount approximately equivalent to 20 liters of fluid for an adult. If formula is reconstituted using lead-contaminated tap water, that infant will receive a significant dose of lead. Similarly, breast milk can be contaminated with lead if the mother's primary source of water is lead-contaminated.

The disease of lead poisoning is also different in children than in adults. (See Figure 1.) In adults, many of the effects are reversible, such as peripheral neuropathies (a loss of sensation or increased sensitivity in the arms or legs); in children, effects persist throughout their life, even after chelation (the drug treatment for severe lead poisoning). Because of these differences, our understanding of lead poisoning in adults cannot be extrapolated to children.

Figure 1. Effects of inorganic lead on children and adults—lowest observable adverse effect levels

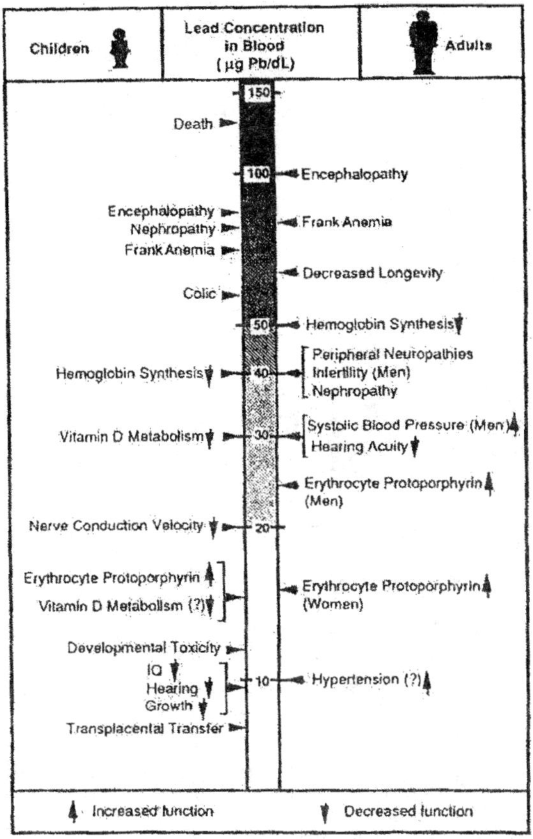

Adapted from ATSDR, *Toxicological Profile for Lead* (1989)

Lead's effects on children. The effects of lead poisoning differ depending on many factors: dose, acuity or chronicity of poisoning, gender, age, nutritional status, the presence or absence of an enriching environment, developmental assets and supports, other toxicants in the body, and genetics. Lead levels typically peak around age 2 years, when normally developing children undergo a major change in dendrite* connections. This time-related association between peak lead levels and major brain development leads to the theory that lead interferes with this critical process.

When studied in the laboratory, lead has been shown to alter basic nervous system functions, such as calcium modulated signaling, even at very low concentrations. Other effects of lead include interference with the synthesis of heme molecules (the oxygen-carrying molecules in red blood cells), leading to anemia, which has also been shown to affect intelligence. One study of lead levels in African American and Mexican American girls suggests that environmental exposure to very small amounts of lead (3 mcg/dl) can delay growth and puberty.[17] This study contributes

*A dendrite is a part of a nerve cell that conducts nerve impulses sent by adjacent nerve cells towards the body of the recipient nerve. During early brain development, many more connections between nerves develop than exist in adult brains. Many of theses connections disappear as the child grows. Many experts believe that this "surplus" of nerve connections makes it possible for the brains of children to develop functions as they are needed, and also lead to the improved recovery of children, compared to adults, from any brain damage that occurs.

to the growing literature on environmental toxins and effects on human endocrine (hormonal) systems. Lead has also been shown to damage kidneys.[18]

The effects of lead poisoning on neurocognitive skills have been identified since at least 1966. Canfield[19] showed that at even very low blood lead levels, children's IQ scores were negatively affected. This study also showed that the effects on IQ were proportionately greater at lower levels than at higher levels. (See Figure 2.)

Figure 2. IQ as a Function of Lifetime Average Blood Lead Concentration.[19]

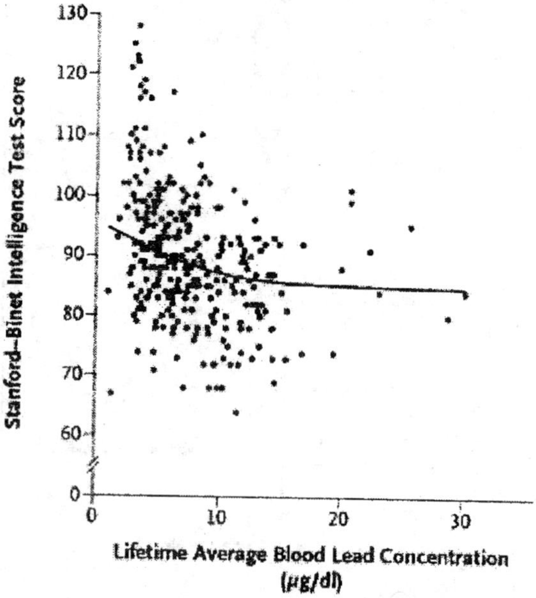

A 4–5 point decrease in IQ can mean the difference between normal and sub-normal intelligence and the ability to function independently; over the long term, it can mean a significant decline in the average intelligence of the affected population. Many other studies have demonstrated similar effects of blood lead levels under 10 mcg/dl; some have shown effects under 5 mcg/dl.[20]

Behavior and psychosocial effects. In addition to effects on IQ, distractibility, decreased reaction time,[21] poor organizational skills, hyperactivity (including ADHD, or Attention Deficit Hyperactivity Disorder), and poor classroom performance[22, 23] have been linked to lead poisoning. These effects have been recognized since at least 1976.[24] The Port Pirie† Cohort Study, a prospective study of the association of lifetime lead levels and emotional, behavioral, and cognitive effects, repeatedly showed significant, permanent, declines in cognition, behavior problems, and emotional problems that persisted throughout childhood to at least age 11–13 years.[25]

Long-term effects. Chronic exposure to lead has been linked to cerebrovascular and kidney disease, more often seen in adults. Lead has been linked to cancers in persons with lifetime lead exposures above 15 ppb in water.[10]

At higher levels.[11] Fortunately, clinical lead toxicity, meaning patients that present with symptoms of lead poisoning such as headaches, abdominal pain, loss of appetite, constipation, clumsiness, agitation, decreased activity, or somnolence is increasingly rare. These symptoms indicate central nervous system involvement that can rapidly proceed to vomiting, stupor, convulsions, encephalopathy, and death. These symptoms typically present in children with blood lead levels higher than 60 mcg/dl. Anyone with these symptoms should be treated for a life-threatening emergency.

† Port Pirie, Australia, is the home of a lead smelter. A group of children born and raised in Port Pirie were studied from birth through age 11–13 years for the effects of lead poisoning.

LEAD IN PREGNANT WOMEN AND NURSING MOTHERS

Because lead is chemically similar to calcium, it is incorporated into bone, which can result in a significant accumulation of lead in bones. If, during pregnancy and breastfeeding, maternal intake of calcium is not sufficient, these stores of lead and calcium are mobilized to supply calcium to the growing fetus and produce human milk.[26, 27] Lead in maternal blood easily crosses the placenta,[28] resulting in lead exposure of the fetus, and is readily incorporated into breast milk, leading to lead-contaminated breast milk.[29, 30] The long-term effects of these exposures are difficult to quantitate in an environment in which many other sources of lead exist. However, one study of breastfed infants linked maternal lead stores to decreased weight gain in the first month of life,[30] and a second concluded that the primary source of lead in infants under age 6 months is dietary, including breast milk and formula.[31] These studies are particularly relevant to the situation in the District of Columbia. At very high levels of maternal lead, pregnancy loss has been reported.[12]

TREATMENT OF LEAD POISONING

The treatment of lead poisoning in children has been described in the CDC's document "Managing Elevated Blood Lead Levels Among Young Children."[32] Unfortunately, no treatment for lead poisoning in children has been shown to reverse the long-term neurocognitive and behavioral effects,[33] and the primary treatment for significant lead poisoning, chelation with succimer,[‡] has been implicated as causing a small decrease in IQ.[34] Many studies have shown persistent cognitive and behavioral effects long after blood lead levels have dropped to levels considered "low."[25, 35, 36] Prevention of lead poisoning is the only solution to this disease.

COMPARING THE RISK OF LEAD POISONING TO OTHER CHILD HEALTH RISKS

The CDC estimated that in 2000, there were 454,000 children in the U.S. with blood lead concentrations higher than 10 mcg/dl. Depending on the quality of these children's environments, we can estimate that each of these children lost at least 4–5 IQ points, and a significant proportion suffer from hyperactivity, behavioral and learning difficulties, and other long-term effects of lead poisoning. Comparing these losses to other child health risks is difficult, for there is no way to measure or place a value on how a person's life would be if they had not been exposed. In terms of the overall health of children in the District of Columbia, the following comparisons can be made:

Preventable Condition	Year of Estimate	Percent of DC Children Affected
Blood lead level greater than or equal to 10 mcg/dl	2002	3.8 percent of children tested at Children's hospital; average level 3 mcg/dl[37]
Exposure to environmental tobacco smoke	2002	46 percent[38] (compared to 38 percent nationwide)[39]
Children living in poverty	2003	29 percent (compared to 17 percent nationwide)[40]
Children without health insurance	2003	12 percent (compared to 12 percent nationwide)[40]
Children living in the Spring Valley/ American University area whose hair had higher levels of arsenic than the general population.	2002	zero[41]

While these figures seem reassuring at first look, with "only" 3.8 percent of District children having a lead level 10 mcg/dl or higher, the reader should remember that even at values of 5 mcg/dl or lower neurocognitive and behavioral effects have been documented. Many of the children whose lead levels are 10 mcg/dl or higher are the same children living in poverty, exposed to environmental tobacco smoke, and/or without health insurance. These conditions add to the effects of lead poisoning; for poverty reduces educational opportunities, environmental tobacco smoke exposure has adverse effects on health, intelligence, and behavior, and lack of health

[‡] Succimer, or dimercaptosuccinic acid, is an oral treatment for chelation of lead in children. It tastes and smells like rotten eggs, making the treatment difficult to administer for children and caregivers alike.

insurance reduces access to the health care that might assist families in reducing lead poisoning and other harmful environmental exposures.

COMPARATIVE RISK BY SOURCE OF LEAD

Since the banning of leaded fuel, lead paint has become the primary source of lead poisoning in the United States. While lead in water has been described, the proportion of lead ingested via water versus lead dust and other sources from lead paint has not been determined. Unfortunately, lead pipes are found in the same older homes in which lead paint is found, making it extremely difficult to separate the contribution of each source. Since there is no level of lead considered to be without negative effect, and since the population most at risk from lead poisoning is the same population that suffers from poor nutrition, inadequate schools, lack of developmental enrichment, and other consequences of poverty, our responsibility is to remove any and all sources of lead poisoning from these children's environments. The harms of lead have been known for thousands of years; with many missed opportunities to remove lead from the environment due to cost concerns. There is no way to place a dollar value on the harm from lead poisoning to children in this city, as well as to children throughout the U.S. and the world, no matter what source, water or paint.

TESTING FOR LEAD

The standard procedure in most laboratories for testing lead in body fluids is the electrothermal atomization atomic absorption spectrophotometry assay. This method replaces less sensitive methods such as the free erythrocyte protoporphyrin, erythrocyte porphyrin, or zinc protoporphyrin tests.[32] There are newer products on the market for testing lead levels in body fluids at this time; the sensitivity, specificity, and validity of these methods have not yet been completely determined, particularly at low levels of lead. Testing of hair, fingernails, and teeth should not be done because they are subject of external contamination, making test results uninterpretable.[32] On occasion, an abdominal radiograph ("X-ray") is useful for determining if a child has a significant amount of chipped paint in his or her digestive tract. If present, the paint chips can be removed. Radiographs of bones looking for "lead lines" are not useful.[32] A new technique, K X-ray fluorescence, is entering the field of lead research. This instrument measures long-term lead deposits as densities in bone, similar to measurements of bone density for the diagnosis of osteoporosis.[42] There are consumer test kits for lead in paint,[43] and many professional lead testing services exist. Reliability of test results varies considerably, so consumers should follow guidelines such as those from the Consumer Product Safety Commission in testing for lead, selection of a method of abatement if lead paint exists, and careful abatement procedures. See Appendix B.

RESEARCH AND LEAD IN DISTRICT OF COLUMBIA WATER

On March 30, 2004, the Centers for Disease Control and Prevention published a report on blood lead levels in residents of homes with elevated lead in tap water in the District of Columbia. This study indicated that a long-term decline in the blood lead levels of children living in homes with lead service lines had halted in 2000, the year chloramines were added to water in the District of Columbia. While there are several limitations to this study, primarily due to the speed with which it was performed, the results are disturbing. The CDC recommends that public health interventions focus on eliminating lead exposures in children, and that lead concentrations in drinking water be below the EPA action level of 15 ppb.[44]

Children's has begun an analysis of the last 10 years of lead test results performed in our laboratory. We will look at the average lead level during the 10-year period, noting any changes in the average. We will also look for associations between lead levels in the children tested and lead levels in household water supply, the presence of lead paint in the home, insurance status, and other potential influences. This study is an extremely high priority; we will inform Congress and the District of Columbia of our results as soon as they are available.

SUMMARY

The children of the District of Columbia deserve a safe environment in which to grow and develop into adults contributing to DC's future. The effect of lead poisoning, even at levels not yet considered to be "poisonous," is to reduce the potential of yet another generation of children. There is no way to place a value on this loss of potential; however, we do know that the loss of IQ points and changes in behavior are measurable and significant. We also know that the resources available to many of the District's children are fragmented, in some instances non-existent, and rarely adequate to the challenges presented by poverty, race and ethnicity, and violence found in this city. This combination sentences the District's children to yet another generation of poverty and poor health. The law says lead levels higher than 15 ppb need to be abated. The children deserve this.

Thank you for this opportunity to inform you about lead and children, pregnant women and breastfeeding women. I am available for questions today or in the future.

APPENDIX A.—NEWS FROM U.S. CONSUMER PRODUCT SAFETY COMMISSION OFFICE OF INFORMATION AND PUBLIC AFFAIRS, WASHINGTON, DC 20207, JUNE 25, 1996, RELEASE 96–150

CPSC FINDS LEAD POISONING HAZARD FOR YOUNG CHILDREN IN IMPORTED VINYL MINIBLINDS

WASHINGTON, DC.—After testing and analyzing imported vinyl miniblinds, the U.S. Consumer Product Safety Commission (CPSC) has determined that some of these blinds can present a lead poisoning hazard for young children. Twenty-five million non-glossy, vinyl miniblinds that have lead added to stabilize the plastic in the blinds are imported each year from China, Taiwan, Mexico, and Indonesia.

CPSC found that over time the plastic deteriorates from exposure to sunlight and heat to form lead dust on the surface of the blind. The amount of lead dust that formed from the deterioration varied from blind to blind.

In homes where children ages 6 and younger may be present, CPSC recommends that consumers remove these vinyl miniblinds. Young children can ingest lead by wiping their hands on the blinds and then putting their hands in their mouths. Adults and families with older children generally are not at risk because they are not likely to ingest lead dust from the blinds. Lead poisoning in children is associated with behavioral problems, learning disabilities, hearing problems, and growth retardation. CPSC found that in some blinds, the levels of lead in the dust was so high that a child ingesting dust from less than one square inch of blind a day for about 15 to 30 days could result in blood levels at or above the 10 microgram per deciliter amount CPSC considers dangerous for young children.

"Some of the vinyl blinds had a level of lead in the dust that would not be considered a health hazard, while others had very high levels," said CPSC Chairman Ann Brown. "Since consumers cannot determine the amount of lead in the dust on their blinds, parents with young children should remove these vinyl miniblinds from their homes."

CPSC asked the Window Covering Safety Council, which represents the industry, to immediately change the way it produces vinyl miniblinds by removing the lead added to stabilize the plastic in these blinds. Manufacturers have made the change and new miniblinds without added lead should appear on store shelves beginning around July 1 and should be widely available over the next 90 days.

Stores will sell the new vinyl blinds packaged in cartons indicating that the blinds are made without added lead. The cartons may have labeling such as "new formulation," "nonleaded formula," "no lead added," or "new! non-leaded vinyl formulation." New blinds without lead should sell in the same price range as the old blinds at about $5 to $10 each. CPSC recommends that consumers with young children remove old vinyl miniblinds from their homes and replace them with new miniblinds made without added lead or with alternative window coverings. Washing the blinds does not prevent the vinyl blinds from deteriorating, which produces lead dust on the surface.

The Arizona and North Carolina Departments of Health first alerted CPSC to the problem of lead in vinyl miniblinds. CPSC tested the imported vinyl miniblinds for lead at its laboratory. The laboratories of NASA's Goddard Space Flight Center and the Army's Aberdeen Test Center used electron microscope technology to confirm that as the plastic in the blinds deteriorated, dust formed on the surface of the blind slats. This testing also established that the dust came from the blinds and not from another source. CPSC laboratory tests confirmed that this dust contained lead.

"This lead poisoning is mainly a hazard for children ages 6 and younger," said Chairman Brown. "Adults and older children generally are not at risk because they are not likely to ingest lead dust from the blinds."

APPENDIX B.—CONSUMER PRODUCT SAFETY COMMISSION—WHAT YOU SHOULD KNOW ABOUT LEAD BASED PAINT IN YOUR HOME: SAFETY ALERT CPSC DOCUMENT 5054

Lead-based paint is hazardous to your health. Lead-based paint is a major source of lead poisoning for children and can also affect adults. In children, lead poisoning can cause irreversible brain damage and can impair mental functioning. It can retard mental and physical development and reduce attention span. It can also retard fetal development even at extremely low levels of lead. In adults, it can cause irritability, poor muscle coordination, and nerve damage to the sense organs and nerves controlling the body. Lead poisoning may also cause problems with reproduction (such as a decreased sperm count). It may also increase blood pressure. Thus, young children, fetuses, infants, and adults with high blood pressure are the most vulnerable to the effects of lead.

Children should be screened for lead poisoning. In communities where the houses are old and deteriorating, take advantage of available screening programs offered by local health departments and have children checked regularly to see if they are suffering from lead poisoning. Because the early symptoms of lead poisoning are easy to confuse with other illnesses, it is difficult to diagnose lead poisoning without medical testing. Early symptoms may include persistent tiredness, irritability, loss of appetite, stomach discomfort, reduced attention span, insomnia, and constipation. Failure to treat children in the early stages can cause long-term or permanent health damage.

The current blood lead level which defines lead poisoning is 10 micrograms of lead per deciliter of blood. However, since poisoning may occur at lower levels than previously thought, various Federal agencies are considering whether this level should be lowered further so that lead poisoning prevention programs will have the latest information on testing children for lead poisoning.

Consumers can be exposed to lead from paint. Eating paint chips is one way young children are exposed to lead. It is not the most common way that consumers, in general, are exposed to lead. Ingesting and inhaling lead dust that is created as lead-based paint "chalks," chips, or peels from deteriorated surfaces can expose consumers to lead. Walking on small paint chips found on the floor, or opening and closing a painted frame window, can also create lead dust. Other sources of lead include deposits that may be present in homes after years of use of leaded gasoline and from industrial sources like smelting. Consumers can also generate lead dust by sanding lead-based paint or by scraping or heating lead-based paint.

Lead dust can settle on floors, walls, and furniture. Under these conditions, children can ingest lead dust from hand-to-mouth contact or in food. Settled lead dust can re-enter the air through cleaning, such as sweeping or vacuuming, or by movement of people throughout the house.

Older homes may contain lead based paint. Lead was used as a pigment and drying agent in "alkyd" oil based paint. "Latex" water based paints generally have not contained lead. About two-thirds of the homes built before 1940 and one-half of the homes built from 1940 to 1960 contain heavily-leaded paint. Some homes built after 1960 also contain heavily-leaded paint. It may be on any interior or exterior surface, particularly on woodwork, doors, and windows. In 1978, the U.S. Consumer Product Safety Commission lowered the legal maximum lead content in most kinds of paint to 0.06 percent (a trace amount). Consider having the paint in homes constructed before the 1980s tested for lead before renovating or if the paint or underlying surface is deteriorating. This is particularly important if infants, children, or pregnant women are present.

Consumers can have paint tested for lead. There are do-it-yourself kits available. However, the U.S. Consumer Product Safety Commission has not evaluated any of these kits. One home test kit uses sodium sulfide solution. This procedure requires you to place a drop of sodium sulfide solution on a paint chip. The paint chip slowly turns darker if lead is present. There are problems with this test, however. Other metals may cause false positive results, and resins in the paint may prevent the sulfide from causing the paint chip to change color. Thus, the presence of lead may not be correctly indicated. In addition the darkening may be detected only on very light-colored paint.

Another in-home test requires a trained professional who can operate the equipment safely. This test uses X-ray fluorescence to determine if the paint contains lead. Although the test can be done in your home, it should be done only by professionals trained by the equipment manufacturer or who have passed a State or local government training course, since the equipment contains radioactive materials. In addition, in some tests, the method has not been reliable.

Consumers may choose to have a testing laboratory test a paint sample for lead. Lab testing is considered more reliable than other methods. Lab tests may cost from $20 to $50 per sample. To have the lab test for lead paint, consumers may:

- Get sample containers from the lab or use re-sealable plastic bags. Label the containers or bags with the consumer's name and the location in the house from which each paint sample was taken. Several samples should be taken from each affected room (see HUD Guidelines discussed below).
- Use a sharp knife to cut through the edges of the sample paint. The lab should tell you the size of the sample needed. It will probably be about 2 inches by 2 inches.
- Lift off the paint with a clean putty knife and put it into the container. Be sure to take a sample of all layers of paint, since only the lower layers may contain lead. Do not include any of the underlying wood, plaster, metal, and brick.
- Wipe the surface and any paint dust with a wet cloth or paper towel and discard the cloth or towel.

The U.S. Department of Housing and Urban Development (HUD) recommends that action to reduce exposure should be taken when the lead in paint is greater than 0.5 percent by lab testing or greater than 1.0 milligrams per square centimeter by X-ray fluorescence. Action is especially important when paint is deteriorating or when infants, children, or pregnant women are present. Consumers can reduce exposure to lead-based paint.

If you have lead-based paint, you should take steps to reduce your exposure to lead. You can:

- Have the painted item replaced. You can replace a door or other easily removed item if you can do it without creating lead dust. Items that are difficult to remove should be replaced by professionals who will control and contain lead dust.
- Cover the lead-based paint. You can spray the surface with a sealant or cover it with gypsum wallboard. However, painting over lead-based paint with non-lead paint is not a long-term solution. Even though the lead-based paint may be covered by non-lead paint, the lead-based paint may continue to loosen from the surface below and create lead dust. The new paint may also partially mix with the lead-based paint, and lead dust will be released when the new paint begins to deteriorate.
- Have the lead-based paint removed. Have professionals trained in removing lead-based paint do this work. Each of the paint-removal methods (sandpaper, scrapers, chemicals, sandblasters, and torches or heat guns) can produce lead fumes or dust. Fumes or dust can become airborne and be inhaled or ingested. Wet methods help reduce the amount of lead dust. Removing moldings, trim, window sills, and other painted surfaces for professional paint stripping outside the home may also create dust. Be sure the professionals contain the lead dust. Wet-wipe all surfaces to remove any dust or paint chips. Wet-clean the area before re-entry.
- You can remove a small amount of lead-based paint if you can avoid creating any dust. Make sure the surface is less than about one square foot (such as a window sill). Any job larger than about one square foot should be done by professionals. Make sure you can use a wet method (such as a liquid paint stripper).
- 4. Reduce lead dust exposure. You can periodically wet mop and wipe surfaces and floors with a high phosphorous (at least 5 percent) cleaning solution. Wear waterproof gloves to prevent skin irritation. Avoid activities that will disturb or damage lead based paint and create dust. This is a preventive measure and is not an alternative to replacement or removal.
- Professionals are available to remove, replace, or cover lead-based paint.
- Contact your State and local health departments lead poisoning prevention programs and housing authorities for information about testing labs and contractors who can safely remove lead-based paint.
- The U.S. Department of Housing and Urban Development (HUD) prepared guidelines for removing lead-based paint which were published in the Federal Register, April 18, 1990, page 1455614614. Ask contractors about their qualifications, experience removing lead-based paint, and plans to follow these guidelines.
- Consumers should keep children and other occupants (especially infants, pregnant women, and adults with high blood pressure) out of the work area until the job is completed.
- Consumers should remove all food and eating utensils from the work area.
- Contractors should remove all furniture, carpets, and drapes and seal the work area from the rest of the house. The contractor also should cover and seal the floor unless lead paint is to be removed from the floor.
- Contractors should assure that workers wear respirators designed to avoid inhaling lead.

• Contractors should not allow eating or drinking in the work area. Contractors should cover and seal all cabinets and food contact surfaces.

• Contractors should dispose of clothing worn in the room after working. Workers should not wear work clothing in other areas of the house. The contractor should launder work clothes separately.

• Contractors should cleanup debris using special vacuum cleaners with HEPA (high efficiency particulate air) filters and should use a wet mop after vacuuming.

• Contractors should dispose of lead-based paint waste and contaminated materials in accordance with State and local regulations.

Government officials and health professionals continue to develop advice about removing lead-based paint. Watch for future publications by government agencies, health departments, and other groups concerned with lead-paint removal and prevention of lead poisoning.

REFERENCES

1. *http://www.webelements.com/webelements/elements/text/Pb/key.html.* Accessed March 31, 2004.

2. *http://itsa.ucsf.edu/snlrc/encyclopaedia*_romana/wine/leadpoisoning.html. Accessed March 31, 2004.

3. Agency for Toxic Substances and Disease Registry PHS, U.S. Department of Health and Human Services, *Toxicological Profile for Lead.* Atlanta, GA: Agency for Toxic Substances and Disease Registry; 1999.

4. Markowitz G, Rosner D. "Cater to the children": the role of the lead industry in a public health tragedy, 1900–1955. *Am J Public Health.* Jan 2000;90(1):36–46.

5. Stewart MD. Notes on some obscure cases of poisoning by lead chromate manifested chiefly by encephalopathy. *Medical News.* 1887;1:676–681.

6. Gibson JL. A plea for painted railings and painted walls of rooms as the source of lead poisoning amongst Queensland children. *Australian Med Gaz.* 1904:149–153.

7. Thomas HM, Blackfan KD. Recurrent meningitis, due to lead in a child of 5 years. *Am J Dis Child.* 1914;8:377–380.

8. Kitman JL. Timeline: 8,500 Years of Lead—79 Years of Leaded Gasoline. *The Nation*; 2000.

9. United State Environmental Protection Agency. *25 Years of the Safe Drinking Water Act: History and Trends*: U.S. Environmental Protection Agency; 1999.

10. *http://www.epa.gov/safewater/dwh/t-ioc/lead.html.* Accessed April 2, 2004.

11. Etzel RA, Balk SJ, eds. *Pediatric Environmental Health.* 2d ed. Elk Grove Village, IL: American Academy of Pediatrics; 2003.

12. Brent R. Environmental causes of human congenital malformations: the pediatrician's role in dealing with these complex clinical problems caused by a multiplicity of environmental and genetic factors. *Pediatrics.* 2004;113(4 (Supplement)):957–968.

13. Sadler TW. *Langman's Medical Embryology.* 6th ed. Baltimore, MD: Williams & Wilkins; 1990.

14. Mendola P, Selevan SG, Gutter S, Rice D. Environmental factors associated with a spectrum of neurodevelopmental deficits. *Ment Retard Dev Disabil Res Rev.* 2002;8(3):188–197.

15. United State Environmental Protection Agency. *Review of the National Ambient Air Quality Standards for Lead: Exposure Analysis Methodology and Validation.* Washington, DC: Air Quality Management Division, Office of Air Quality Planning and Standards, U.S. Environmental Protection Agency.; 1989.

16. Plunkett LM TD, Rodricks JV,. Differences between adults and children affecting exposure assessment. In: Guzelian PS HC, Olin SS, ed. *Similarities and Differences Between Children and Adults: Implications for Risk Assessment.* Washington, DC: ILSI Press; 1992:79–94.

17. Selevan SG, Rice DC, Hogan KA, Euling SY, Pfahles-Hutchens A, Bethel J. Blood lead concentration and delayed puberty in girls. *N Engl J Med.* Apr 17 2003;348(16):1527–1536.

18. Solhaug M, Bolger P, Jose P. The developing kidney and environmental toxins. *Pediatrics.* 2004;113(4 (Supplement)):1084–1091.

19. Canfield RL, Henderson CR, Jr., Cory-Slechta DA, Cox C, Jusko TA, Lanphear BP. Intellectual impairment in children with blood lead concentrations below 10 microg per deciliter. *N Engl J Med.* Apr 17 2003;348(16):1517–1526.

20. Bellinger D. Lead. *Pediatrics.* 2004;113(4 (Supplement)):1016–1022.

21. Hunter J, Urbanowicz MA, Yule W, Lansdown R. Automated testing of reaction time and its association with lead in children. *Int Arch Occup Environ Health.* 1985;57(1):27–34.

22. Thomson GO, Raab GM, Hepburn WS, Hunter R, Fulton M, Laxen DP. Blood-lead levels and children's behaviour—results from the Edinburgh Lead Study. *J Child Psychol Psychiatry*. Jul 1989;30(4):515–528.

23. Fulton M, Raab G, Thomson G, Laxen D, Hunter R, Hepburn W. Influence of blood lead on the ability and attainment of children in Edinburgh. *Lancet*. May 30 1987;1(8544):1221–1226.

24. Needleman HL, Gunnoe C, Leviton A, et al. Deficits in psychologic and classroom performance of children with elevated dentine lead levels. *N Engl J Med*. Mar 29 1979;300(13):689–695.

25. Burns JM, Baghurst PA, Sawyer MG, McMichael AJ, Tong SL. Lifetime low-level exposure to environmental lead and children's emotional and behavioral development at ages 11–13 years. The Port Pirie Cohort Study. *Am J Epidemiol*. Apr 15 1999;149(8):740–749.

26. Gulson BL, Jameson CW, Mahaffey KR, Mizon KJ, Korsch MJ, Vimpani G. Pregnancy increases mobilization of lead from maternal skeleton. *J Lab Clin Med*. Jul 1997;130(1):51–62.

27. Gulson BL, Mahaffey KR, Jameson CW, et al. Mobilization of lead from the skeleton during the postnatal period is larger than during pregnancy. *J Lab Clin Med*. Apr 1998;131(4):324–329.

28. Goyer RA. Transplacental transport of lead. *Environ Health Perspect*. Nov 1990;89:101–105.

29. Gulson BL, Jameson CW, Mahaffey KR, et al. Relationships of lead in breast milk to lead in blood, urine, and diet of the infant and mother. *Environ Health Perspect*. Oct 1998;106(10):667–674.

30. Sanin LH, Gonzalez-Cossio T, Romieu I, et al. Effect of Maternal Lead Burden on Infant Weight and Weight Gain at One Month of Age Among Breastfed Infants. *Pediatrics*. May 1, 2001 2001;107(5):1016–1023.

31. Gulson BL, Mizon KJ, Palmer JM, et al. Longitudinal study of daily intake and excretion of lead in newly born infants. *Environ Res*. Mar 2001;85(3):232–245.

32. Centers for Disease Control and Prevention. *Managing Elevated Blood Lead Levels Among Young Children: Recommendations from the Advisory Committee on Childhood Lead Poisoning Prevention*. Atlanta, GA: Centers for Disease Control and Prevention.; 2002.

33. Liu X, Dietrich KN, Radcliffe J, Ragan NB, Rhoads GG, Rogan WJ. Do children with falling blood lead levels have improved cognition? *Pediatrics*. Oct 2002;110(4):787–791.

34. Rogan WJ, Dietrich KN, Ware JH, et al. The effect of chelation therapy with succimer on neuropsychological development in children exposed to lead. *N Engl J Med*. May 10 2001;344(19):1421–1426.

35. Tong S, Baghurst PA, Sawyer MG, Burns J, McMichael AJ. Declining blood lead levels and changes in cognitive function during childhood: the Port Pirie Cohort Study. *Jama*. Dec 9 1998;280(22):1915–1919.

36. Lanphear BP, Dietrich K, Auinger P, Cox C. Cognitive deficits associated with blood lead concentrations <10 microg/dL in U.S. children and adolescents. *Public Health Rep*. Nov–Dec 2000;115(6):521–529.

37. Soldin O, Pezzullo J, Hanak B, Miller M, Soldin S. Changing trends in the epidemiology of pediatric lead exposure: interrelationship of blood lead and ZPP concentrations and a comparison to the U.S. population. *Ther. Drug Monitor*. 2003;25:415–420.

38. Moon RY. Personal communication. May 18, 2002.

39. Gergen P, Fowler J, Maurer K, Davis W, Overpeck M. The burden of environmental tobacco smoke exposure on the respiratory health of children 2 months through 5 years of age in the United States: Third National Health and Nutrition Examination Survey, 1988 to 1994. *Pediatrics*. 1998;101(2):*http://www.pediatrics.org/cgi/content/full/101/102/e108*.

40. *http://aecf.org/cgi-bin/kc.cgi?action=profile&area=District+of+Columbia*. Accessed April 4, 2004.

41. *http://www.atsdr.cdc.goc/statefactsheets/sfs-dc.pdf*. Accessed April 4, 2004.

42. Todd AC, McNeill FE, Fowler BA. In vivo X-ray fluorescence of lead in bone. *Environ Res*. Dec 1992;59(2):326–335.

43. Meyers P. Cranky Consumer: Scouring the Home for Lead Paint. *The Wall Street Journal*. Tuesday, March 30, 2004 2004:D2.

44. Centers for Disease Control and Prevention. Blood lead levels in residents of homes with elevated lead in tap water—District of Columbia, 2004. *MMWR*. March 30, 2004 2004;53:1–3.

RESPONSE BY DANA BEST TO ADDITIONAL QUESTION FROM SENATOR JEFFORDS

Question. Do you believe that lead poisoning in children can be completely eliminated in this country without addressing lead in drinking water?

Response. No. However, lead in drinking water is not the primary source of lead—lead paint is. Abatement of lead paint in homes and other settings in which children spend time should be our first priority. Abatement of lead in drinking water should be pursued simultaneously because of the potential scope of impact on entire populations. Information on lead abatement in homes can be found at:

• The Office of Lead Hazard Control of the U.S. Department of Housing and Urban Development (HUD) publishes Lead Paint Safety—A Field Guide for Painting, Home Maintenance and Renovation Work. This booklet can be ordered by calling 1-800-424-5323 or by downloading from *www.hud.gov/lea/leahome.html.* HUD also offers a one-hour, web-based training course on visual inspection of paint at *www.hud.gov/lea/lbptraining.html.*

• The Center for National Lead-Safe Housing provides information about safe home repair at *www.leadsafehousing.org/html/tech*_assistance.htm.

• The Alliance to End Childhood Lead Poisoning provides information about safe home repair at *www.aeclp.org/painting/index.html.*

RESPONSES BY DANA BEST TO ADDITIONAL QUESTIONS FROM SENATOR CRAPO

Question 1. When a patient at Children's Hospital is found to have lead in his or her bloodstream, what typically happens to address this problem? Do organizations or agencies other than the Hospital become involved?

Response. All lead levels are reported to the DC Department of Health by Children's laboratory staff. The DC Department of Health typically makes a home visit when a blood lead level is 15 mcg/dl or higher, to suggest abatement, determine the source of the exposure, etc. For further details about these visits, please contact the DC Department of Health.

If a child has a lead level higher than 10 mcg/dl, the recommendations of the Advisory Committee on Childhood Lead Poisoning Prevention are usually followed. These recommendations can be found at: *http://www.cdc.gov/nceh/lead/CaseManagement/caseManage*_main.htm.

The following table is from page 41 of that document:

Blood Lead Level (BLL) (mcg/dL)

10–14	15–19	20–44	45–69	>70
Lead education .... Dietary .... Environmental .... Follow-up blood lead monitoring.	Lead education .... Dietary .... Environmental .... Follow-up blood lead monitoring. Proceed according to actions for 20–44 mcg/dl if:. A follow-up Bll is in this range at least 3 months after initial venous test. .... or BLLs increase	Lead education .... Dietary .... Environmental .... Follow-up blood lead monitoring. Complete history and physical exam. Lab work; .... Hemoglobin or hematocrit. Iron status .... Environmental investigation. Lead hazard reduction Neurodevelopmental monitoring. Abdominal X-ray (if particulate lead ingestion is suspected) with bowel decontamination if indicated.	Lead education .... Dietary .... Environmental .... Follow-up blood lead monitoring. Complete history and physical exam. Lab work: .... Hemaoglobin or hematocrit. Iron status .... FEP or ZPP .... Environmental investigation. Lead hazard reduction Neurodevelopmental monitoring. Abdominal X-ray with bowel decontamination if indicated Chelation therapy	Hospitalize and commence chelation therapy. Proceed according to actions for 20-44 mcg/dL.

The following actions are *NOT* recommended at any blood lead level:

• Searching for gingival lead lines

- Testing of neurophysiologic function
- Evaluation of renal function (except during chelation with EDTA)
- Testing of hair, teeth, or fingernails for lead
- Radiographic imaging of long bones
- X-ray fluorescence of long bones

Question 2. Recognizing that cases vary in their particulars, what important examples can you provide of atypical treatment situations?

Response. There are bizarre cases in which the recommendations do not apply. For instance, there was a patient here at Children's who had been shot by a "stray" bullet at age 3. Since the bullet lodged near the child's spinal column, it could not be removed. At age 10 the child was still undergoing regular chelation due to chronically elevated lead levels. (This case illustrates two of our city's major problems.)

STATEMENT OF ERIK D. OLSON, SENIOR ATTORNEY, NATURAL RESOURCES DEFENSE COUNCIL

Thank you for the opportunity to submit this testimony. I am Erik D. Olson, a Senior Attorney with the Natural Resources Defense Council (NRDC), a national non-profit public interest organization dedicated to protecting public health and the environment, with over 500,000 members. I am Chair of the Campaign for Safe and Affordable Drinking Water, an alliance of over 300 medical, public health, nursing, consumer, environmental, and other groups working to improve drinking water protection. I also serve on the steering committee of a new organization called Lead Emergency Action for the District (LEAD), a coalition of local and national civic groups, environmental, consumer, medical, and other organizations and citizens urging a stronger public response to the DC lead crisis. I testify today only on behalf of NRDC.

The drinking water lead crisis in Washington DC poses serious public health risks to thousands of residents of the national capital area, and casts a dark shadow of doubt over the ability, resources, or will of Federal and local officials to fulfill their duty to protect our health. Preliminary data released by the Centers for Disease Control and Prevention recently found that there are reasons to be concerned about lead in DC tap water. While severe acute lead poisoning due to drinking water was not found, blood lead levels in DC children who drink water in homes served with lead lines did not decrease, whereas they did decrease in children served by non-lead lines. This suggested to health experts that lead in tap water is likely contributing to higher blood lead levels in some children in the District. Because of deficiencies in the DC blood lead monitoring program design, and because blood lead levels begin to drop fairly shortly after exposure is stopped (with time much of the lead deposits in bone and tissues), it is quite possible that more serious problems were not detected. Mary Jean Brown, the lead poison prevention chief at the CDC and a co-author of the report said in releasing the report that,

> "there is no safe level of lead . . . Even a small contribution, especially in small children, is not something that we want to happen. . . . We don't want to increase the blood lead levels of those individuals by even 1 microgram if it can be prevented."

See Avram Goldstein, "Blood Lead Levels Affected by Disinfectant," Washington Post, March 31, 2004, available online at *http://www.washingtonpost.com/wp-dyn/articles/A37404–2004Mar30.html* and CDC study at *http://www.cdc.gov/mmwr/preview/mmwrhtml/mm53d330al.htm.*

It is important to note that new data published in major medical journals the past few years show that even at levels below 10 μg/dl in blood, lead has been linked to reduced cognitive function in children, and surprisingly, the most significant effects are seen at levels below 10 μg/dl. See CDC Advisory Committee on Childhood Lead Poisoning Prevention, Evidence of Health Effects of Blood Lead Level <10 μg/dl, available online at *http://www.cdc.gov/nceh/lead/ACCLPP/meetingMinutes/lessThan 10MtgMAR04.pdf.*

The U.S. Environmental Protection Agency (EPA) has the primary responsibility for protecting drinking water only in Washington DC, Wyoming, and a few U.S. territories. EPA has failed to fulfill its obligation to aggressively oversee the safety of DC's water supply, to ensure that the public is fully apprised of the health threats posed by our drinking water, and to enforce the law.

This raises important questions about the adequacy of EPA's drinking water program not only for the Nation's Capital, but also for the whole Nation. The U.S. Army Corps of Engineers' Washington Aqueduct Division (the Corps) has failed to treat the water it delivers to DC and neighboring Northern Virginia communities sufficiently to assure that the water is not corrosive, in order to reduce lead contamination. The DC Water and Sewer Authority (WASA) has failed to act promptly or adequately on the lead contamination crisis, and has repeatedly confused and mislead the public about the lead problem. To date, the local and Federal response has been far too slow and manifestly inadequate. The Nation's capital's water supply should be the best in the world, an international model. Instead, it is among the worst big city supplies in the Nation.

It should not be assumed that Washington is the only city in the U.S. affected by lead or other important tap water problems. We are now learning of lead problems in Northern Virginia, and there are several other cities have struggled with lead contamination in recent years, including:

• *Seattle, Washington* (19 ppb 90th percentile lead according to Annual Consumer Confidence Report (CCR) issued in 2003, citing 1997 data).

• *Portland, Oregon* (17 ppb 90th percentile according to CCR issued in 2003)

• *St. Paul, Minnesota* (45 ppb 90th percentile in 1996, reportedly brought down through treatment to 25 ppb in 1999, 20 ppb in 2000, and I 1 ppb in 2003).

• *Bangor, Maine* (6–8 ppb 90th percentile from 1993–1999, increased to "15 ppb" in 2001 after switch to chloramines and subsequent nitrification problem; reportedly reduced since then after additional treatment-compliance issue boiled down to 1 ppb in one home out of 38 tested).

• *Madison, Wisconsin* (22.2 ppb 90th percentile lead level, city is now doing lead service line replacement, according to February 2004 report available online at *http://www.ci.madison.wi.us/water/Report%20PhaseII%2OS.pdf*).

• *Greater Boston, Massachusetts* communities (most recent Mass. Water Resource Authority's 2003 CCR reports system-wide (consolidated) 90"' percentile lead level has dropped from 65 ppb in 1992 to 11 ppb in 2002, but MWRA's 2001 CCR reported, on a community-by-community basis, that 13 Boston area communities substantially exceeded the 90th percentile level. It is unclear why MWRA stopped reporting these community-by community data in its annual CCRs thereafter.)

• *Newark, New Jersey* (2001 90th percentile in Wanaque system was 24 ppb and 13 ppb in Pequannock system; 2002 level reportedly changed to 12 and 14, respectively, with new corrosion treatment).

• *New York City* (2002 and 2001 90th percentile levels reported in CCR at 15 ppb, with levels up to 3,555 ppb in 2001; compliance issue boils down to 1 ppb in one home out of 107 tested in all of NYC).

• *Oneida, New York* (2002 CCR reported 19 ppb 90th percentile level, reportedly doing lead service line replacement).

As several of these examples highlight, there are opportunities to "game" the system by slightly altering the monitoring program. Though there is evidence that this may have happened in Washington DC, we are not aware of any evidence of this elsewhere, though the temptation could be large, and the lack of serious EPA oversight makes detection of such problems unlikely. If the compliance of a system serving millions of people boils down to less than 1 ppb measured at one or a few homes out of about 100 tested, this raises significant issues.

Many other cities have had similar lead problems to those noted above. However, incredibly EPA maintains no accurate up-to-date national information on this issue. Some of these cities will assert that they are now in compliance with EPA's lead action level despite recent documented problems, but EPA has done little to aggressively ensure that this is correct.

School systems in many cities across the country including in Seattle, Boston, Baltimore, Philadelphia, Montgomery County, Maryland, and many others have found serious lead contamination problems, but often have been slow to inform parents and resolve the problem. Many other States and school systems have entirely failed to comply with the Lead Contamination Control Act of 1988s' mandate to test school water for lead and replace coolers that serve lead-contaminated water. EPA and many States have done a poor job of assuring that the EPA lead rule, and the school testing and cooler programs are fully implemented.

The EPA Inspector General has recently issued a stinging report finding that EPA's national drinking water data base mandated by Congress and EPA rules is woefully incomplete and out of date, and that EPA has repeatedly mislead the public about drinking water quality and compliance because violations are seriously underreported.[1] EPA has acknowledged that there are major problems with State reporting of all violations and specific lead levels to EPA—indeed, NRDC has learned that fully 20 States have not been reporting any required information on

[1] EPA Inspector General, "EPA Claims to Meet Drinking Water Quality Goals Despite Persistent Data Quality Shortcomings," Report 2004–P–0008, available online at *www.epa.gov/reports/2004/20040305–2004-P–0008.pdf*.

lead rule compliance, contrary to EPA rules. Yet EPA has has failed to crack down on States that are not complying with Federal reporting rules, making effective Federal tracking, oversight, and enforcement impossible. Moreover, the Washington crisis and experience in other cities highlight that the EPA lead rule and public education requirements are manifestly inadequate and almost designed to be difficult to enforce.

Below, we summarize some key problems with the response to the lead crisis, and the actions that need to be taken to resolve the problem locally and to avoid possible repetition of the problem nationally:

EPA. The EPA bears a special responsibility for addressing the DC water crisis, since EPA has primary responsibility for drinking water protection only in Washington, DC and Wyoming. EPA must take emergency enforcement action against WASA and the Corps. EPA's recent notice of violation issued to WASA was extremely long in coming. EPA's deals with WASA and the Corps lack the clarity, detail, and enforceability that are needed to assure this problem is promptly resolved. Only years after the alleged violations, of which EPA was well aware, and only after a barrage of 2 months of adverse publicity, did EPA take this feeble action of issuing an NOV. An emergency enforcement order should be issued that would not only mandate immediate actions to deal in the short-term with the lead crisis, but should also require a comprehensive top-to-bottom third party review of both WASA and Corps operations.

EPA has failed to ensure prompt and accurate public education and reporting on lead problems, and there are substantial questions about whether EPA adequately oversaw WASA's lead monitoring and sample invalidations. EPA failed to promptly and adequately review, or to insist upon the updating the Corps' corrosion control program. It is unclear whether EPA insisted upon an adequate and accurate materials survey, and EPA reportedly allowed WASA to avoid lead service line replacement by taking advantage of a regulatory loophole.

EPA has been slow to force WASA to redo its manifestly invalid and misleading school testing, or to mandate testing of day care centers or private schools. The EPA lead rule itself, which is drafted in a way that makes it extremely difficult to enforce, needs to be substantially strengthened. In addition, as noted above, EPA's data reporting systems are woefully inadequate, to the point that EPA management cannot accurately and timely answer simple questions such as "which public water systems are above the lead action level and which are replacing lead service lines?" EPA also has done little to ensure that school testing for lead has been carried out nationally, perhaps in part due to a court ruling casting doubt on the program (*Acorn* v. *Edwards*, 81 F.3d 1387 (5h Cir. 1996)). EPA's inspection and enforcement program for drinking water has always been weak, but has gotten demonstrably worse during the Bush Administration, as is shown in graphs a the end of this testimony.

Army Corps of Engineers. The Corps has failed to ensure that its water is adequately treated to reduce its corrosivity and to thereby reduce lead levels in Washington and the Northern Virginia suburbs that it serves. The Corps has repeatedly responded to water quality problems by adopting the cheapest and often least effective band-aid solutions. Instead of using orthophosphate or other sophisticated corrosion inhibiters as recommended as best by its consultants, the Corps chose to simply adjust water pH with lime, a cheaper and apparently less effective alternative.

Instead of moving toward advanced treatment such as granular activated carbon filters and UV light or ozone disinfection, or membranes to reduce cancer-causing (and possibly miscarriage and birth defect-inducing) disinfection byproducts, and to more effectively remove the dangerous parasite *Cryptosporidium* and other contaminants, the Corps opted for the cheapest and least effective choice. It simply added ammonia to its chlorine to make chloramines. The switch to chloramines did slightly reduce chlorination byproduct levels, but also appears to have increased corrosivity of the water and therefore increased lead problems. It should be noted that contrary to the inaccurate assertions of some critics, the EPA rules setting new limits on disinfection byproducts were not the result of wild environmental extremists, but were negotiated by a diverse regulatory negotiating committee over a several-year period. The committee included major water utility trade associations, chlorine manufacturers, health departments, public health.experts, States, local officials, and environmentalists (see 1998 agreement in principle at *http://www.ena.gov/safewater/mdbp/mdbpagre.html*).

WASA. WASA's response to the lead crisis has been slow, plagued by misleading statements to the public and even to senior DC officials, and often characterized by missteps and at best grudging compliance with EPA rules. Whether it is the alleged firing of a WASA employee for reporting lead problems to EPA, or the failure to notify customers with high lead levels for many months after samples were taken, or the failure to effectively notify the Mayor, City Council, and all city residents of the extensive and serious lead problem until the Washington Post broke the story, WASA has a lot to answer for. EPA has recently listed six alleged violations of Federal regulations that may have contributed to the lack of public knowledge. See EPA Non Compliance Letter to WASA, dated March 31, 2004, available online at *http://www.epa.gov/dclead/johnson-letter2.htm.*

WASA's conflicting advice to customers (such as a February 9 letter to all customers telling them to flush their water for 15–30 seconds, followed by a public announcement a few days later to flush lead lines for 10 minutes, followed a few days later by a recommendation that pregnant women and children under six served by lead service lines should use a filter) has confused and justifiably outraged citizens. WASA's invalid and misleading testing of city schools, in which virtually all samples were taken after water was flushed for 10 minutes (with the likely effect of reducing or eliminating lead levels), necessitates a re-conducting of a valid school and day care testing program. At the mayor's and EPA's insistence, WASA has now said it will do additional school testing.

In addition, it appears that WASA's *partial* lead service line replacement program may be making matters worse, increasing lead levels in some homes' water. Since local and Federal authorities have approved and encouraged the use of lead service lines in DC for over 100 years, we believe that WASA should fully remove all of the lead service lines at its expense (with Federal assistance, see "Congress" below), instead of stopping at the property line. A comprehensive third-party public review of WASA's lead program and all water quality operations also is desperately needed.

Congress. We urge Congress to help DC and EPA to fund the response to the lead crisis, including lead service line replacement and upgrades to the DC and Corps water infrastructure. Congress also should respond to the national water infrastructure problem through national legislation and increased appropriations. In addition, Congress should vigorously oversee EPA's drinking water program, including its national implementation of the lead rule and its enforcement and data collection programs. Members of this Committee should urge their colleagues on the Appropriations Committee to increase funding for EPA drinking water programs, and particularly for drinking water enforcement. We also urge Congress to insist that EPA take emergency enforcement action against WASA and the Corps, as discussed below.

Specifically, among the actions that we believe Congress should take to address problems raised by the lead crisis are:

- Water Infrastructure or Grants/Trust Fund Legislation
  - Congress should substantially increase the Safe Drinking Water State Revolving Fund authorization and appropriations (now funded at $850M; authorization of $1B expired in 2003).
  - Congress should adopt broad water infrastructure bill and/or water infrastructure trust fund legislation.
  - Congress should adopt targeted legislation for lead rule compliance/lead service line replacement and filters for DC residents at least, since the Federal Government approved and oversaw the installation of the lead lines.
  - The Corps of Engineers should pay for DC lead service line replacement since Corps built the system, and operates the treatment plant that is providing corrosive water. Also, Federal agents (federally appointed Commissioners and engineers) approved and sometimes required lead service lines in DC.
  - Congress should adopt new legislation that provides grants to needy water systems, like Reid-Ensign bill (S. 503, 107th Congress).
- Fix Lead Pipe and Fixtures provision in the SDWA
  - Congress should redefine "Lead Free" in SDWA § 1417(d) to mean really lead free (i.e. no lead added, and no more that 0.1 or 0.25 percent incidental lead—as required by L.A., Bangor, Maine, etc.).
  - Congress should fix the public notice provisions in SDWA § 1417(a)(2), which clearly have been inadequate (as shown by the DC experience).
- Fix the SDWA lead in schools and day care provisions (SDWA §§ 1461–1463)
  - Congress should redefine lead free in the Lead Contamination Control Act (LCCA), which added SDWA § 1461, to mean really lead free (0.1 percent or 0.25 percent, see above).
  - Congress should order an EPA review of § 1462 implementation and effectiveness of lead fountain recall provision in all States.
  - Congress should clarify §§ 1461–63 to eliminate any constitutionality doubts raised by *Acorn* v. *Edwards*, 81 F.3d 1387 (5th Cir. 1996).
  - Congress should require ongoing retesting of all schools and day care centers in light of *Acorn* and widespread non-compliance, and new info on lead leaching.
- Fix the EPA Lead Rule & Associated Regulations

- Adopt a 10 or 15 ppb MCL at the tap. There was an MCL (50 ppb) until 1991.
- As a clearly second-best alternative, the rule needs serious overhaul:
  - Require immediate review of corrosion control programs for systems that make treatment changes, and also require review periodically;
  - Change monitoring requirements so systems cannot go for years without testing, and to clarify and strengthen test methods, site selection, and number of tests (50 or 100 per city are not enough);
  - Strengthen/overhaul public education and public notice requirements in 40 CFR 141.85 which are obviously inadequate;
  - Require full lead service line replacement, or at a minimum require that water systems that approved, authorized, or required use of lead service lines to replace those lines if they are contributing to lead over action level;
  - Require in-home certified filters to be provided to high-risk people who have high lead levels, with water system-supplied maintenance in accordance with 40 CFR 141.100;
  - Eliminate the loophole that allows systems to count homes tested at below 15 ppb as is their lead service lines were replaced in implementing the 7 percent/year lead service line replacement provision;
  - Require an overhaul/upgrade of EPA's compliance & data tracking.
- Fix the Consumer Confidence Report & Right to Know Requirements
  - WASA's report said on the cover "Your Drinking Water is Safe" and buried the facts. No one knew of the problem. Similar problems have been ocumented for water systems across the country. EPA's right to know and consumer confidence report rules need to be overhauled & strengthened.
- Fix SDWA Standards Provisions
  - Congress should require that standards to protect pregnant women, children, vulnerable people.
  - Congress should overhaul the new contaminant selection & 6 year standard review provisions. These provisions have been complete failures since 1996.

EPA'S RESPONSIBILITIES

EPA has known, at least since the mid-1990s, that lead contamination of tap water is a significant issue in Washington, and that the public was ill-informed about the problem. In 1995–1996, in response to a Freedom of Information Act request, NRDC learned that many homes across the city had lead levels well in excess of the EPA Action Level, and that those homeowners had not been informed of the contamination. The *Washington Post* ran a story about the issue in April 1996. Meanwhile, the Corps' filed its corrosion control plan with EPA, and EPA substantially delayed in its approval, well beyond the legal deadline. Finally, EPA apparently simply accepted the Corps' plan to use only pH adjustment, rather than requiring the Corps to further study-or use orthophosphate or other more sophisticated corrosion inhibiters recommended by some consultants. When the Corps later switched to chloramines as a disinfectant, EPA made the serious mistake of not insisting upon a full review of the corrosion control plan in light of the apparently more corrosive disinfectant.

Even when the lead Action Level was exceeded in Washington in 2001, EPA required no changes in corrosion control, went along with WASA's plan to replace only a small number of lead service lines, and did not insist that WASA conduct an effective public education program. There also are substantial unresolved questions about whether EPA allowed WASA to "invalidate" lead samples and avoid an exceedence of the Action Level, as alleged by a former WASA employee who was reportedly fired for informing EPA of the lead problem. Additionally, while EPA has issued a notice of violation recently to WASA for failing to comply with public notification and public education rules, EPA has never challenged the adequacy of WASA's water quality reports sent to all consumers in June 2003 boldly proclaiming that "YOUR DRINKING WATER IS SAFE," despite the exceedence of the lead Action Level.

Moreover, while EPA enforcement of the Safe Drinking Water Act (SDWA) has never been strong, this testimony documents that nationally, it has substantially dropped off since President Bush took office (see Figures at the end of this testimony). EPA's drinking water inspections, administrative penalty orders, administrative penalties, and other measures of enforcement activity generally have taken a substantial downturn in the past 3 years. We understand there is only one EPA staffer in EPA's Washington enforcement office dedicated to drinking water enforcement (though there are pieces of a few others who spend small amounts of time on drinking water enforcement), and that the dedicated drinking water enforcement staffing in the EPA's regions is small and dwindling. This enforcement downturn may have contributed to the lack of action in this case, compared to a far more vigorous EPA enforcement response to previous DC water crises in 1993–1994 and 1995–1996. There is a serious need for a major infusion of resources and a will to enforce in EPA's drinking water and enforcement programs.

The only solution to the DC water crisis is for EPA to initiate a full civil and criminal investigation, and to immediately issue emergency administrative orders to WASA and the Corps. The orders should mandate that they address the multitude of problems with their response to the lead crisis and other water quality problems, including enforceable deadlines for:

(1) expedited, valid testing of all schools and day care centers;

(2) expanded testing of multiple family and single family homes and apartments beyond those with lead service lines;

(3) reissued accurate, understandable notices to consumers of lead levels, health risks, and options to avoid lead;

(4) professional installation *and maintenance* of certified filters for homes with lead service lines or high lead levels in their water, and that have young children, pregnant women, women who expect they may become pregnant, and other high risk individuals;

(5) an aggressive, honest, ongoing public education campaign developed with public input;

(6) a comprehensive third-party review of all available records and archives to determine whether the DC materials survey correctly identifies all locations where lead components were used;

(7) an expedited third-party review of the Corps' corrosion control and disinfection byproduct control strategy, with mandatory implementation of solutions by specified dates certain; and

(8) a top-to-bottom third party expert review of WASA and the Corps' water quality, source water, and overall performance, including a detailed review of their implementation of past consultant recommendations, Comprehensive Performance Evaluations, and sanitary surveys, and recommendations for long-term compliance with current and upcoming rules and water quality objectives. The review should seek public input and should be published.

(See LEAD coalition recommendations below for a more detailed discussion of the terms of possible orders). Finally, EPA must overhaul its lead rule, and its overall and substantially better fund its drinking water and enforcement program's oversight, sampling, data collection, and legal enforcement to ensure that this or other similar problems are not repeated in other cities around the country.

THE ARMY CORPS OF ENGINEERS' RESPONSIBILITIES

The Corps has repeatedly opted for the cheapest, easiest way out of water quality problems, even if the "solution" is manifestly inadequate. Thus, instead of following consultants' advice to consider aggressive and sophisticated corrosion inhibiters such as orthophosphates to reduce lead problems, the Corps chose merely to adjust pH. Instead of addressing the underlying problem creating the high chlorination byproduct contamination of city water by installing advanced treatment such as activated carbon and ozone or UV disinfection, or membranes, the Corps opted for a cheap "band-aid" solution of using chloramines alone, apparently exacerbating the corrosion problem with our water. As noted above, EPA should immediately issue an emergency order to the Corps requiring: (1) a comprehensive and public third party expert review of the Corps' corrosion control and water treatment problems; (2) enforceable deadlines for completion of the review and implementation of recommend solutions; and (3) a longer-term top-to-bottom third party review, with public input, of the Corps' water quality and treatment.

DC WATER AND SEWER AUTHORITY'S (WASA) RESPONSIBILITIES

WASA has bungled its response to the DC lead problem. In addition to violating EPA rules, WASA's public education and public notice efforts have been conflicting, confusing, misleading, and manifestly woefully inadequate. The direct notices provided to customers whose water was tested and confirmed to be highly contaminated was misleading and failed to provide any sense of health risk or urgency. The WASA water quality reports issued to the public proclaiming that "YOUR DRINKING WATER IS SAFE," despite evidence to the contrary, was highly misleading, as were a variety of other WASA public communications. WASA's changing advice on how long and whether to flush tap water, and whether filters are necessary, has confused the public.

WASA's program testing about 750 samples from over 150 city schools' fountains and faucets was fundamentally flawed and either completely inept or intentionally misleading. WASA admits that contrary to standard EPA regulatory protocol and standard scientific practice, they ran the water for 10 minutes before taking school samples, thereby likely substantially reducing lead levels in the samples. No child runs water for 10 minutes before drinking it. WASA's press conference portraying the results as demonstrating that there is no lead problem in DC schools was highly misleading and likely false. The Mayor and EPA have now told WASA to redo this testing. It should be done for all school and day care center faucets and fountains used by children for consumption.

In addition, there are serious unanswered questions about when WASA first learned of the lead problem, whether WASA "invalidated" lead samples to avoid exceeding the Action Level, and whether WASA fired an employee allegedly for notifying EPA of water quality problems (as has been found by a U.S. Department of Labor whistleblower review). It is also unclear whether the city's materials survey (intended to identify lead components in the system) adequately documents where lead service lines and high risk homes are located. The WASA lead sampling plan and monitoring program clearly are inadequate, since to date they have not sought to document the extent of the lead problem in homes not served by lead service lines.

WASA's lead service line replacement program is insufficiently aggressive and will not promptly resolve the city's lead problems. In addition to the slow pace of replacement (at WASA's current rate, it will take about 15 years to complete), it also is becoming apparent that *partial* lead service line replacement (leaving the lead line on the homeowner's property in place) may actually make lead problems worse. Partial service line replacement can exacerbate lead problems by shaking loose lead particles during and after the replacement process, and by creating galvanic corrosion (similar to a battery) caused when two pipes made of different metal are connected. We believe that WASA should pay for—with Federal assistance—full lead service line replacements.

A long history of problems with the operation and maintenance of the DC water distribution system, including past city-wide boil water alerts during the microbial crises in 1993–1994 and 1995–1996, and WASA's inability or unwillingness to candidly inform customers and apparently even senior city officials about water quality problems makes clear the need for EPA to issue an emergency order mandating a comprehensive top-to-bottom third party expert review of WASA's water quality and operations, with public input and public release of the fmdings, and a schedule for implementation of the recommendations.

HISTORY OF RECENT LEAD CRISIS IN DC

On Saturday January 31, 2004, residents of the Nation's Capitol picked up their morning papers and were stunned to learn that thousands of homes' drinking water in the District was seriously contaminated with lead. Officials at the DC Water and Sewer Authority (WASA) and at the U.S. Environmental Protection Agency (EPA) had known about the lead problem for over a year, and probably longer, but had failed to effectively notify the public about the problem. The Mayor, City Council, Members of Congress, and the general public were caught by surprise that over 4,000 of 6,000 homes whose water WASA tested was contaminated with lead at levels above EPA's action level-the safety level at which Federal rules require prompt action to reduce lead levels. There has been over a month of front-page stories, saturation TV and radio coverage, hostile City Council hearings, public outrage, and repeated (albeit often conflicting) WASA public statements that there was no serious health threat. Finally, WASA recommended on February 25 that pregnant women and children under age six whose homes were served by lead service lines should not drink city water, fueling further public concern, confusion, and outrage that WASA and EPA had known about the health threat for so long and never previously told pregnant women and parents of young children not to drink the water.

WASA also held a press conference in late February to announce that school drinking water was safe, based upon testing of over 750 fountains and faucets in DC schools. It then came out that the results were seriously misleading because in almost all cases, WASA flushed the water lines for 10 minutes, likely removing most lead from the water, contrary to EPA rules and all scientific protocols for lead testing. No child stands at a fountain flushing water for 10 minutes before taking a drink. WASA refused to retest DC school drinking water, or to comprehensively test day care centers, posing a serious health risk to DC school and preschool children, until ordered to do so by the Mayor and EPA.

Now we are learning that it appears that similar problems may be plaguing Northern Virginia communities that also receive their water from the U.S. Army Corps of Engineers' Washington Aqueduct Division (the Corps). The Corps changed its disinfection practice to use chloramines in 2000, a switch many experts believe may account for increased corrosivity of the water and therefore more lead leaching into tap water. Chloramines are a "band-aid" that modestly reduce cancer-causing chlorination byproducts, but only a switch to modern water treatment technologies such as granular activated carbon plus UV light or ozone disinfection will actually solve both the chlorination byproduct problem.

The February 25 "don't drink the water" advice, though necessary, is woefully inadequate. Citizens are infuriated that they have been mislead and given conflicting advice. District leaders announced, as this scandal erupted in early February, that they would name an "independent" blue ribbon panel to investigate. However, this was followed days later by an announcement of a panel consisting entirely of WASA and other District government officials, with no independent experts and no citizens, environmentalists, or consumer representatives. The District government's retreat from its promise that there would be an independent review showed a lack of commitment to swiftly resolve this serious health problem or to get to the bottom of why WASA continues to fail in its duty to protect the public.

The decisions to approve the use of lead service lines were made with the explicit approval and oversight of Federal officials, who were overseeing the construction of the city's water lines and supply. There had been a vigorous public debate about the safety of lead service lines stretching back to the 1890s, yet Federal officials who ran the city supply decided to use lead lines. Thus, the Federal Government bears some culpability for the problem.

CONCLUSION

We urge members of this Committee to consider the legislative and oversight recommendations noted above. Without changes in applicable statutory provisions, and aggressive Congressional oversight, it is likely that problems like those in Washington, DC could happen in many cities and towns across the country. Public health protection requires increased vigilance by EPA, Congress, health authorities, and water utility professionals, and increased public awareness.

APPENDIX A

LEAD COALITION'S RECOMMENDATIONS

Lead Emergency Action for the District (LEAD), a coalition of local and national health, environmental, and other citizen organizations of which NRDC is a member, recommended the following actions in February; only part of a few of these recommendations have been carried out:

1. *The U.S. Environmental Protection Agency (EPA) has the responsibility to immediately take enforcement action against WASA to ensure our health is protected, and should initiate a full criminal and civil enforcement investigation.*

The EPA has primary responsibility for overseeing the safety of the District's drinking water supply. Unlike its vigorous actions to resolve microbiological threats a decade ago, the agency has shirked its responsibility in response to the recent lead problem. The EPA should immediately initiate an enforcement action under its emergency order authority (which allows the EPA to enforce when there is an imminent health threat, requiring no finding of a violation of law), and should initiate a parallel criminal and civil enforcement investigation. The EPA order should mandate several specific actions, *including enforceable deadlines* for:

(1) *Expedited, valid testing of all schools and day care centers*, both first draw and flush samples.

(2) *Expanded testing of homes beyond those with lead service lines.* WASA should arrange free water lead tests for all DC residents. (This is what the New York City Department of Environmental Protection has been doing for more than 10 years.) Notice of these free lead tests should be drafted in consultation with EPA and the public, and should note the health implications of elevated lead levels in water and the threat from lead paint in DC

(3) *Reissued accurate, understandable notices to consumers* of lead levels, health risks, and options to avoid lead, by mail and through broadcast media. WASA should be required to immediately notify all DC households whether they are believed to have lead service lines or not, what the risks are, and should arrange for free lead testing of any tap water on request. Notices similar to those recently sent to lead service line customers should be sent to customers who are not believed to

have lead service lines noting that there still may be a risk of lead contamination, and offering to arrange for free lead testing.

(4) *Professional installation and maintenance of certified filters* for homes with lead service lines or high lead levels in their water, and that have young children, pregnant women, women who expect they may become pregnant, and other high risk individuals.

(5) *An aggressive, honest, ongoing public education campaign* developed with public input. This should include several specific requirements, such as:

a. WASA should send all DC residents a *detailed* city-wide map of all areas with known or suspected lead service lines with accompanying health and other explanations.

b. WASA must acknowledge the public's right to know and issue a city-wide map of lead levels detected on a detailed map, and should provide *real time* monitoring results for lead and all contaminants found in its water.

c. WASA must notify any home with a lead service line that has been found to have excessive lead in an appropriate water test that it is eligible for free lead service line replacement, and the schedule for replacement. The notice should also note whether WASA is responsible for only part of the service line replacement or full service line replacement under DC law.

d. EPA and WASA must issue notices that publicly recommend that those pregnant women, or parents of young children, with lead service lines or whose water lead levels are in excess of EPA's Action Level (or some other reasonable safety level), should obtain blood screening for lead for their children. This is not an emergency that would require going to the emergency room, but it is a matter of importance, and blood tests for lead levels should be provided by the DC Department of Health.

(6) *A comprehensive third-party review of all available records and archives to determine whether the DC materials survey* correctly identifies all locations where lead components were used;

(7) *An expedited third-party review of the Corps' corrosion control and disinfection byproduct control strategy*, with mandatory implementation of solutions by specified dates certain; and

(8) *A top-to-bottom third party expert review of WASA and the Corps' water quality, source water, and overall performance*, including a detailed review of their implementation of past consultant recommendations, Comprehensive Performance Evaluations, and sanitary surveys, and recommendations for long-term compliance with current and upcoming rules and water quality objectives. The review should seek public input and should be published.

2. *EPA should immediately take enforcement action against the Army Corps of Engineers' Washington Aqueduct and order it to aggressively treat the water to reduce lead leaching.*

The EPA's 1991 lead and copper regulations require the Washington Aqueduct to treat our water in order to reduce its corrosivity; less corrosive water should mean less lead leaching from pipes. While the Corps and WASA do have a corrosion control program (albeit one that reportedly was reviewed by the EPA far later than envisioned by the 1991 rules), it is obvious that it must be critically examined and improved. Recent changes in water treatment at the Washington Aqueduct (apparently made after the corrosion control plan went into effect), aimed at reducing disinfection byproducts, may have altered the chemistry of the city's water. An urgent independent review of the corrosion control plan is warranted, with EPA-ordered steps to implement recommended actions. Deadlines should be established for completion of the review and implementation of its recommendations, and the results should be made public as soon as they are completed. When WASA was constituted, it entered into a governance agreement with the city of Falls Church and Arlington County over Washington Aqueduct, with oversight over expenses and actions. WASA and other customers should long ago have insisted upon improvements in the Washington Aqueduct's corrosion control program.

3. *WASA must re-conduct its testing of District school water to be sure that all drinking water fountains and all faucets used for consumption in District schools and day care centers are tested—both first draw and flushed samples—within 2 weeks.*

WASA's recent water test results were highly misleading because more than 97 percent of the samples taken were from faucets and fountains flushed for 10 minutes. Since no student flushes a fountain for 10 minutes before taking a drink, flushing water for a test sample would create misleading samples and test results. (Flushing often will reduce or eliminate lead levels in large buildings.) Since infants and young children are most vulnerable to lead poisoning, schools and day care centers should be top priorities for testing.

4. *EPA and Congress should help WASA and the DC government fund home treatment units or bottled water for pregnant women and infants under age 6 in households that have lead service lines or lead in the drinking water at levels above the EPA action level.*

There are likely thousands of pregnant women and young children under the age of 6 who are drinking tap water that contains lead at levels higher than 15 parts per billion, EPA's action level. These people need a safe alternative water supply until the problem has been resolved. The DC government, EPA and Congress should fund alternative water supplies for high-risk water drinkers. Bottled water is not necessarily any safer than tap water unless it is independently tested and confirmed to be pure, and many filters are not independently certified to remove the levels of lead found in many DC homes' water. Therefore, EPA should assist residents by assuring that any alternative water supply (such as bottled water) is indeed free of lead and other harmful contaminants, or that a filter is independently certified (see *www.nsf.org*) to take care of lead. It should be noted that NSF certifies only that lead levels up to 150 ppb will be reduced to below 10 ppb; there is no guarantee for reducing levels above 150 ppb. Finally, it is critical that WASA and other officials involved ensure that there is a followup program for maintenance of filters, since poorly maintained filters can fail to remove lead or even make contamination worse.

5. *WASA should expedite replacement of lead service lines, and the City Council should review policies on replacement of the homeowner's portion of the line.*

Under EPA's lead and copper rule, WASA reportedly has begun to implement its obligation to replace 7 percent of the District's lead service lines (or to test and clear homes served by lead service lines as containing less than 15 ppb lead in their water) each year. At this pace it will take nearly 15 years—until about 2018—for WASA to replace all the city's lead service lines. In the meantime, thousands of pregnant women, infants and children could be consuming water with excessive lead levels. We strongly urge that the lead service line replacement program be aggressively expedited. A schedule should be published, with objective criteria for which lines will be replaced first (presumably based primarily upon replacement of those lines posing the greatest public health risk first). Federal and city general funds should be set aside for this program to augment promised rate increases on our water bills. WASA customers should not foot the entire bill, since the decisions to approve the use of lead service lines were made with the explicit approval and oversight of Federal officials who were overseeing the construction of the city's water lines and supply. There was a vigorous public debate about the safety of lead service lines stretching back to the 1890s, yet Federal officials who ran the city supply decided to use lead lines. District officials also should consider using the city's multimillion dollar rainy-day fund to help pay for service line replacements.

In addition, the City Council should review WASA's and the city's policy about lead service line replacement for the portions of the line that are supposedly owned by homeowners. Evidence is mounting that partial lead service line replacement often will not solve the problem, and actually can make lead levels worse by shaking loose lead in the pipes and causing galvanic corrosion that may exacerbate lead problems.

Under recent EPA rule changes, it is apparently up to the City Council to determine how much of the service line should be replaced by WASA. In 1991, EPA originally required full lead service line replacement unless the water utility could prove that it did not control part of the line, in which case it was to replace only that portion that the utility controlled. After being sued successfully by a water industry group, the EPA changed the rules to provide that it is largely a question of local law what portion of the lead service line is the responsibility of the water utility. We believe that it is only fair that since many of the lead service lines were installed from the 1890s through the 1940s under the direction, approval and control of the District and Federal officials, those authorities should be responsible for replacing them, not homeowners. The cost to homeowners of their portion of lead service line replacement could be thousands of dollars, but it is far more efficient and cost-effective to replace the entire service line at once, rather than digging up yards twice. This is a question that deserves a full public airing by the City Council.

6. *The City Council should create a permanent citizen water board for water to oversee WASA and the Washington Aqueduct, to address longstanding problems with DC's water supply.*

In 1996, the Natural Resources Defense Council (NRDC), Clean Water Action (CWA), and the DC Area Water Consumers Organized for Protection (DC Water COPs) issued a report, based in large part on city and Federal records obtained under the Freedom of Information Act. That report found serious ongoing problems with the District's water, and identified likely problems that could occur in the future. Among the current and future problems noted were lead contamination, bacteria and parasites, cloudiness (turbidity) in the water—which may indicate poor filtration and can interfere with disinfection—and disinfection byproducts that cause cancer and may cause birth defects and miscarriages. The report also noted that the Washington Aqueduct's water treatment plants need a major infusion of funds to modernize and upgrade treatment, and that the District has ancient and deteriorating water pipes leading to water main breaks, regrowth of bacteria, and lead problems. Those pipes must be replaced. In addition, the WASA-operated sewage collection and treatment systems have serious inadequacies, including major problems whenever stormwater runoff overloads the treatment plant's capacity, causing raw sewage to flow into the Anacostia and Potomac rivers.

In the wake of the DC citywide boil-water alerts in 1993 and 1996 due to turbidity and bacteria problems, and EPA's enforcement orders issued thereafter, comprehensive sanitary surveys and engineering reviews by outside contractors found a series of serious problems with our water treatment and distribution system. These reviews recommended hundreds of millions of dollars in improvements in the city's water supply system.

While the city has addressed some of the most pressing problems, it has not made many of the important investments needed to repair local water infrastructure. We strongly recommend that the City Council establish a citizen water board to oversee the city's water supply and sewer system. The board should oversee not only steps to improve our drinking water system, but also WASA's storm water and sewer obligations, because of the overall competition for water infrastructure dollars and need to focus on whole watershed and "sewer shed" solutions. This board—like those created by some States to oversee electric and other utilities—should be funded with a small surcharge on water and sewer bills, and should be wholly independent of WASA and the Washington Aqueduct. It should include independent engineering and public health experts and citizen activists interested in drinking water, and should issue an annual progress report on WASA's and the Washington Aqueduct's performance, progress and problems.

7. *The City Council must improve its oversight of WASA.*

The District's City Council is responsible for overseeing WASA's day-to-day activities, and has failed to do its job over recent years to make sure that WASA is carrying out its responsibilities to deliver safe drinking water and to safely collect and fully treat city sewage. More aggressive City Council oversight is needed to avoid continued problems with WASA.

8. *The mayor should make tap water and all environmental protection a high priority.*

The mayor should make drinking water safety, sewage collection and treatment and environmental protection a high priority. The mayor bears some responsibility for ensuring that WASA is doing its job. He has many ways to influence WASA's board and daily operations, and should insist on regular briefings and updates on how the city is fulfilling its obligations to provide these most basic city services.

9. *Consumers, health, and citizens groups should be on the blue ribbon commission, and should recommend people to serve on the panel.*

The announced "independent" panel to review WASA's embarrassing performance in addressing the lead problem has instead morphed into an internal review panel of city officials, including two of the WASA officials who so obviously have failed to do their jobs. In order to avoid a panel that merely papers over the problems and whitewashes the lead crisis, LEAD is calling upon city officials to name independent experts, consumers, citizen groups and environmentalists to the panel.

10. *The EPA, CDC, the DC Dept of Health and the City Council should establish a joint task force with citizen participation, to evaluate the extent of lead poisoning from all sources in the District, and its environmental justice implications, particularly for low-income AfricanAmerican and Latino households.*

According to expert estimates, the District has widespread lead poisoning, affecting perhaps tens of thousands of District children. Because of the city's demographic and economic realities, most of these children are African American and Latino. The District and Federal officials should establish a joint task force, with citizens and medical experts, to evaluate the extent of the problem and its environmental justice implications, and to recommend actions to remedy it.

APPENDIX B
RECENT TRENDS IN EPA DRINKING WATER ENFORCEMENT

Source: EPA Data, 2004[2]

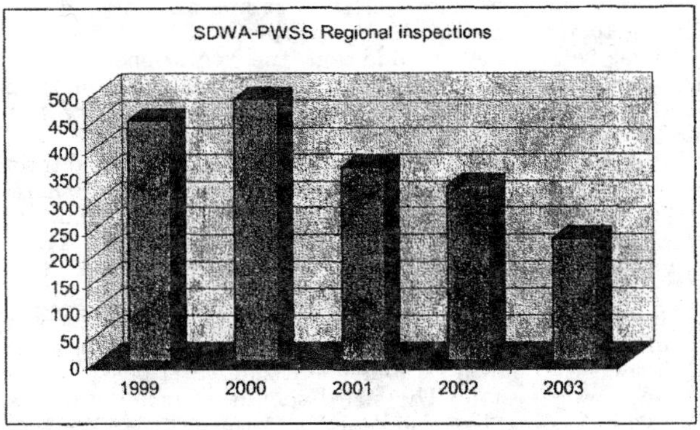

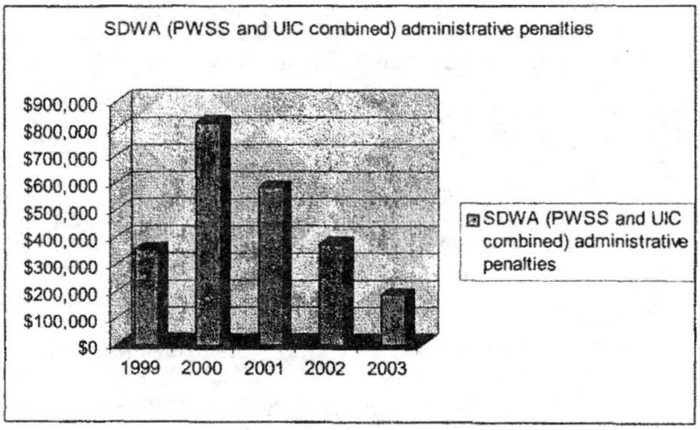

2

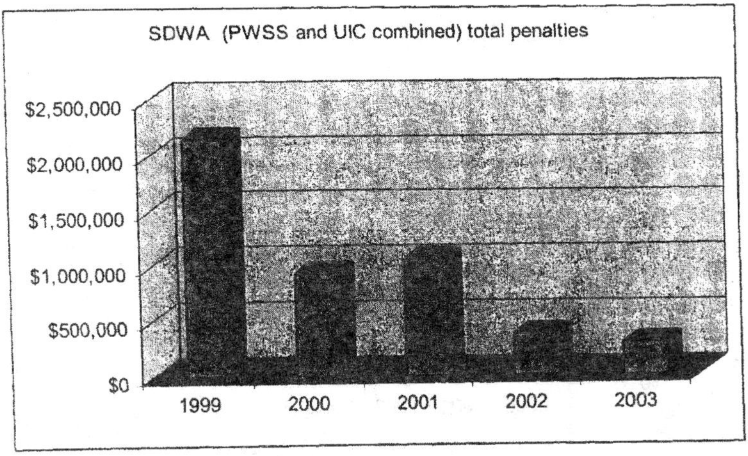
SDWA (PWSS and UIC combined) total penalties
$2,500,000
$2,000,000
$1,500,000
$1,000,000
$500,000
$0
1999
2000
2001
2002
2003

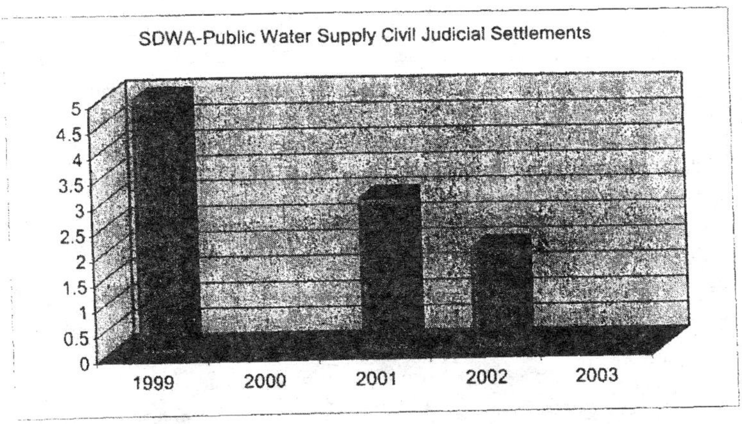
SDWA-Public Water Supply Civil Judicial Settlements
5
4.5
4
3.5
3
2.5
2
1.5
1
0.5
0
1999
2000
2001
2002
2003

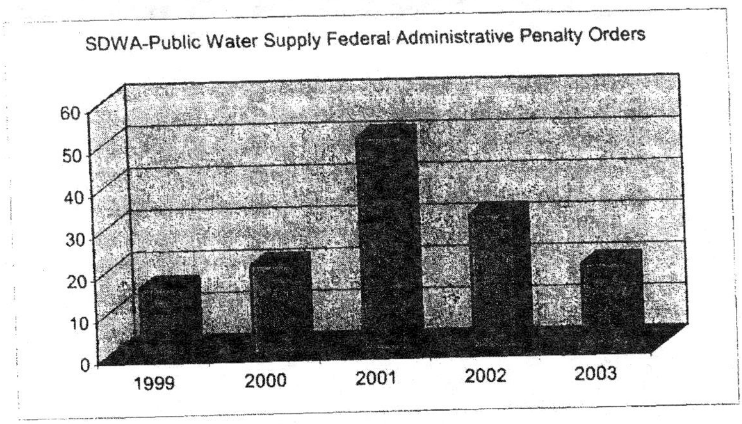
SDWA-Public Water Supply Federal Administrative Penalty Orders
60
50
40
30
20
10
0
1999
2000
2001
2002
2003

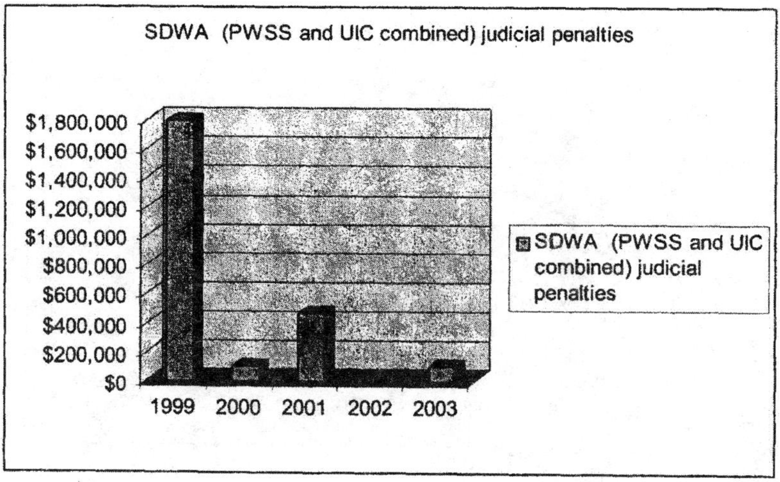
SDWA (PWSS and UIC combined) judicial penalties
$1,800,000
$1,600,000
$1,400,000
$1,200,000
$1,000,000
$800,000
$600,000
$400,000
$200,000
$0
1999
2000
2001
2002
2003
SDWA (PWSS and UIC combined) judicial penalties

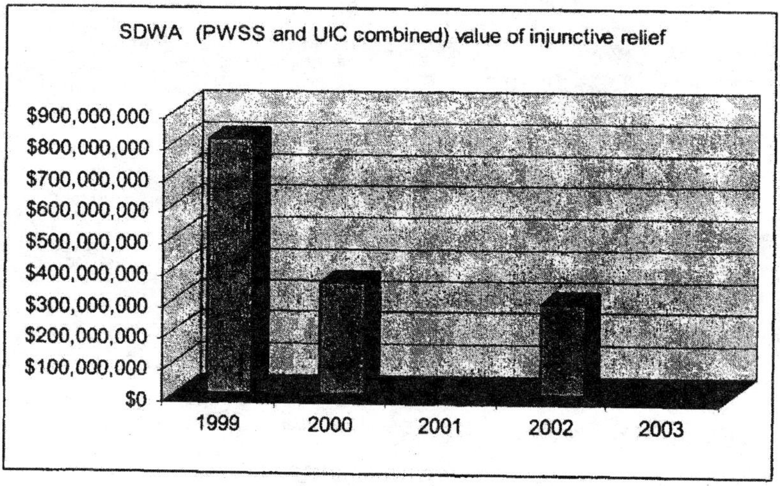
SDWA (PWSS and UIC combined) value of injunctive relief
$900,000,000
$800,000,000
$700,000,000
$600,000,000
$500,000,000
$400,000,000
$300,000,000
$200,000,000
$100,000,000
$0
1999
2000
2001
2002
2003

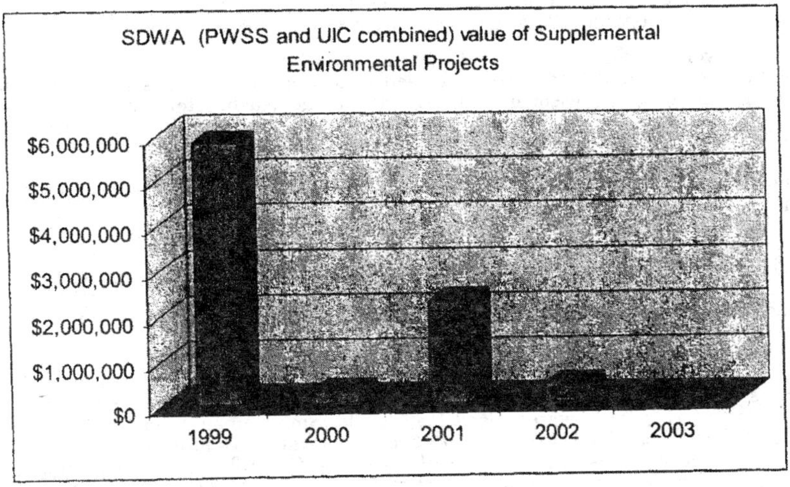
SDWA (PWSS and UIC combined) value of Supplemental Environmental Projects
$6,000,000
$5,000,000
$4,000,000
$3,000,000
$2,000,000
$1,000,000
$0
1999
2000
2001
2002
2003

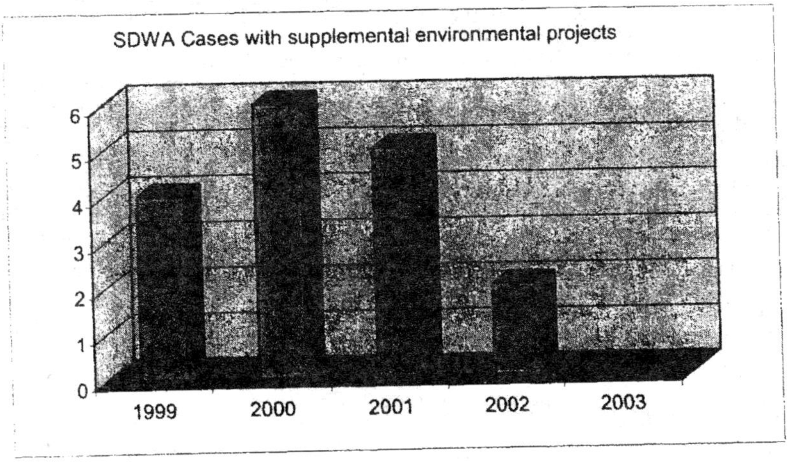
SDWA Cases with supplemental environmental projects
6
5
4
3
2
1
0
1999
2000
2001
2002
2003

STATEMENT OF MURIEL WOLF, M.D., CHILDREN'S MEDICAL CENTER, GEORGE WASHINGTON UNIVERSITY MEDICAL SCHOOL

Thank you for the opportunity to provide this testimony to you today regarding the effect of elevated lead levels and lead poisoning. My name is Dr. Muriel Wolf. I am an Associate Professor of Pediatrics at George Washington University Medical School, and Attending in Pediatrics and Cardiology and Senior Pediatrician at Children's National Medical Center in Washington, DC. I have taken care of children with elevated lead levels and lead poisoning for over 30 years.

Fortunately, the health care landscape has changed significantly since the 1970s when we admitted over 100 children per year with elevated lead levels of over 60 mcg/dl. Now, we admit fewer than 5 children per year with elevated lead levels at this number.

The problem of lead in the water in the District of Columbia has alerted all of us about the possibility of elevated blood lead levels. But as of this writing, there is no strong evidence that the lead in the water has caused any serious elevation of blood lead levels.

While the issue of lead in the District water supply is an important one, let it not be lost that most elevated lead levels in children are due to lead paint in old houses. Children exposed to an environment where there is peeling or flaking lead paint are at risk for elevated blood lead levels. Small amounts of lead paint chips or dust can cause blood lead levels to become elevated. Because of the presence of lead paint in old houses, the Centers for Disease Control (CDC) has recommended in cities where more than 20 percent of the houses were built before 1970, children should be tested for elevated blood lead at age one, and then again at 2 years of age. In DC, over 50 percent of the housing was built before 1970. Accordingly, all District children should have blood lead tests at 1 and 2.

Lead can cause significant health problems. Currently the acceptable blood lead level according to CDC guidelines is 9mcg/dl or less. Very high blood lead levels (over 50 to 60 mcg/dl) may cause serious health problems such as marked learning disability and mental retardation. Even higher blood lead levels can be associated with brain swelling and seizures. Elevated blood lead levels may cause significant anemia and kidney damage.

Children with blood lead levels above 20 mcg/dl may have learning disabilities and attention deficit disorder problems, and hearing and growth may even be affected. More recent studies have shown that even mild elevations of blood lead levels between 10 and 20 mcg/dl may minimally lower the IQ by 1 to 3 points.

The goal is to prevent elevated blood lead level, and currently those elevated levels almost always come from exposure to lead paint and dust. Homes with lead paint should be screened for lead hazards where there is peeling and flaking paint. Windows and doors should be wiped with high phosphate soap. Floors should be. wet-mopped rather than vacuumed so that the lead in the dust is not spread throughout the room. Children and adults should frequently wash their hands to prevent environmental exposure to lead dust. Finally, the paint causing the problem should be sealed or removed.

Lead paint remains the most serious source of lead problems at this time. But lead in the water is a significant issue as well—especially if the CDC decides to lower the acceptable level of lead for children. Lead in the water may contribute to elevated lead levels, but nobody knows for sure. It has not been shown so far in DC to be the cause of elevated lead levels beyond the 10 mcg/dl level. Of the 14 patients identified in the District with elevated lead levels above 10 mcg/dl, all lived in environments where there was lead paint and tested positive on dust wipes.

Current research reported in the New England Journal of Medicine in April 2003 suggested that lead levels of 5–9 mcg/dl can indeed lower IQ by 5–7 points. If this research can be corroborated by other studies, then we should be significantly concerned that lead in the water in the District of Columbia may be contributing to elevated lead levels.

It is appropriate to study the issue of lead in the water in the District of Columbia, but the lead found in housing is the major problem at the current time.

STATEMENT OF ANDY BRESSLER, RESIDENT, DISTRICT OF COLUMBIA

Chairman Crapo, Ranking Member Graham, Members of the Committee, I want to thank you for holding this important hearing.

I would like to tell you about my family. My name is Andy Bressler. My wife Shellie and I have lived on Capitol Hill for the past 7 years. Some of you may recognize our sons Adam and Casey. Back in February, our family was profiled in the

Washington Post. I ask that the article[1] be included in the record. You may have also seen them in the halls of these buildings when they come to visit and to have lunch with their Mommy who works in the Hart Building.

Three years ago, we had the great fortune of adding twin boys to our household. Despite some of the obstacles urban living entails, we have enjoyed living in the city and have looked forward to being able to enjoy everything that city life offers. Little did we know that we were exposing our children to potential health hazards through lead-poisoned drinking water.

Imagine our shock a little over a year ago to hear from our pediatrician that our healthy, thriving recently turned 2 year olds had tested over the CDC recommended level for lead. We immediately contacted the DC Department of Health's Office of Lead and requested an inspection of our home. A test was conducted on our 125 year old house it was determined there was no lead paint exposure on the interior, but, there was a limited area of lead paint on an exterior door. We then proceeded to have that door replaced. When we inquired about the possibility of lead in our water, the inspector reassured us that that DC water was fine and safe to drink, and that could not be the source of the lead problem.

Months later, my wife and I took the boys in for their 3-year checkup. The next afternoon, my wife received a call from the doctor's office to say that once again the boys lead levels had not declined, and in fact had risen. Casey tested at 14; Adam tested at 12 (both at levels above the 10 mg threshold). In knowing we had done work to remediate the problem months early, the doctor asked if the boys drink water. When we replied that we use it to water down their juice and to cook with, he advised us to stay away from unfiltered tap water and to solely use bottled water for their cups and in preparing their meals.

The next day, we contacted the DC Lead Hotline at WASA to request a water test for our home—this was in mid-January 2004. Through the Moms on the Hill group, we had learned that there had been some concern about lead in DC water, and that there had been some testing going back 6 months or more. A few days later, the *Washington Post* broke the story about the extreme lead levels found in DC water. We waited over 2 weeks for someone from WASA to return our multiple phone messages. We finally had our water tested by WASA, and it does show significant elevated lead levels (24 ppb).

Between our own research and speaking with experts, we have learned that that there is no cure or antidote for our sons' exposure level. By eliminating the exposure, it would eventually leave their bodies. But, we understand that it will likely take years before the lead is out of their system. We have taken all possible steps to rid our house of the lead, and little did we know that every time we gave our children something to drink, we were exposing them once again to the lead.

Another uncertainty is the long-term effects these levels of exposure will have on our children. Experts have testified that at their level of exposure, minimally they will lose precious IQ points. Other problems could include learning difficulties, attention disorders, and/or general behavior problems. These symptoms would not present themselves until years later once they are in elementary and middle school.

As parents of twins, we have been cautioned not to compare developmental milestones with other children their age. We were told that our children would reach these steps at their own pace. As a part of human nature, it is very difficult not to compare and wonder. At this time, our greatest concern that each time one of them has difficulty in grasping a subject matter in school or an unexplained emotional outburst, we will question if it is long-term effect of being exposed to leaden water up until their third birthday.

Some of the issues that we would like to see addressed by Congress, EPA, WASA, and the City:

• Let's move quickly to a solution—if it means replacing the lead pipes, then let's get on with it!

• We need much better oversight from both Congress and the EPA—It is obvious that there have been failures over the last several years, as both the EPA, WASA, and the WASA board have failed the public by not coming forward sooner, and moving toward a solution sooner.

• We are also disappointed that the "Task force" working on this issue is not open to the public, nor does it have representation from citizens affected by this severe problem.

• There needs to be real accountability for the lack of leadership and management oversight at WASA, EPA, and especially the Board of WASA, who were appointed to represent the public.

[1] The article referred to may be found in Committee files.

• From what we understand, this is not a new issue, as DC's water had a lead problem back in the late 1980s, and early 1990s. At that time, WASA undertook steps and developed a plan to fix the lead pipe problem. However, it appears that since that time WASA and its Board abandoned those efforts, and we would very much like to understand why they did not follow through on those plans.

• We also are concerned about the continually changing advice that we have received from WASA regarding how to reduce the lead levels in our water (such as how long to run the water—1 minute, 5 minutes, 10 minutes??). In addition, WASA's hotline has given us conflicting information regarding whether the service line leading to our house is lead or not (it appears as though it is).

• Scientists have stated incidences leaching are higher in warmer weather. Now that summer is approaching, what can be done in the immediate future to help alleviate the high lead exposure levels in the water.

• Finally, we understand there is a plan to begin replacing lead service lines, and we would like to have a better understanding and guidance as to how WASA is prioritizing these replacements.

Thank you for holding this hearing, and we would be pleased to speak with you or your staff regarding any of these issues, or our experiences with the DC government and WASA.

Thank you.

TIMELINE FOR THE IMPLEMENTATION OF THE LEAD AND COPPER RULE IN THE DISTRICT OF COLUMBIA 2000 TO 2004

Nov. 1, 2000: The Aqueduct replaced its secondary disinfection treatment by converting from free chlorine to chloramines.

2001–2002: DCWASA initiates massive water meter replacement program; some 18,000 water meters were replaced with new, remote read capability units.

July 30, 2002: EPA first receives preliminary information that DCWASA exceeded the lead action level; final report submitted August 27, 2002. EPA advises DCWASA that it has to return to regular sampling frequency and sample site numbers (100 samples every 6 months), beginning with next compliance period (January 1, 2003), conduct lead public education program and initiate lead service line replacement program.

October 2002: DCWASA conducts lead in drinking water public education program within required timeframe. On October 30 and 31, 2002, DCWASA sent:

- lead brochure mailed to every residence via *Washington Post* circulation department;
- brochure and public service announcement delivered to *Washington Post's* and *Washington Times'* editorial boards;
- brochure and PSA sent to 8 television stations; 18 local radio stations and the AP;
- brochure to city libraries, DC public schools and Board of Education, DC Department of Health, and to 12 hospitals and clinics;
- DC Council members Patterson and Cropp; and
- brochures to all lead sampling volunteer homeowners (hand delivered).

  From November 4–December 14, 2002 brochures were delivered to remaining hospitals, libraries and schools.

Jan. 1, 2003: DCWASA begins routine tap sampling program with increased monitoring frequency and increased sample site numbers.

Jan. 24, 2003: EPA receives first report on DCWASA's public education program: due November 10, 2002; report received January 24, 2003).

May 5, 2003: EPA obtains the services of corrosion expert, Marc Edwards of Virginia Tech, through a HQ level of effort contractor to assist EPA Region III in analyzing data and to make recommendations to EPA and to DCWASA for potential treatment options—assistance to WASA that is above and beyond EPA's responsibilities.

May 21, 2003: Lead Service Line Replacement (LSLR) program plan submitted showing DCWASA's intended implementation of the LSLR program. Due September 30, 2003; Received May 21, 2003 (four months early).

June 27, 2003: EPA Region III approved DC WASA's LSL Replacement program plan on June 27, 2003.

July 30, 2003: Lead and copper tap monitoring shows action levels exceeded for monitoring period January–June 2003. EPA advises DCWASA that Lead service line replacement and public education program must continue.

Sept 2003: The public education program was conducted by DCWASA (program report was due October 10, 2003; report received October 14, 2003).

- brochure insert in water bills sent to all billing units in August 2003 bills, with highlighted message on bill face.
- pamphlet and brochure mailed Sept 30, 2003 to Mayors' office of Latino Affairs, Dept. of Health, all ANC Chairpersons, DC Public schools, libraries, hospitals and clinics.
- Newspaper ad placed in *Washington Post* Sept 30, 2003.
- PSA's faxed to TV and radio stations Sept 30, 2003.

Oct. 2003: The public education program was conducted by DCWASA (program report was due November 10, 2003; report received early on October 14, 2003).

Sept. 30, 2003: EPA receives LSLR preliminary report; official report received October 27, 2003. (program report was due September 30, 2003).

Oct. 17, 2003: Region III's contractor, Marc Edwards, submits his draft report on his research which recommends that DCWASA conduct specific research in several areas.

Nov. 19, 2003: EPA Region III completed the initial review of DC WASA's LSLR report.

Jan 26, 2003: DC WASA conducted full lead and copper tap monitoring during July–December 2003 compliance period. EPA instructs DCWASA to continue LSL replacement and public education programs.

Nov 11, 2003: Marc Edwards presents findings and recommendations to DCWASA, Washington Aqueduct, and EPA Region III.

January 21, 2004: DCWASA presents to Washington Aqueduct and Virginia wholesale customers preliminary research plan containing actions recommended by Marc Edwards and study results to date.

Feb. 5, 2004: Technical Expert Working Group (TEWG) formed on conference call.

Feb. 9, 2004: First face to face meeting of the TEWG conducted at the Washington Aqueduct offices, outline of the research action plan developed.

Feb. 11, 2004: EPA ORD in Cincinnati begins analyzing lead service line pipe scale with X-ray defraction techniques to begin preliminary analysis of pipe conditions.

Feb 16, 2004: DCWASA staff and their contractors meeting with Steve Reiber at University of Washington in Seattle, who now is EPA's external corrosion expert through contract. Purpose of meeting is to learn set up for electro-chemical testing of pipe loops. Equipment to be shipped back to District of Columbia to run tests on pipe specimens with actual District tap water.

Feb 25, 2004: First draft action plan due to be ready for briefing to DC City Council.

March 10, 2004: Final Research Action Plan due (30 days past initial meeting).

Statement of Robert Vinson Brannum, Parent

Good afternoon, Mr. Chairman and members of the Subcommittee. Mr. Chairman let me introduce myself to you and other members of the Subcommittee. My name is Robert Vinson Brannum. I am a parent of an eleven-year old son and a proud native Washingtonian. I am also a resident of Ward 5 and the historic Bloomingdale community along the North Capitol Corridor. I have served three terms as president of the Bloomingdale Civic Association. As a member of this African American and greater Washington community, I believe in the spirit of volunteerism and community service. I am happy to have been invited by the Subcommittee to submit this statement for its official record and review.

In the aftermath of recent disclosures regarding high levels of lead discovered in the DC water system, this afternoon the Subcommittee on Fisheries, Wildlife, and Water will hold an oversight hearing on the detection of lead in DC drinking water, focusing on needed improvements in public communications and the status of short- and long-term solutions.

It is disturbing in the year 2004 to have questions raised about the quality of the drinking water in my home city—the Nation's Capitol. I would rather have a conversation about full voting rights rather than talking about how long to let my faucet run each morning when I rise and each evening when I return from work. Good and safe drinking water not only sounds good, but also it serves as an indicator of a healthy society.

On 22 March 2004, I submitted water samples for testing. I have not received the results. There are many District of Columbia residents who wonder what happened, how did it happened and why it took so long for the information to become public? What did our Federal and local officials know about the lead levels, when did they know it, and what did they do when they learned about it?

Officials from the U.S. Environmental Protection Agency (EPA), the Washington Aqueduct, the DC Water And Sewer Authority (WASA), and the DC Department of Health must answer these questions and many more. However, in the search for answers I do hope we do lose focus on the critical issue of solving the problem while directing our sights to responsibility and accountability. As a parent and a teacher, I am naturally concerned about the impact of high lead levels on the physical and cognitive development of our children, particularly African American children. As a community activist, I am concerned about the increased cost of maintaining safe water for daily consumption by those who are on low and fixed incomes.

There are many who do not trust WASA to perform ongoing tests. Like so many other concerned parents, I am not a scientist or a chemist. Yet, from all I have read it appears the issue is not the actual tests conducted, but rather the apparent delay in the notification of the public by WASA and EPA. From what I have been able to read and have been told, lead is not being exposed to the water at the Aqueduct and distributed by WASA. By most accounts, the water becomes exposed to lead via the lead service lines or lead soldered joints.

Other residents of this community have expressed to me their concerns about the water. Several residents have paid to have the lead pipes from the service line to their homes and they are still not comfortable about the safety of the water. Some residents feel WASA has the records to determine where all the lead service lines are and should pay for the replacement. Still others believe there should be a moratorium on water bill payments until the problem is resolved. There was almost a unanimous view the Army Corps of Engineers should review the current chemical make-up of the water to make certain that change was not the principal cause of the lead.

If it becomes necessary to repair or replace all the lead service lines, I do not feel the costs should be borne by the residents of the District of Columbia or the Government of the District. Historically, Presidential appointed commissioners governed the District of Columbia and the U.S. Congress directed all municipal functions of the District. Even to this day, the District of Columbia does not have complete control of its own municipal operations let alone its own water system. It is my expectation the Federal Government would bear the complete cost of the service line replacement. I do hope the subcommittee will support an appropriation to cover the Federal Government's responsibility in this crisis.

Many who have voiced questions about the lead have expressed concerns because of its negative affect on children, particularly African American children. If the statement of the DC Department of Health are correct, none of the children tested and found to have high led levels were exposed to lead from water. All of children were exposed to lead because of lead paint and lead dust in the home.

To this day no one is able to state with presumptive certainty how the lead got into the water. However, this fact has not diminished the critics of the District of Columbia government. In addition, I, personally find it incredulous for EPA to assert the lead notification problem has been the sole responsibility of WASA and EPA bears no accountability.

The water distribution in the District of Columbia is the responsibility of WASA. There are some, as an expression of their disappointment in the conduct of WASA during this lead crisis who feel the governance of WASA has to be restructured to exclude any local District government involvement. As a resident of the District, I cannot support the idea of a Federal takeover of WASA to the agency that may have contributed to the current crisis and may be seeking to recreate history to absolve itself of any responsibility or accountability.

In a discussion with students, their questions regarding the lead issue are simple. How did the lead get into the water? Why did it take so long to the people? Can there be a double filtration system of the water? Is this just a ploy for the water companies (commercial) to get more money? How can you ever know if the water is safe?

Mr. Chairman, this committee, along with WASA, EPA, and the DC Department of Health has a duty to assure the public and our young people the water is safe to drink. This can only happen when all entities stop finger-pointing to one another and worked closely to solve the lead in the water problem. I do not shrink to no one on the urgency to solve this matter. However, I am troubled by the rush to judgment and the push for the concept of a "Federal takeover" of WASA. I feel District officials; particularly Deputy Mayor and City Administrator Robert Bobb should be given the opportunity to respond and to determine the best course of city action. Deputy Mayor Bobb has assembled District agency directors and he has taken a "clear hands" on approach to lead the District government response. The District Government is engaged at the highest levels.

Lead service lines are not unique to the District of Columbia. It is my understanding major cities such as New York City, Chicago, and Richmond have lead service lines in their city systems. Just as in the District of Columbia, high lead levels have been found in the water of suburban jurisdictions and schools in Maryland and Virginia.

Mr. Chairman, everyone wants to have safe drinkable water for all who live, work and visit the Nation's Capitol. Unfortunately, no one is able to identify the source of the lead. All our immediate efforts should be to find the cause and solve the problem before casting blame.

OFFICE OF INSPECTOR GENERAL

Catalyst for Improving the Environment

Memorandum Report

EPA Claims to Meet Drinking Water Goals Despite Persistent Data Quality Shortcomings

Report No. 2004-P-0008

March 05, 2004

Report Contributors:

Jill Ferguson
Linda Pettit-Waldner
Tim Roach
Dan Engelberg

Abbreviations

EPA	Environmental Protection Agency
GPRA	Government Performance and Results Act
M/R	Monitoring and Reporting
OIG	Office of Inspector General
PWSS	Public Water Supervision System
SDWIS/FED	Safe Drinking Water Information System/Federal Version

UNITED STATES ENVIRONMENTAL PROTECTION AGENCY
WASHINGTON, D.C. 20460

OFFICE OF
INSPECTOR GENERAL

March 05, 2004

MEMORANDUM

SUBJECT: EPA Claims to Meet Drinking Water Goals Despite Persistent Data Quality Shortcomings
Report No. 2004-P-0008

FROM: Kwai Chan /s/
Assistant Inspector General
Office of Program Evaluation

TO: Benjamin Grumbles
Acting Assistant Administrator
Office of Water

In each of the past 4 years, the Environmental Protection Agency (EPA) incorrectly reported meeting its drinking water goal under the Government Performance and Results Act (GPRA). The Agency reported meeting its annual performance goal for drinking water quality even though it concurrently reported that the data used to draw those conclusions were flawed and incomplete. In each of those years, EPA reported that it met its annual goal of 91 percent of the population drinking water that met health-based standards. However, EPA's own analysis, supported by our review, indicated the correct number was unknown but less than what was reported. We must note that this inaccuracy in reporting does not necessarily indicate a direct or immediate threat to human health.

Purpose

We initiated this review to evaluate the "drinking water performance measure," a key component of EPA's GPRA goal of "Clean and Safe Water." The evaluation questions were: (1) how do incomplete or inaccurate drinking water data affect the drinking water GPRA calculation; and (2) what actions have EPA undertaken to ensure that drinking water data collected and distributed to the public are reliable and valid?

During the preliminary research phase, we learned that the Office of Water was conducting analyses that largely overlapped our own, and was working with States and other stakeholders to address data quality problems. Since we already completed work on our first question but not the second, we are reporting the results on the first and suspending our work on the second.

Details on our scope and methodology, as well as background on GPRA and drinking water reporting, are in Appendix A.

Results

EPA has reported just meeting its annual performance goal for drinking water for fiscal years 1999 through 2002. In each of those years, EPA reported performance equaling the 91 percent GPRA annual performance goal. However, because EPA and the Office of Inspector General (OIG) reviews indicated that performance is less than what EPA reported, due to missing data on violations of drinking water standards, the Agency did not in fact meet its drinking water performance goals for these 4 years. Our assessment also mirrors statements made by the Agency in its performance reports and elsewhere.

EPA Consistently Reported Meeting Drinking Water Goals

EPA officials and reports consistently noted that national drinking water performance goals were being achieved. Annual performance reports, the 2003 Draft Report on the Environment, and statements by Agency officials indicated that national drinking water quality was high and EPA was progressing toward its goal of having 95 percent of the population drinking water that meets health-based standards. This was also repeated by the media. Figure 1 summarizes EPA's most recent annual performance report about drinking water quality, showing Agency claims that it just met its performance goal for each of the last 4 fiscal years:

Figure 1: EPA Reports Meeting Drinking Water Performance Goals

APG 8	Safe Drinking Water	Planned	Actual
FY 2002	**91% of the population served by community water systems will receive drinking water meeting all health-based standards, up from 83% in 1994.** Goal Met. ➡Corresponds with FY 2002 NEPPS Core Performance Measure (CPM).	91%	91%
FY 2001	*Same Goal, different targets.* *Goal Met.*		
	Performance Measures		
	- *Population served by community drinking water systems with no violations during the year of any federally enforceable health-based standards that were in place by 1994.*	91%	91%
	- *Population served by non-community, non-transient drinking water systems with no violations during the year of any federally enforceable health-based standards that were in place by 1994.*	96%	92%
FY 2000	*Same Goal.* *Goal Met*	91%	91%
FY 1999	*Same Goal, different targets.* *Goal Met*	91%	91%

FY 2002 Result: In FY 2002, 244 million people were served by community water systems meeting all health-based standards. This result is 91% of the 268 million people served by 53,437 community water systems in FY 2002.

Source: *EPA 2002 Annual Performance Report, page II-22.*

In addition to annual performance reports, EPA portrayed its success at improving drinking water quality through other reports and through statements by Agency officials. For example, the 2003 Draft Report on the Environment stated, "In 2002, (S)tates reported that 94 percent of the population served by community water systems were served by systems that met all health-based standards, up from 79 percent in 1993."[1] A July 20, 2003, statement from the Assistant Administrator for Research and Development repeated this conclusion that, in 2002, 94 percent of Americans were served by drinking water that met health-based standards. Using Agency reports, the media communicate such information to the public. For example, a June 23, 2003, New York Times editorial, "An Environmental Report Card," used similar language in the press release cited in Figure 2 to report that, "Fully 94 percent of Americans are served by drinking water systems that meet federal health standards, as opposed to 79 percent 10 years ago." Most recently, the new Administrator's Draft 500-Day Water Quality Plan continued using drinking water quality projections that remained unchanged from previous claims and press releases. The Plan stated, "(i)n 2002, 93.6 percent of the population received drinking water that met all health-based standards. By 2015, all people served by community water systems will receive drinking water that meets standards."

Figure 2: Statement on Drinking Water Quality

> **"Our drinking water is purer. In 2002, 94 percent of Americans were served by drinking water systems that meet our health-based standards – an increase of 15 percent in the last decade."**

Source: *EPA's June 23, 2003, Press Release*

EPA Reports Data Quality Problems While Reporting Performance Goals Met

EPA's recent Performance Reports contain statements about the quality of the drinking water data used to report under this performance measure. In the last three Annual Performance Reports the Agency's message was consistent: there are problems with the quality of data in the Safe Drinking Water Information System/Federal Version (SDWIS/FED). These statements were echoed in other reports, such as the 2003-2008 Strategic Plan. In that Plan, EPA reported, "the baseline statistic of national compliance with health-based drinking water standards likely is lower than reported." However, EPA continued to report that drinking water performance goals were being met. The following tables contain disclaimers related to data quality problems that EPA has included in reports:

---

[1] This statistic is based on the 2002 calendar year, while the 91 percent in EPA's 2002 annual performance is based on the Federal fiscal year.

Statements in EPA Annual Reports	
Report	**Data Quality Description**
1999 Annual Performance Report	There is no indication of data quality in the discussion of performance for this Annual Performance Goal.
2000 Annual Performance Report	"There are recurrent reports of discrepancies between national and state data bases . . . Given the particular need for confidence in the completeness and accuracy of data about drinking water quality, EPA designated SDWIS content as an Agency material weakness in 1999, under the Federal Managers' Financial Integrity Act."
2001 Annual Performance Report	A technical appendix noted under-reporting of monitoring and violations data to EPA and that "failures to monitor could mask treatment technique and MCL violations."
2002 Annual Performance Report	"The most significant data quality problem is under reporting to EPA of both monitoring and reporting violations and incomplete inventory characteristics . . . failures to monitor could mask treatment technique and MCL violations. Such underreporting of violations limits EPA's ability to precisely quantify the population served that are meeting health based standards."

Statements in Other EPA Reports	
Report	**Data Quality Description**
Fiscal Year 2001 Annual Plan	"SDWIS data quality has been problematic. It has been demonstrated that there are discrepancies between SDWIS data and state databases. In addition, utilities have pointed out specific data quality problems."
2003 Draft Report on the Environment *(Technical Document)*	"Underreporting and late reporting of CWS violations data by states to EPA affect the ability to accurately report the quality of our nation's drinking water... Based on this analysis, the agency estimated that states were not reporting 40 percent of all health-based violations to EPA."
2003-2008 Strategic Plan	"Routine data analyses of the Safe Drinking Water Information System (SDWIS) have revealed a degree of nonreporting of violations of health-based drinking water standards... As a result of these data quality problems, the baseline statistic of national compliance with health-based drinking water standards likely is lower than reported."
Data Reliability Analysis of the EPA SDWIS/FED and Plan *(Draft)*	"If the quality of the data measured and reported to SDWIS-FED, the source of the data for the [GPRA performance] calculation, are less than 100 percent as defined by this analysis, then the progress toward meeting the strategic goal may not be as great as reported."

The information in the previous Figures and Tables indicate that while the Agency consistently reported meeting its drinking water performance goals, the Agency also consistently acknowledged problems with drinking water data quality. Therefore, we believe the Agency has wrongly reported that it met its 91 percent performance goal for the years 1999 to 2002 – the actual number is lower by some unknown amount.

EPA and OIG Reviews Indicated GPRA Measure Less Than Reported

Since 2000, EPA has developed two drinking water data quality reports (with the second still in draft form). Both reports noted problems with under-reporting, in that States did not transmit all health-based violations and all monitoring and reporting violations into SDWIS/FED, which houses compliance information about drinking water systems. Our analysis confirmed these data quality problems of under-reporting of violations to health-based drinking water standards.

EPA Reports Indicate Drinking Water Data Quality Improved But Still "Low"

EPA has been conducting "data verifications" of drinking water data since 1991. In 2000, EPA issued the "Data Reliability Analysis of the EPA Safe Drinking Water Information System/Federal Version (SDWIS/FED)" report from data verifications conducted between 1996 and 1998. This report identified problems with the accuracy and completeness of SDWIS/FED data. The followup 2003 report, currently under internal draft review, includes information from data verifications from 1999 to 2001.

These two reports distinguish between"health-based" and "monitoring and reporting" (M/R) violations. As shown in Figure 3, the first report indicated that "data quality" for health-based violations at the community water systems (or "systems") whose records were examined during the 1996-1998 data verification time period was 40 percent.[2] That is, 40 out of every 100 health-based violations that should have been in SDWIS/FED were in SDWIS/FED, and 60 out of 100 were not. In the second time period (1999-2001), data quality improved to 65 percent. According to EPA's own assessment, this is still in the "low quality" range.

Figure 3: Despite Improvements "Data Quality" Remains a Problem

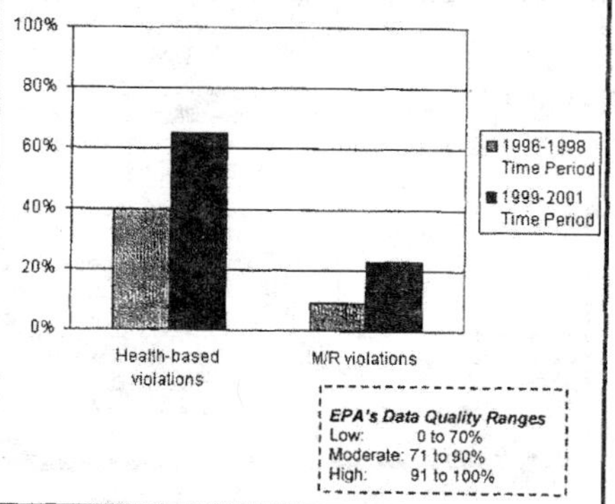

Data quality for the monitoring and reporting component of drinking water compliance also improved among the communities audited – from 9 percent to 23 percent between the two time periods. In other words, the data quality problem for M/R was at 91 percent and has reduced to 77 percent for an overall improvement of 14 percent. Because the enforcement portion of SDWIS/FED tracks only non-compliance with drinking water regulations, the

[2] EPA has defined SDWIS/FED data quality as the percent of data that should be in SDWIS/FED that are in the database with no discrepancies or errors.

monitoring and reporting violations indicate instances in which information about water quality was not reviewed by the State to make compliance determinations. The reported levels of health-based and monitoring and reporting data quality indicate that EPA performance reports reflected a best-case scenario and that in all likelihood performance was lower than reported.

EPA has recognized the data quality limitations of SDWIS/FED and its impact on the Agency's ability to manage the drinking water program, as well as to accurately report to Congress and the public. In the draft 2003 Data Reliability Analysis, EPA points out that, "Overall, the violations data that are reported to and accepted by SDWIS/FED are highly accurate. The weak link in data quality continues to be the large number of violations that are not reported to SDWIS/FED (as estimated by Completeness), with monitoring and reporting data being the least complete and of very low quality."

Recent improvements in drinking water data quality are attributed to EPA, State, and third party efforts to identify data quality deficiencies and implement activities to remedy those deficiencies. This has been a long-term effort, signified by the publication of reports outlining problems with drinking water data quality and activities to fix those problems. The 2003 Draft Data Reliability Report of SDWIS/FED noted that in 1999, a workgroup of EPA, State, and stakeholder representatives developed a data quality action plan. Some of the components of that action plan included setting data quality goals for SDWIS/FED, quantify and qualify the quality of SDWIS/FED data, and take interim steps to improve data quality.

The Draft Data Reliability Report also noted that EPA and States undertook and completed a number of activities to reach their data quality improvement goals. Some of those actions taken by EPA and the States included: (1) improved data entry processes, tools, and training for regions and States; (2) improved and simplified data retrieval and reporting tools; (3) improved data verification audit procedures; and (4) accelerated ongoing data quality improvement activities (development of SDWIS/STATE, and electronic reporting between utilities, labs, and States). A second data quality action plan is being developed and implemented. When we met with EPA officials to discuss our draft report, they told us that data quality continues to improve, as measured by the most recent data verifications.

EPA also noted in March 2003 in the Draft Strategic Plan for 2003-2008 that it would consider how to best classify water systems that experienced monitoring and reporting violations. Options included (1) classifying systems with monitoring and reporting violations as not being in compliance with health-based standards, and (2) excluding these systems from the GPRA calculation. By doing so, the Agency would remove from its performance reporting the systems that it cannot determine provided water that met all health-based standards. However, the final Strategic Plan issued in September 2003 stated, "(The) Agency is currently engaged in statistical analysis to more accurately quantify the impact of these data quality problems."

During our meeting with Agency officials to discuss the draft report and these two options, they explained that there is a potential for water systems with no health-based violations to be eliminated from the GPRA calculation because of one M/R violation. They believe that this would distort their reporting under GPRA, and they are studying the issue. While we understand that this potential exists, we also believe that EPA's current policy of treating systems with M/R violations as being in compliance with health-based standards can also distort the GPRA measure. The Agency's 2001 and 2002 annual reports (see page 4) note that there is a potential for M/R violations to mask violations to drinking water standards.

OIG's Review of EPA's Database Confirmed Agency's Assessment

Our review of the EPA data verification database confirmed the Agency's conclusion that States did not report all health-based violations into SDWIS/FED. The actual percentage of people drinking water that met health-based standards in this sample was likely to be lower, but we cannot use the database to determine a range for the nation as a whole. This is because the methods used to select the 761 water systems do not support estimates for the nation's approximately 54,000 community water systems.[3]

In the data verification database, we observed that the error rate was high for systems with health-based violations to the contaminants reviewed during data verifications. Of the 71 systems with violations for the drinking water standards that were reviewed during the data verifications, 17 had not been reported into SDWIS/FED.[4]

Overall, the direction of errors in the reporting of health-based violations caused a downward bias in the drinking water performance measure among the 761 systems. This suggests that by utilizing incomplete results from SDWIS to report performance under GPRA, EPA portrayed an incorrect picture of the percentage of people drinking water that met all health-based standards.

There is always potential for errors when collecting any type of information. In Appendix A, Figure 5 illustrates the flow of information from the water system, through the laboratories, and to the State and EPA. State data verifications identify errors related to data analysis and reporting of violations.

Conclusions

Congress, the public, and the media rely on transparent and accurate reports about drinking water quality. The significance of this measure is that each percent of the population served by community water systems receiving drinking water meeting all health-based standards represents more than 2.6 million people in the United States. As EPA persistently reports meeting its drinking water performance goal while acknowledging drinking water data quality problems, it in fact has not accurately reported its performance to the approximately 268 million people drinking

---

[3] A Community Water System serves at least 25 people or 15 service connections on a year-round basis.
[4] Our review also indicated that 2 of the 761 systems were identified to have a health-based violation in SDWIS/FED when no violation actually occurred.

water from community water systems. EPA's increasing candor about the limitations associated with basing performance measures on its compliance database (SDWIS) and the fact that it identified possible corrective actions in its 2003-2008 Draft Strategic Plan for addressing the problems of water system monitoring and reporting violations indicate the Agency's willingness to consider alternative approaches for how it reports performance.

We suggest that while EPA and States continue moving forward to correct data deficiencies, the Agency should also identify methods to better account for the impact of the "large number" (as described in the Draft 2003 Data Reliability Report) of violations that are not reported to SDWIS/FED. It should determine how best to account for community water systems with monitoring and reporting violations when reporting into GPRA and adjust the measure to reflect this. Options include those mentioned above that were described in the Draft Strategic Plan. In order to address broader concerns over this measure, given the inherent problems utilizing SDWIS for reporting on performance, we also suggest that in the future the Agency move toward employing an altogether different methodology for reporting performance for this Annual Performance Goal. One approach would be for EPA to base its future reporting on a stratified sample of the nation's 54,000 community water systems and audit those systems for compliance with health-based drinking water standards. This has the potential to provide a more accurate and transparent accounting of the nation's drinking water quality.

Agency Comments and OIG Response

In the Agency's February 2, 2004, response to our draft report, the Agency did not directly acknowledge our principal finding concerning the incorrect conclusions about drinking water performance contained in recent annual performance reports. In addition, while the Agency agreed to continue to improve how EPA communicates health risks associated with drinking water, no commitment to specific steps to correct the inconsistencies we had pointed out were agreed to. Appendix B contains Agency comments, and Appendix C contains some of our specific responses to those comments.

Based on the Agency's comments, we made several revisions and clarifications to our report. However, insofar as data quality within SDWIS is not the principal focus of this report, the comments did not address our principal concern: that for 4 years, EPA has reported to Congress and the public that it met an important annual performance goal when available evidence indicates it did not. After reviewing EPA's comments, we continue to believe that the Agency inappropriately claimed to have met performance goals for its drinking water program for the past 4 years. Steps to account for missing and inaccurate data when reporting performance under GPRA are being considered by EPA, but no decisions have been made. We reiterate our suggestion that EPA change how it reports under GPRA to compensate for known concerns over the reliability of this measure.

If you or your staff have any questions, please contact me at (202) 566-0827, or Dan Engelberg, Director of Water Issues, at (202) 566-0830.

Background, Scope, and Methodology

Background

In 1993, Congress enacted GPRA to shift Federal planning, management, and decision-making away from a traditional focus on resources and activities to a focus on results and outcomes. The Office of the Chief Financial Officer and program offices produce an annual report to Congress on the Agency's progress toward achieving annual strategic goals. In response to GPRA, EPA established major goals, including "Clean and Safe Water." One of the sub-objectives under this goal was "Water [that is] Safe to Drink," which is designed to reflect the quality of the drinking water supplied to the population.

In 1999, EPA established a measure of progress toward meeting this sub-objective: by the year 2005, 95 percent of the population served by community water systems would have water that is safe to drink, meaning that the water meets all applicable health-based standards. As is shown in Figure 4, over the past 9 years, EPA has reported an increasing percentage of the population drinking water that meets health-based standards. In its new strategic plan, EPA has retained the performance sub-objective of water that is safe to drink and the measure of percent of population, but has extended the timetable to accomplish the objective by 3 years, to 2008.

Figure 4: Population Reported by EPA Meeting Health Based Standards

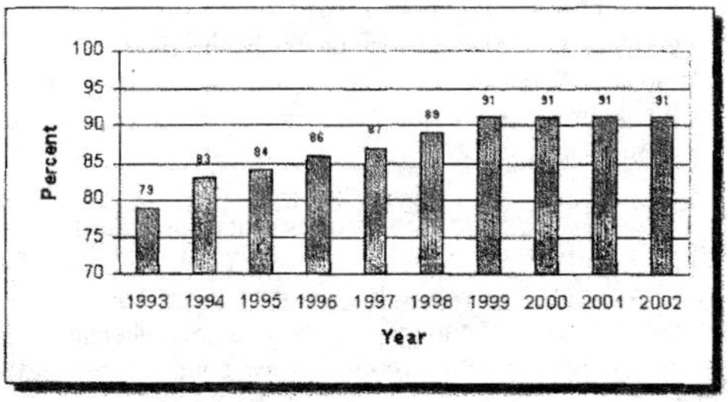

Source: *EPA 2002 Drinking Water Factoids. This data is based on the Federal fiscal year, which ends September 30. The Draft Report on the Environment reported 94 percent of the population served by community water systems drank water that met all health-based standards for the 2002 calendar year.*

The drinking water performance measure is based on compliance information contained in SDWIS/FED. The information utilized from this database is derived by sampling and analyses of drinking water, and assessments of treatment techniques from the approximately 54,000 community water systems that supply water to 268 million Americans. SDWIS/FED is designed to support many program management functions, including storing basic water system information, enforcement actions, and sampling results for unregulated contaminants.

Public water systems are responsible for monitoring their own systems and collecting and reporting sampling results to a primacy agent (typically State or Tribal drinking water programs). All States have primacy except for Wyoming. Primacy agents determine compliance with drinking water regulations and report violations to EPA (via SDWIS/FED). This process is illustrated in Figure 5. SDWIS/FED contains data when violations of drinking water standards and mandated treatment techniques are reported into it. To measure program performance, EPA aggregates the SDWIS/FED data into a national measure of overall compliance with health-based drinking water standards, which it reports as a percentage.

Figure 5: SDWIS Data Flow

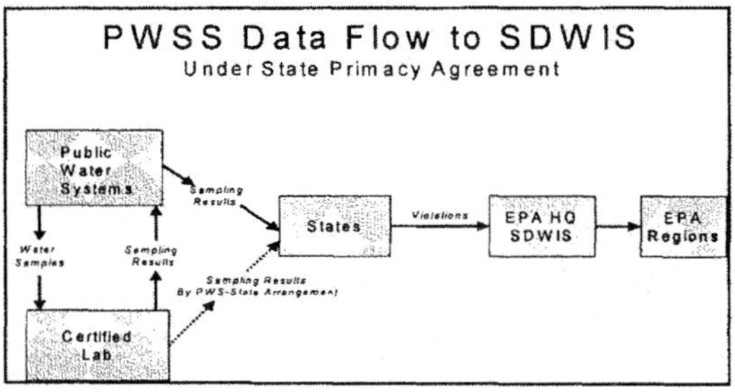

Source: *EPA Office of Water*

Scope and Methodology

We reviewed a database containing the results of a series of contractor-conducted audits (known as data verifications) of public water system data in State files.[5] We used this database to discuss two sources of errors that affect the precision of the data used in the drinking water quality measure: (1) errors in the process of reporting drinking water data from States to EPA, and (2) problems associated with under-reporting of drinking water information because of a reliance upon an exceptions-based database (meaning that only violations are recorded) for tracking drinking water violations.

We followed all but one of the applicable *Government Auditing Standards*, issued by the Comptroller General of the United States. We were unable to adhere to the standard that we assess management controls. Specifically, there were no written procedures to document the flow of information from data verifications into the data verification database. Based on our discussions with Office of Water staff, we believe that the data in this database were sufficient for the purposes of our review. We followed all other standards for performance audits or evaluations. All work was completed between August 2002 and September 2003.

[5] We focused on data from community water systems because compliance data from this group are used for the drinking water GPRA measure.

Appendix B

Agency Comments

MEMORANDUM

SUBJECT: Comments on the Draft Report *EPA Reports Meeting Drinking Water Goals Despite Persistent Data Quality Shortcomings*

FROM: Benjamin H. Grumbles
Acting Assistant Administrator

TO: Nikki L. Tinsley
Inspector General

Dear Ms. Tinsley:

Thank you for the opportunity to comment on your office's draft report, "EPA Reports Meeting Drinking Water Goals Despite Persistent Data Quality Shortcomings." I appreciate your general interest in this issue. The Office of Water recognizes the importance of high quality data and is committed to continue to make improvements in this area in the drinking water program as well as across our other activities

EPA's data verification audits and associated analyses, which are also the basis for your draft report, indicate that the data in SDWIS–FED are highly accurate with very few errors, but are still incomplete. EPA and the states have made significant progress in improving the quality of SDWIS-FED data since we first became aware of this issue. We acknowledge, however, that more work remains to be done. EPA's report, *Drinking Water Data Reliability Analysis and Action Plan (2003)*, to be released shortly, will highlight our continuing efforts and additional steps that we intend to take in partnership with states to further improve drinking water data quality.

While the Agency does use SDWIS-FED data to meet reporting requirements under the GPRA, we are aware of the data's shortcomings and have been diligent in flagging those to key audiences as well as to the general public. EPA in its GPRA reporting is using the data that is available to us through the national reporting system. We will continue to explore ways to communicate the range of issues associated with the nature and quality of SDWIS-FED data and the relationships to public health risk with your suggestions in mind. We will also continue to engage in discussions with states regarding potential new approaches for reporting drinking water data (e.g., electronic transfer of monitoring results from the laboratory to the federal database).

In a broader context, I would like to note that in the vast majority of instances where states make compliance determinations these determinations are correct. Most of the determinations correctly find that public water systems are meeting health-based standards and thus do not require an entry to be made in SDWIS-FED, which is a violations-only database. I mention this not to diminish the very real need to improve data quality, but as an important reminder that SDWIS data quality and drinking water quality are far from synonymous.

I have attached a more detailed set of comments on specific aspects of the draft to assist your office as you prepare a final report. Please call Cynthia Dougherty, Director of the Office of Ground Water and Drinking Water, at (202) 564-3750 if you would like further clarification on any of these issues.

Attachments

cc: Kwai-Cheung Chan
Mike Shapiro
Cynthia C. Dougherty
Elizabeth Corr
Dan Engelberg
Jill Ferguson
Linda Pettit-Waldner
Tim Roach
Michael Mason

Office of Ground Water and Drinking Water Specific Comments on Draft Inspector General Report,
EPA Reports Meeting Drinking Water Quality Goals Despite Persistent Data Quality Shortcomings (12/23/03)

1) In general, the IG draft report addresses two topics EPA is actively engaged in analyzing:

- The quality of data reported by states to EPA's database of record, the Safe Drinking Water Information System/Federal Version (SDWIS/FED) for public water system inventory, violations and enforcement actions, and
- The implications of that quality on a Government Performance and Results Act (GPRA) measure for populations receiving drinking water from community water systems that are in compliance with all health-based standards.

We are addressing both of these issues through our work internally and with states to continue to improve data quality and to identify new ways of communicating its significance to the public. We are also at this time near completion of our second comprehensive report on the quality of data in SDWIS. We expect to release this final report in the near future and will provide a copy to the IG at that time.

2) The IG draft report characterizes EPA as having "mistakenly" reported meeting its drinking water goal under the Government Performance and Results Act (GPRA). In actuality, however, we use the best available data reported to us by the states under their primacy agreements with the Agency. In using this data to describe results under the GPRA, we have tried to be clear that the results are based on the data as reported in SDWIS and that our audits indicate that there is incomplete reporting of this data. We have also made our first triennial report on SDWIS data quality, published in October 2000, available to the public on our website. We will soon be making our second report, noted above, on data quality available on the web.

3) EPA's data verifications indicate that the data the states report to the Agency are very accurate, although incomplete.

4) Several of the IG's comments, including an incorrect flow chart, indicate the need for improved understanding of data flow from public water systems to states to EPA. We have prepared a revised flow chart at the same level of detail (attached). We would suggest a meeting between OGWDW and the IG's staff before you finalize your draft report to ensure an accurate understanding of key details related to data flow which are beyond the depictions in the chart. Such a meeting would also serve as an opportunity for us to provide, and discuss where necessary, other detailed edits to the draft report for purposes of accuracy.

5) The draft report includes a chart that depicts data quality improvements, but the draft report includes only brief discussion on this point. Specifically, we believe that the draft

report should recognize that data quality has improved based on actions taken jointly by EPA and states since 1998. We are concerned that the absence of elaboration on this point undercuts the concerted efforts as well as the progress that EPA and states have made and are continuing to make.

6) To achieve a balanced examination, we suggest that the IG evaluate factors that could affect the results of data quality calculation in either direction, rather than emphasizing only factors that might appear to reduce the reported levels of the population receiving safe water all the time. We would like to take this opportunity to draw certain key points to the IG's attention for discussion in the draft report.

- In developing our own report, we have been examining the issue of over-reported violations. We looked at all the large water systems (over 50,000 population served) that had been identified as being in violation and found that one-third or more had corrected the violations and should not have been reported as being in violation in 2001. If considered, this could have the effect of increasing the GPRA percentage.

- Another noteworthy factor is that for many large water systems (which have the greatest effect on the GPRA number) a violation may not affect water quality throughout the entire water system, even though for GPRA accounting purposes the entire water system is credited with a violation. Quantitative consideration of this factor, while challenging to do, would likely contribute additional and substantial population to be counted as receiving drinking water which had no violation.

- A further set of potentially relevant factors is the impact of violation timing, frequency and duration on the significance and potential public health consequences of violations.

- Similarly, the varying nature of violations (e.g., a one-time violation for a chronic contaminant versus for an acute contaminant, or a violation significantly above the standard versus one that is close to the standard) may have differing public health implications.

We are very interested in finding ways to communicate these complexities succinctly to enhance public understanding of the GPRA measure and what it means. We would welcome the IG's comments on these factors.

7) The draft report characterizes the success of improved drinking water quality (distinct from data quality) as the Agency's result when in fact it is the result of a broad partnership that includes EPA, the states that are the primary implementers of the national drinking water program and public water systems that carry out the regulatory requirements.

8) An area that the IG touches on in the draft report and which resurfaces in the context of the conclusions is the potential impact of monitoring and reporting violations on GPRA

reporting. Monitoring and reporting violations, however, may have no link to whether a water system met health-based standards. For instance, where monitoring and reporting violations are scattered among numerous water systems that otherwise routinely demonstrate that they meet health-based standards, the likelihood of a significant impact on EPA's GPRA reporting is less than if monitoring and reporting violations occur repeatedly within the same water systems. We suggest that the IG reconsider whether to emphasize this complex issue as part of its conclusions in the absence of further analysis.

9) In the first paragraph on page 7, it is unclear to us whether the IG is discussing the utility of the data verification database or the SDWIS database and also unclear as to whether the estimates under discussion are the data quality estimates or the GPRA number.

In this and the following paragraph, there also appears to be a misunderstanding about the regulatory framework of the SDWA. Under the drinking water program's regulatory structure, health standards for multiple contaminants are addressed within single rulemakings. The draft report indicates that data verifications only evaluate eight drinking water standards. This is incorrect and affects the IG's conclusion about extrapolating data verification results for data quality purposes and possible "larger discrepancies" in the Agency's GPRA calculation. In fact, EPA's data verifications examined all 87 contaminants that were regulated under the SDWA in 2001 and EPA will continue to evaluate all regulated contaminants in future data verifications.

10) We believe there are a wider range of circumstances affecting shortcomings in the data that should be considered in an evaluation, including but not limited to:

- Relationship of waivers, variances and exemptions to violation data points;
- Conditions in state programs that result in non-reporting of violations;
- Methodologies that would improve evaluation and understanding of state program processes to determine compliance;
- Development of methodologies for estimating the proportion of populations in larger water systems actually affected by violations, rather than charging the entire water system with a violation that only affects a portion of its population.

11) Concerning the listing "Stakeholders Identified Other Potential Sources of Error" in Appendix B, the draft report does not present an evaluation of these potential sources of error nor indications of which may be more problematic. In general, the reference to "error" is inappropriate. Some of these appear to be undocumented and/or unevaluated opinions about potential sources of error and others are not potential sources of error, but rather process vulnerabilities. This particularly applies to the GPRA measure section.

Appendix C

Additional OIG Responses

Note to Agency Comment #2:

We agree that the Agency currently uses the best available data reported by the States and has moved in recent years to be more transparent in the presentation of problems with SDWIS data. For this reason, we have changed "mistakenly" to "incorrectly" in the report. We realize that this isn't an error of oversight on the part of EPA. However, EPA continues reporting that GPRA goals are met while warning about the implications of missing data. In our view, correctly reporting whether it has met a performance goal is at least as important as disclosing the existence of errors.

Note to Agency Comment #4:

The final report contains the data flow chart provided by the Office of Water.

Note to Agency Comment #5:

Protecting and improving the nation's drinking water quality and drinking water data quality is a collective effort with credit for success attributable to many parties. We did not intend to reduce the share of credit to any one group by briefly noting data quality improvement efforts undertaken in previous years. We did intend to highlight that drinking water data quality improvements are a result of activities to improve data management systems and processes. The issues regarding the transparency of GPRA reporting are the focus of our report, which is why we did not further elaborate on data quality improvement activities. However, we have changed our presentation in the final report to better reflect the shared responsibility and accomplishments of EPA and its partners.

Note to Agency Comment #6, first bullet:

We agree that the GPRA percentage is affected by water systems with incorrectly reported health-based violations. Our review of the data verification database factored in 2 such systems out of the 71 that experienced health-based violations. We were aware of other drinking water databases, but chose to review only the data verification database because of the semi-random sampling methodology used for selecting the 761 community water systems.

General Note to Agency Comment #6:

These factors all contribute to the complexity of presenting a picture of the nation's drinking water quality for the purposes of GPRA reporting. We suggest that while EPA works to address data issues such as those described here, the Agency also more clearly report that the absence of drinking water data in SDWIS/FED have an effect on the accuracy of the annual GPRA reports.

Note to Agency Comment #7:

See our response to EPA's Comment 5.

Note to Agency Comment #8:

EPA's 2001 and 2002 Annual Performance Reports noted that failures to monitor could mask violations of health-based standards (see page 4). We agree with this position. We feel that EPA is mistaken in asserting that the impact of GPRA reporting is less if M/R violations occur repeatedly in a single water system than if the same number of M/R violations are spread among several different systems.

Note to Agency Comment #9

We clarified language and corrected errors in the draft report.

Note to Agency Comment #10

We agree that these are important factors that affect the accuracy and validity of the GPRA measure. For the purposes of this report, our focus was on the implications of reporting success at meeting the drinking water GPRA goal while concurrently reporting problems with the completeness of drinking water data.

Note to Agency Comment #11

We removed references to potential sources of error based on Agency comments.

Appendix D

Report Distribution

Assistant Administrator, Office of Water (4101)
Director, Office of Ground Water and Drinking Water (4607)
Comptroller (2731A)
Agency Followup Official (the CFO) (2710A)
Agency Audit Followup Coordinator (2724A)
Associate Administrator for Congressional and Intergovernmental Relations (1301A)
Associate Administrator, Office of Public Affairs (1101A)
Inspector General (2410)

STATEMENT OF DAVID C. BELLINGER, PH.D., M.SC., FROM THE CHILDREN'S HOSPITAL BOSTON, HARVARD MEDICAL SCHOOL, BOSTON, MASSACHUSETTS

LEAD (SUPPLEMENTAL ARTICLE)

ABSTRACT

Children differ from adults in the relative importance of lead sources and pathways, lead metabolism, and the toxicities expressed. The central nervous system effects of lead on children seem not to be reversible. Periods of enhanced vulnerability within childhood have not consistently been identified. The period of greatest vulnerability might be endpoint specific, perhaps accounting for the failure to identify a coherent "behavioral signature" for lead toxicity. The bases for the substantial individual variability in vulnerability to lead are uncertain, although they might include genetic polymorphisms and contextual factors. The current Centers for Disease Control and Prevention screening guideline of 10 µg/dL is a risk management tool and should not be interpreted as a threshold for toxicity. No threshold has been identified, and some data are consistent with effects well below 10. Historically, most studies have concentrated on neurocognitive effects of lead, but higher exposures have recently been associated with morbidities such as antisocial behavior and delinquency. Studies of lead toxicity in experimental animal models are critical to the interpretation of nonexperimental human studies, particularly in addressing the likelihood that associations observed in the latter studies can be attributed to residual confounding. Animal models are also helpful in investigating the behavioral and neurobiological mechanisms of the functional deficits observed in lead-exposed humans. Studies of adults who have been exposed to lead are of limited use in understanding childhood lead toxicity because developmental and acquired lead exposure differ in terms of the maturity of the organs affected, the presumed mechanisms of toxicity, and the forms in which toxicities are expressed.

Key Words: lead toxicity • children • toxicology • epidemiology

Abbreviations: CDC, Centers for Disease Control and Prevention

Although children are viewed as the most vulnerable segment of the population with regard to lead poisoning, recognition of lead as an adult toxicant preceded by thousands of years the first description of childhood lead poisoning.[1] For millennia, exposure to lead was primarily via occupation, but the introduction of leaded paint for residential use in the 19th century brought large amounts of this metal within easy reach of children.[2] The later use of lead as a gasoline additive, begun in the 1920s and lasting into the 1990s in the United States,[3] contributed further to the contamination of environmental media with which children have intimate daily commerce, including air, dusts, and soils.

DIFFERENCES BETWEEN CHILDREN AND ADULTS IN LEAD SOURCES, METABOLISM, AND TOXICITIES

Children and adults differ somewhat in the relative importance of different lead exposure sources and pathways, in aspects of lead metabolism, and in the specific ways in which toxicities are expressed. To a greater extent than adults, young children normally explore their environment via hand-to-mouth activity, behaviors that are likely to increase the lead intake of a child who lives in an environment with hazards such as leaded paint in poor repair or elevated levels of lead in house dust or yard soils.[4,5] The average fractional gastrointestinal absorption of lead is much greater in infants and young children than in adults,[6] and absorption is increased in the presence of nutritional deficiencies that are more common in children than in adults (eg, iron, calcium).[7,8]

In both children and adults, lead toxicity can be expressed as derangements of function in many or an systems. Although lead causes central nervous system abnormalities in adults,[9,10,11] peripheral neuropathies tend to be more prominent. In the developing nervous system, in contrast, central effects are more prominent than peripheral effects.[12] Moreover, peripheral nervous system effects in adults tend to reverse after cessation of exposure,[13,14] whereas the central effects in children seem not to do so,[15,16,17,18] perhaps because lead perturbs the complex processes by which synaptic connections are selected and modified.[19] Even pharmacotherapy, at least succimer administered to young children who present with blood lead levels of 20 to 44 µg/dL, does not seem to reduce or reverse cognitive injury.[20] An important exception to these generalizations is that neurobehavioral deficits associated with modest elevations of prenatal lead levels, if ever present, seem largely to attenuate by the time children reach school age.[21]

CRITICAL WINDOWS OF VULNERABILITY AND INTERINDIVIDUAL DIFFERENCES IN SUSCEPTIBILITY

It is difficult to identify discrete windows of enhanced developmental vulnerability to lead exposure. The intraindividual stability of blood lead level over time is substantial, particularly in lead-rich environments such as the inner city or areas around lead smelters, where many of the major epidemiologic studies have been conducted.[22, 23] In addition, under many exposure scenarios, the half-life of lead in blood is greater in children than in adults.[5, 24] When blood lead levels do vary over time, age at exposure and magnitude of exposure are often highly confounded, with blood lead level peaking in the age range of 1 to 3 years.[25] This is probably because this period encompasses both the onset of independent ambulation and the time when a child's oral exploration of the environment is greatest. As a result, if a study were to find that blood lead level measured at age 2 is most predictive of some critical neurocognitive outcome at school age, then it would be difficult to ascertain whether this reflects a special vulnerability of the central nervous system at age 2 or that blood lead levels tend to be highest during this period. Some studies support the former hypothesis,[26] whereas others have found that school-age neurocognitive outcomes are most strongly related to recent or concurrent blood lead levels.[22, 27] The findings of yet other studies fail to provide evidence for the existence of any critical periods of vulnerability.[28]

Another reason that it is difficult to identify a single critical period of heightened vulnerability to lead toxicity is that there might be many such periods, depending on the particular endpoint of interest. Using primate models, Rice [29, 30] demonstrated that the timing of developmental lead exposure affected the nature and the severity of deficit on a variety of tasks (spatial discrimination reversal, nonspatial discrimination reversal, and a fixed-interval response operant task). In contrast, performance on a spatial delayed alternation task was not affected by age at exposure.[31] Morgan et al[32] observed different expressions of attentional dysfunction in rats depending on the timing of lead exposure. It seems eminently plausible that this fundamental principle of toxicology applies to children as well, although the evidence is meager. This is likely to be attributable, in no small measure, to the absence in most human epidemiologic studies of sufficiently detailed exposure data that capture, at least, features such as timing, duration, and dose. If the specific effects of lead do differ according to exposure scenario, then this lacunae in exposure data would account, at least in part, for the general lack of success in discerning a coherent "behavioral signature" of lead exposure in children.[33] This suggests, however, that we should not necessarily expect strict consistency across studies in the patterns of neurocognitive impairment associated with lead.[33, 34]

An individual's vulnerability to neurodevelopmental injury is also likely to vary according to host characteristics that are, at present, largely unknown. Individuals differ widely in the blood lead level at which signs of clinical intoxication appear, with some individuals seeming well at a blood lead level that in others results in encephalopathy or even death. Plots of "subclinical" blood lead level and endpoints such as covariateadjusted IQ reveal tremendous scatter of observations around the regression lines (eg,[35]), with low R^2 values associated with the regressions, suggesting that children are variable in their responses to lower levels of exposure, as well. An important implication is that children with the same blood lead level should not be considered to be at equivalent developmental risk.

The potential sources of individual variability in lead-associated neurodevelopmental risk are legion, although none has been confirmed with even a modest degree of certainty. One type of explanation focuses on toxicokinetic and toxicodynamic factors. It is assumed that blood lead level, the biomarker of internal dose that is most often used, is a valid index of the biologically effective dose at the brain, the critical target organ for neurotoxicity. The many intervening steps that link the internal dose and the response in the brain, however, provide many opportunities for interindividual differences in sensitivity to arise.[36] Certain genetic polymorphisms involved in lead metabolism are thought to affect individual vulnerability, including those for the vitamin D receptors[11] and for lead-binding red blood cell proteins such as amino levulinic acid dehydratase.[37, 38] Supportive evidence is sparse, however.[39] Gender differences have been reported in the immunotoxicity of gestational lead exposure in rats.[40] In humans, gender differences in neurotoxicity have been reported,[41, 42, 43, 44, 45] although in some studies, it is male individuals who seem to be more vulnerable, whereas in others it is female individuals. Co-exposure to other toxicants is another candidate explanation for individual differences in susceptibility, although greater attention has been paid to the potential of co-exposures to be confounders than to be effect modifiers. In a rodent model, the effect of lead on mortality, spatial learning, and the N-methyl-D-aspartate receptors differed depending on whether pups were exposed to lead alone or in combination with magnesium and zinc.[46] Finally, characteristics of a child's rearing environment might influence the toxicity of a given lead dose.[47] Lead seems to be similar to other biological risks, such as low birth weight, in that children from environments that offer fewer developmental resources and supports express deficits at a lower blood lead level than do children from more optimal environments[45, 48] and show less recovery after exposure.[43]

FUNCTIONAL FORM OF THE DOSE-EFFECT RELATIONSHIP: A THRESHOLD?

A threshold value below which lead has no apparent adverse developmental effect has not been identified. The 1991 Centers for Disease Control and Prevention (CDC) statement on childhood lead poisoning[49] set 10 μg/dL as the screening action guideline. Although this blood lead level was intended to serve as a risk guidance and management tool at the community level, it has been widely—and incorrectly—imbued with biological significance for the individual child. Indeed, it often seems to be interpreted as a threshold, such that a level <10 μg/dL is viewed as "safe" and a higher level as "toxic." The truth is unlikely to be so simple, however. No single number can be cited as a threshold, divorced from a context that specifies factors such as the endpoint of interest, the age at exposure and at assessment, the duration of blood lead elevation, and characteristics of the child's rearing environment. Although few data were available at the time on putative effects below 10 μg/dL, the 1991 CDC statement stated that adverse effects are likely to occur in this range (p. 9). This should not be surprising given that even after 2 decades of steady decline in population blood lead levels,[50] the mean still lies between 1[51] and 2[52] orders of magnitude greater than estimates of natural background levels in humans. It strains credulity to conclude that the threshold for neurotoxicity lies within the narrow and, in an evolutionary sense, still quite elevated range of present-day blood lead levels. Data reported since the 1991 CDC statement support this position. Among children in the Boston prospective study, for whom the mean blood lead level at age 2 years was 7 μg/dL (90th percentile, 13 μg/dL), a significant inverse association was found between blood lead level and both IQ and academic achievement at 10 years of age.[26] No point of inflection in this relationship was identified when nonparametric regression models were fitted, and the data were most consistent with a linear (ie, nonthreshold) model extending to the lowest blood lead levels represented in the cohort (<1 μg/dL).[53] In the Third National Health and Nutrition Examination Survey sample, among 4,853 6- to 16-year-old children, current blood lead was inversely associated with 4 measures of cognitive function, even when the sample was restricted to children with blood lead levels <5 μg/dL.[54] Blood lead histories of the children were not available, however, so it is possible that their levels had been much higher at younger ages and that it was those levels that were responsible for the later performance deficits of the children with higher concurrent blood lead levels. Moreover, measures of key potential confounders such as parent IQ and home environment were not available, although strong confounding by these factors within such a narrow range of blood lead levels is unlikely. These limitations were addressed in the study of Canfield et al.[55] In the subgroup of 101 children whose blood lead levels were <10 μg/dL at 6, 12, 18, 24, 36, 48, and 60 months of age, significant covariate-adjusted associations were observed between blood lead level and IQ at ages 3 and 5. Chiodo et al[56] also reported significant inverse associations between neuropsychological function and blood lead levels <10 μg/dL. In the Canfield et al[55] study, moreover, the slope of the association was greater in the subgroup of children whose peak blood lead was <10 μg/dL than it was in the complete study sample that included children whose teak blood lead levels exceeded 10 μg/dL. Reanalyses of the Boston prospective study[57] suggested the same pattern. Collectively, these new studies provide compelling evidence that 10 μg/dL should not be viewed as a threshold. The precise shape of the dose-effect relationship in the lower portion of the exposure range remains uncertain, however. Although the data are consistent with the slope being steeper below 10 μg/dL than above 10 μg/dL, a convincing mechanism has not been proposed.

NONCOGNITIVE EFFECTS OF LEAD: BROADENING OUTCOME ASSESSMENTS BEYOND IQ

The neurocognitive effects of pediatric lead toxicity have garnered the greatest attention from both researchers and regulators, perhaps for reasons of ease of measurement by the former and ease of interpretation by the latter. Indeed, enough studies provide data on endpoints such as IQ to make meta-analyses feasible,[21, 58, 59, 60] with all such efforts reaching similar conclusions, viz, that an IQ decline of 1 to 5 points is associated with a 10-µg/dL increase in blood lead (eg, from 10 to 20 µg/dL). Many studies have identified distractibility, poor organizational skills, and hyperactivity as possible reasons for the reduced global cognitive function of more highly exposed children.[61, 62, 63, 64, 65, 66]

Recently, the range of outcomes examined in relation to childhood lead exposure has been expanded, building on older reports of serious behavioral pathologies in case series of children with subencephalopathic lead poisoning. In 1 of these early reports, Byers and Lord[67] noted that poor school progress among children who were previously treated for lead poisoning was attributable not only to their cognitive deficits but also to their aggression and explosive tempers. Within the past decade, several studies have suggested that even "subclinical" lead exposure is a risk factor for antisocial, delinquent behaviors. For example, a history of childhood lead poisoning was the strongest predictor of adult criminality among male individuals in the Philadelphia subsample of the Collaborative Perinatal Project.[68] Needleman et al[69] found that male adolescents with increased bone lead levels self-reported more delinquent acts and were rated by both their parents and teachers as having scores that exceeded clinical cutoffs on the Attention, Aggression, and Delinquent Behavior scales of the Child Behavior Checklist. Furthermore, between ages 7 and 12, the behaviors of boys with higher bone lead levels deteriorated more than did the behaviors of boys with lower bone lead levels. Among adolescents in the Cincinnati Lead Study, the frequencies of self-reported delinquent and antisocial behaviors were significantly associated with both prenatal and early postnatal blood lead levels.[70] In a case-control study, adjudicated delinquents had significantly higher bone lead levels than did community control youths and were 4 times more likely to have a bone lead level at the 80th percentile of the distribution (approximately the detection limit).[71] Finally, in a set of historical analyses, Nevin[72] reported striking, provocative concordances between temporal trends in the amount of lead used commercially and in violent crime and unwed pregnancies. Although such ecologic analyses provide a weak basis for causal inference, they do suggest hypotheses that should be evaluated in settings in which information is available on exposure, outcome, and potential confounders at the individual rather than the community level. Much work remains to be done to clarify the potential contributions of lead, as well as other environmental pollutants, to child psychiatric morbidity.[73, 74]

UTILITY OF ANIMAL STUDIES

Because studies of children's environmental lead exposure must necessarily be observational rather than experimental (apart from randomized clinical trials comparing alternative treatment modalities), much of the controversy surrounding their interpretation has focused on the possibility that residual confounding, rather than lead toxicity itself, explains the associations between higher body burdens and reduced function. Such discussions are difficult to conclude to everyone's satisfaction because there is no logical conclusion to the line of argument that posits a succession of unmeasured factors that might be responsible for creating such spurious associations. Moreover, errors in model specification can result in bias toward the null hypothesis in the estimate of lead's neurotoxicity, if statistical adjustments are made for factors that are in the causal pathway between lead and poor outcome. For this reason, animal behavioral models of lead toxicity, in which the possibility of confounding (in either direction) is reduced by random assignment to exposure groups and by active control of relevant (known) genetic and environmental factors, are crucial elements of the total data base to which regulators can and should appeal in setting exposure standards. The inference that low-level lead exposure causes human behavioral morbidity becomes more plausible when behavioral changes are also observed after lead is administered to animals under experimental conditions. Indeed, the striking similarities between the general pattern of behavioral abnormalities in lead-exposed animals and in "free range" lead-exposed children provides support, albeit indirect, for the inference that the relationships observed in humans are causal.[33, 75]

The converse is true, as well, in that sometimes the results of animal studies suggest that an association observed in humans might not reflect a causal influence. For instance, analyses of the Second National Health and Nutrition Examination Survey data set suggested that very modest elevations in current blood lead level, well within the range of community exposures, were associated with increased hearing threshold in children.[76, 77] Although some studies of animal models provide limited evidence of a modest effect at high blood lead levels,[78] the results of a recent study in 31 rhesus monkeys with blood lead levels of 35 to 40 µg/dL for the first 2 postnatal years cast doubt on the validity of the conclusion that low-level lead exposure causes hearing deficits in children. In this study, no lead-associated effects were detected on any level of auditory processing using tympanometry (middle ear function), otoacoustic emissions (cochlear function), or auditory brainstem-evoked potentials (auditory nerve, brainstem pathways).[79] This might explain why recent studies of 2 cohorts of Ecuadorian children with substantially elevated blood lead levels (means of 40 and 52 µg/dL) failed to find a significant association between blood lead level and hearing threshold.[80, 81]

One reason that animal models of lead toxicity are so useful in understanding childhood lead toxicity is the deep level of analysis that they allow in the effort to identify the behavioral mechanism(s) of functional deficit. The assessments included in most human epidemiologic studies tend to be global or apical tests of cognition and achievement rather than experimental, laboratory tests. One reason for this is that exposure-associated decrements on such tests are more highly valued by risk analysts and regulators as bases for exposure standards. Although poor performance on global tests is often strongly predictive of adaptive difficulties in school or the workplace,[82] the mere fact of poor performance provides relatively little insight into the reasons for it, i.e., about the underlying "behavioral lesion." For example, in many studies, higher lead levels are associated with reduced scores on a design-copying task. A child might perform poorly on such a task for many reasons, however, including poor visual-perceptual skills, poor fine motor control, metacognitive or organizational deficits, poor impulse control, anxiety, or a depressed mood. In a diagnostic clinical evaluation, the relative merits of these various hypotheses can be explored using a test battery tailored to the child's presentation and modified on the basis of the tester's observations as the evaluation proceeds. In a field epidemiologic study, an investigator might have 1-time access to a child for perhaps 3 hours, needing to administer a fixed battery to all children to ensure comparability of the data and the circumstances of its collection. Under such constraints, dissection of a behavioral deficit by means of a detailed process analysis is not feasible, and an exposure-associated decrement in performance on apical tests tends to be "explained," inappropriately, in terms of a deficit in a complex construct such as "attention" or "memory." Limited efforts to deconstruct such global constructs have been conducted in lead-exposed children. Application of an assessment battery based on a neuropsychological model of attention[83] revealed that elevated dentine lead levels were associated with deficits in 2 of the 4 elements of attention in this model: the ability to select a focus and carry out operations on it, and the ability to shift focus in a flexible and adaptive manner.[39] The continuous accessibility of experimental animals makes them an ideal resource for explicating the bases for the global deficits observed in human subjects. They are literally a captive audience from whom cooperation and consent for repeated testing is not required and who do not need to miss work or school to participate. Fine-grained process analyses of the behaviors of lead-exposed primates, for instance, are consistent across, laboratories and with the limited human data available,[39] in identifying several specific aspects of the global construct "attention" that are sensitive to lead: a tendency to be distracted by irrelevant stimuli, to respond in a perseverative manner, an inability to inhibit inappropriate responses, difficulty changing response strategies when reinforcement contingencies shift, and difficulty abstracting general rules (i.e., "learning how to learn").[33] No substantial obstacles stand in the way of efforts to administer to children batteries that would allow similarly fine-grained dissection of behavior, and investigators are currently working toward this goal.[84, 85]

Animal models are also better suited than human studies to the task of testing limits to evaluate the effects of lead on the ability to weather "periods of behavioral transition,"[86] as well as to identify factors that exacerbate or reduce lead toxicity (ie, effect modification). In the laboratory, one can "program" life histories to explore the impact of different factors on the severity and nature of lead-associated deficits and to see whether the point at which and the way in which an animal's behavior breaks down over time or under stress are affected by previous lead exposure. Animal models can also be helpful in probing the nature and bases of individual differences in sensitivity to lead toxicity.[29]

Animal models are of relatively little help, however, in evaluating lead's effects on the ability to manipulate symbolic or abstract systems, such as reading or mathematics, that have no compelling nonhuman analogues. In addition, studies of lead's effects on behavioral systems that tend to be species specific (eg, communication, affect, reproduction, social behaviors) are less relevant to understanding childhood lead toxicity than are nonhuman models of systems with strong cross-species parallels in the morphology of behavior, such as problem solving and learning.[87]

UTILITY OF ADULT STUDIES

Studies conducted on adults are likely to be of limited relevance in understanding lead toxicity in children, particularly with regard to nervous system effects. This organ continues to undergo substantial changes well into the second decade of postnatal life, involving the establishment of hemispheric dominance, the completion of myelination (particularly in the frontal lobes), synaptic pruning, and synaptic reorganization. As a result, the impact of an acquired brain lesion in an adult can differ dramatically from the impact of a similar lesion incurred during development.[88] Even in the absence of an insult, the brain-behavior relationships underlying complex cognitive processes can differ substantially between adults and children. For example, lesions that spare language in proficient speakers can impair language acquisition, suggesting that the neural substrate for language processing is not as highly localized in children as in adults.[89] A functional magnetic resonance imaging study of performance on a verbal fluency task identified the expected regions of activation in both children and adults (left inferior frontal cortex, left middle frontal gyrus) but more widespread cortical activation among children than in adults, particularly in the right hemisphere (right inferior frontal gyrus).[90] This seems not to be attributable simply to age-related differences in competence but to age-related differences in functional neuroanatomy. In another functional magnetic resonance imaging study, comparing visual lexical processing in adults and 7- to 10-year-olds, different patterns of activation were found in children and adults, even when the 2 age groups were matched in terms of accuracy on the task.[91] This suggests that, to some extent, the specific regions of the brain enlisted to solve a particular problem change with age. Thus, it seems that the adult and the developing child differ in so many critical respects that few lessons about pediatric lead neurotoxicity can be gleaned from studying adult lead neurotoxicity.

CONCLUSION

Conceptually, excessive lead exposure in children poses a relatively simple problem. We know where the most important hazards are in the environments of young children, the major pathways of exposure, the range of effects (to a level of detail far greater than for any other environmental pollutant), and at least the general features of the dose-effect relationships for the most intensively studied endpoints. Studies continue to describe apparent effects that were previously unknown, as well as show that known effects can be detected at lower and lower levels of exposure. Fortunately, even as these advances in knowledge were being achieved, children's exposures to lead were in dramatic decline, with the mean blood lead level now barely >2 μg/dL.[92] Although much is known about the effects of lead on brain chemistry and physiology, we nevertheless lack a unifying model of the mechanisms of lead neurotoxicity. It is not obvious, however, that additional evidence on the health effects of lead or the mechanisms of its protean toxicities is needed to motivate public health interventions to reduce children's lead exposure. In terms of housing and community interventions, apart from the obvious immediate and long-term benefits of complete residential lead abatement, if conducted properly, as a way to reduce childhood exposures, we know relatively little about other environmental, nutritional, or social interventions that are effective (including cost-effective). Given the apparent absence of commitment at a societal level to eradicate this entirely preventable childhood disease even in the face of economic analysis that demonstrates it to be cost-effective,[93] it seems that the answer to the question posed 10 years ago, "Lead toxicity in the 21st century: will we still be treating it?"[94] is, sadly, "Yes."

FOOTNOTES

Received for publication Oct. 7, 2003; accepted Oct. 20, 2003.

Reprint requests to (D.C.B.) Children's Hospital, Farley Basement Box 127, 300 Longwood Ave, Boston MA 02115. E-mail: *david.bellinger@childrens.harvard.edu.*

REFERENCES

1. Gibson JL, Love W, Hardine D, Bancroft P, Turner AJ. Note on lead poisoning as observed among children in Brisbane. In: Huxtable LR, ed. *Transactions of the Third Intercolonial Medical Congress of Australasia.* Sydney, Australia: Charles Potter; 1982:76–83.

2. Markowitz G, Rosner D. "Cater to the children": The role of the lead industry in a public health tragedy, 1900–1955. *Am J Public Health.* 2000; 90:36–46 *[Abstract/Free Full Text].*

3. Rosner D, Markowitz G. A "gift of God"? The public health controversy over leaded gasoline during the 1920s. *Am J Public Health.* 1985; 75:344–352 *[ISI][Medline].*

4. Calabrese EJ, Stanek EJ, James RC, Roberts SM. Soil ingestion: a concern for acute toxicity in children. *Environ Health Perspect.* 1997; 105:1354–1358 *[ISI][Medline].*

5. Manton WI, Angle CR, Stanek KL, Reese YR, Kuehnemann TJ. Acquisition and retention of lead by young children. *Environ Res.* 2000; 82:60–80 *[ISI][Medline].*

6. Ziegler EE, Edwards BB, Jensen RL, Mahaffey KR, Fomon SJ. Absorption and retention of lead by infants. *Pediatr Res.* 1978; 12:29–34 *[ISI][Medline].*

7. Wright RO, Shannon MW, Wright RJ, Hu H. Association between iron deficiency and low-level lead poisoning in an urban primary care clinic. *Am J Public Health.* 1999; 89:1049–1053 *[Abstract].*

8. Bradman A, Eskenazi B, Sutton P, Athanasoulis M, Goldman LR. Iron deficiency associated with higher blood lead in children living in contaminated environments. *Environ Health Perspect.* 2001; 109:1079–1084 *[ISI][Medline].*

9. Stollery BT. Reaction time changes in workers exposed to lead. *Neurotoxicol Teratol.* 1996; 18:477–483 *[ISI][Medline].*

10. Muldoon SB, Cauley JA, Kuller LH, et al. Effects of blood lead levels on cognitive function of older women. *Neuroepidemiology.* 1996; 15:62–72 *[ISI][Medline].*

11. Schwartz BS, Stewart WF, Bolla KI, et al. Past adult lead exposure is associated with longitudinal decline in cognitive function. *Neurology.* 2000; 55:1144–1150 *[Abstract/Free Full Text].*

12. National Research Council. *Measuring Lead Exposure in Infants, Children, and Other Sensitive Populations.* Washington, DC: National Academy Press; 1993.

13. Baker EL, White RF, Pothier LJ, et al. Occupational lead neurotoxicity: improvement in behavioral effects after reduction of exposure. *Br J Ind Med.* 1985; 42:507–516 *[ISI][Medline].*

14. Yokoyama K, Araki S, Aono H. Reversibility of psychological performance in subclinical lead absorption. *Neurotoxicology.* 1988; 9:405–410 *[ISI][Medline].*

15. Needleman HL, Schell A, Bellinger D, Leviton A, Allred E. The long-term effects of exposure to low doses of lead in childhood: an 11-year follow-up report. *N Engl J Med.* 1990; 322:83–88 *[Abstract].*

16. White RF, Diamond R, Proctor S, Morey C, Hu H. Residual cognitive deficits 50 years after lead poisoning during childhood. *Br J Ind Med.* 1993; 50:613–622 *[ISI][Medline].*

17. Tong S, Baghurst PA, Sawyer MG, Bums J, McMichael AJ. Declining blood lead levels and changes in cognitive functioning during childhood: the Port Pirie Cohort Study. *JAMA.* 1998; 280:1915–1919 *[Abstract/Free Full Text].*

18. Stokes L, Letz R, Gerr F, et al. Neurotoxicity in young adults 20 years after childhood exposure to lead: The Bunker Hill experience. Occup *Environ Med.* 1998; 55:507–516 *[Abstract].*

19. Johnston MV, Goldstein GW. Selective vulnerability of the developing brain to lead. *Curr Opin Neurol.* 1998; 11:689–693 *[ISI][Medline].*

20. Rogan WJ, Dietrich KN, Ware JH, et al. The effect of chelation therapy with succimer on neuropsychological development in children exposed to lead. *N Engl J Med.* 2001; 344:1421–1426 [Abstract/Free Full Text].

21. Pocock S, Smith M, Baghurst PA. Environmental lead and children's intelligence: a systematic review of the epidemiological evidence. *BMJ.* 1994; 309:1189–1197 *[Abstract/Free Full Text].*

22. Dietrich KN, Berger O, Succop P. Lead exposure and the motor developmental status of urban 6-year-old children in the Cincinnati Prospective Study. *Pediatrics.* 1993; 91:301–307 *[Abstract].*

23. McMichael AJ, Baghurst PA, Robertson EF, Vimpani GV, Wigg NR. The Port Pirie cohort study: blood lead concentrations in early childhood. *Med J Aust.* 1985; 143:499–503 *[ISI][Medline].*

24. Succop PA, O'Flaherty EJ, Bomschein RL, et al. A kinetic model for estimating changes in the concentration of lead in the blood of young children. In: Lindberg SE, Hutchinson TC, eds. *International Conference on Heavy Metals in the Environment.* Vol 2. Edinburgh: CEP Consultants, Ltd; 1987:289–291.

25. Brody DJ, Pirkle JL, Kramer RA, et al. Blood lead levels in the U.S. population. Phase 1 of the Third National Health and Nutrition Examination Survey (NHANES III, 1988 to 1991). *JAMA.* 1994; 272:277–283 *[Abstract].*

26. Bellinger DC, Stiles KM, Needleman HL. Low-level lead exposure, intelligence and academic achievement: a long-term follow-up study. *Pediatrics.* 1992; 90:855–861 *[Abstract].*

27. Tong S, Baghurst P, McMichael A, Sawyer M, Mudge J. Lifetime exposure to environmental lead and children's intelligence at 11–13 years: the Port Pirie cohort study. *BMJ.* 1996; 312:1569–1575 *[Abstract/Free Full Text].*

28. Wasserman GA, Liu X, Popovac D, et al. The Yugoslavia Prospective Lead Study: contributions of prenatal and postnatal lead exposure to early intelligence. *Neurotoxicol Teratol.* 2000; 22:811–818 *[ISI][Medline].*

29. Rice DC. Lead-induced behavioral impairment on a spatial discrimination reversal task in monkeys exposed during different periods of development. *Toxicol Appl Phannacol.* 1990; 106:327–333 *[ISI][Medline].*

30. Rice D. Lead exposure during different developmental periods produces different effects on FI performance in monkeys tested as juveniles and adults. *Neurotoxicology.* 1992; 13:757–770 *[ISI][Medline].*

31. Rice DC, Gilbert SG. Sensitive periods for lead-induced behavioral impairment (nonspatial discrimination reversal) in monkeys. *Toxicol Appl Pharmacol.* 1990; 102:101–109 *[ISI][Medline].*

32. Morgan RE, Garavan H, Smith EG, Driscoll LL, Levitsky DA, Strupp BJ. Early lead exposure produces lasting changes in sustained attention, response initiation, and reactivity to errors. *Neurotoxicol Teratol.* 2001; 23:519–531 *[ISI][Medline].*

33. Bellinger D. Interpreting the literature on lead and child development: the neglected role of the "experimental system." *Neurotoxicol Teratol.* 1995; 17:201–212 *[ISI][Medline].*

34. Rice D. Behavioral effects of lead: commonalities between experimental and epidemiologic data. *Environ Health Perspect.* 1996; 104(suppl 2):337–351 *[Medline].*

35. Wasserman GA, Liu X, Lolacono NJ, et al. Lead exposure and intelligence in 7-year-old children: the Yugoslavia Prospective Study. *Environ Health Perspect.* 1997; 105:956–962 *[ISI][Medline].*

36. Links JM, Schwartz BS, Simon D, Bandeen-Roche K, Stewart WF. Characterization of toxicokinetics and toxicodynamics with linear systems theory: application to lead-associated cognitive decline. *Environ Health Perspect.* 2001; 109:361–368 *[ISI][Medline].*

37. Smith C, Wang X, Hu H, Kelsey K. A polymorphism in the delta-aminolevulinic acid dehydratase gene may modify the pharmacokinetics and toxicity of lead. *Environ Health Perspect.* 1995; 103:248–253 *[ISI][Medline].*

38. Schwartz BS, Lee BK, Lee GS, et al. Associations of blood lead, dimercaptosuccinic acid-chelatable lead, and tibia lead with polymorphisms in the vitamin D receptor and delta-aminolevulinic acid dehydratase gene. *Environ Health Perspect.* 2000; 108:949–954.

39. Bellinger D, Hu H, Titlebaum L, Needleman HL. Attentional correlates of dentin and bone lead levels in adolescents. *Arch Environ Health.* 1994; 49:98–105 *[ISI][Medline].*

40. Bunn TL, Parsons PJ, Kao E, Dietert RR. Exposure to lead during critical windows of embryonic development: differential immunotoxic outcome based on stage of exposure and gender. *Toxicol Sci.* 2001; 64:57–66 *[Abstract/Free Full Text].*

41. Dietrich KN, Krafft KM, Bornschein RL, et al. Low-level fetal lead exposure effect on neurobehavioral development in early infancy. *Pediatrics.* 1987; 80:721–730 *[Abstract].*

42. Pocock S, Ashby D, Smith MA. Lead exposure and children's intellectual performance. *Int J Epidemiol.* 1987; 16:57–76 *[Abstract].*

43. Bellinger D, Leviton A, Sloman J. Antecedents and correlates of improved cognitive performance in children exposed in utero to low levels of lead. *Environ Health Perspect.* 1990; 89:5–11 *[ISI][Medline].*

44. McMichael AJ, Baghurst P, Vimpani G, Robertson E, Wiig N, Tong S-L. Sociodemographic factors modifying the effect of environmental lead on neuropsychological development in early childhood. Neurotoxicol Teratol. 1992; 14:321–327 *[ISI][Medline].*

45. Tong S, McMichael AJ, Baghurst PA. Interactions between environmental lead exposure and sociodemographic factors on cognitive development. *Arch Environ Health.* 2000; 55:330–335 *[ISI][Medline].*

46. Newman HM, Yang RSH, Magnusson KR. Effects of developmental exposure to lead, magnesium and zinc mixtures on spatial learning and expression of NMDA receptor subunit mRNA in Fischer 344 rats. *Toxicol Lett.* 2002; 126:107–119 *[ISI][Medline].*

47. Bellinger D. Effect modification in epidemiological studies of low-level neurotoxicant exposures and adverse health outcomes. *Neurotoxicol Teratol.* 2000; 22:133–140 *[ISI][Medline].*

48. Bellinger DC, Leviton A, Waternaux C, Needleman HL, Rabinowitz M. Low-level lead exposure, social class, and infant development. *Neurotoxicol Teratol.* 1988; 10:497–503 *[ISI][Medline].*

49. U.S. Centers for Disease Control. *Preventing Lead Poisoning in Young Children*. Atlanta, GA: U.S. CDC; 1991.

50. Pirkle JL, Brody DJ, Gunter EW, et al. The decline in blood lead levels in the United States. The National Health and Nutrition Examination Survey (NHANES). *JAMA*. 1994; 272:284–291 *[Abstract]*.

51. Mushak P. New directions in the toxicokinetics of human lead exposure. *Neurotoxicology*. 1993; 14:29–42.

52. Flegal AR, Smith DR. Lead levels in preindustrial humans. *N Engl J Med*. 1992; 326:1293–1294 *[ISI][Medline]*.

53. Schwartz J. Low-level lead exposure and children's IQ: a meta-analysis and search for a threshold. *Environ Res*. 1994; 65:42–55 *[ISI][Medline]*.

54. Lanphear BP, Dietrich K, Auinger P, Cox C. Cognitive deficits associated with blood lead concentrations <10 µg/dL in U.S. children and adolescents. *Public Health Rep*. 2000; 115:521–529 *[ISI][Medline]*.

55. Canfield RC, Henderson CR, Cory-Slechta DA, Cox C, Jusko TA, Lanphear BP. Intellectual impairment in children with blood lead concentrations below 10 µg per deciliter. *N Engl J Med*. 2003; 348:1517–1526 *[Abstract/Free Full Text]*.

56. Chiodo LM., Jacobson SW, Jacobson JL. Neurodevelopmental effects of postnatal lead exposure at very low levels. *Neurotoxicol Teratol*. 2004, in press.

57. Bellinger DC, Needleman HL. Intellectual impairment and blood lead levels. *N Engl J Med*. 2003; 349:500 *[Free Full Text]*.

58. Needleman HL, Gatsonis C. Low-level lead exposure and the IQ of children. *JAMA*. 1990; 263:673–678 *[Abstract]*.

59. Schwartz J. Beyond LOEL's, p values, and vote counting: methods for looking at the shapes and strengths of associations. *Neurotoxicology*. 1993; 14:237–246 *[ISI][Medline]*.

60. World Health Organization/International Programme on Chemical Safety. *Environmental Health Criteria 165. Inorganic Lead*. Geneva, Switzerland: World Health Organization; 1995.

61. Needleman HL, Gunnoe C, Leviton A, et al. Deficits in psychologic and classroom performance of children with elevated dentine lead levels. *N Engl J Med*. 1979; 300:689–695 *[Abstract]*.

62. Yule W, Urbanowicz M-A, Lansdown R, Millar I. Teachers' ratings of children's behaviour in relation to blood lead levels. *Br J Dev Psychol*. 1984; 2:295–305 *[ISI]*.

63. Thomson GO, Raab GM, Hepburn WS, Hunter R, Fulton M, Laxen DP. Bloodlead levels and children's behaviour—results from the Edinburgh lead study. *J Child Psychol Psychiatry*. 1989; 30:515–528 *[ISI][Medline]*.

64. Bellinger D, Leviton A, Allred E, Rabinowitz M. Pre- and postnatal lead exposure and behavior problems in school-aged children. *Environ Res*. 1994; 66:12–30 *[ISI][Medline]*.

65. Wasserman G, Staghezza-Jaramillo B, Shrout P, Popovac D, Graziano J. The effect of lead exposure on behavior problems in preschool children. *Am J Public Health*. 1998; 88:481–486 *[Abstract]*.

66. Burns J, Baghurst P, Sawyer M, McMichael A, Tong S. Lifetime low-level exposure to environmental lead and children's emotional and behavioral development at ages 11–13 years: the Port Pirie Cohort Study. *Am J Epidemiol*. 1999; 149:740–749 *[Abstract]*.

67. Byers RK, Lord EE. Late effects of lead poisoning on mental development. *Am J Dis Child*. 1943; 66:471–494.

68. Denno DW. *Biology and Violence*. New York, NY: Cambridge University Press; 1990.

69. Needleman HL, Riess JA, Tobin MJ, Biesecker GE, Greenhouse JB. Bone lead levels and delinquent behavior. *JAMA*. 1996; 275:363–369 *[Abstract]*.

70. Dietrich KN, Ris MD, Succop PA, Berger OG, Bornschein RL. Early exposure to lead and juvenile delinquency. *Neurotoxicol Teratol*. 2001; 23:511–518 *[ISI][Medline]*.

71. Needleman HL, McFarland C, Ness RB, Fineberg SE, Tobin MJ. Bone lead levels in adjudicated delinquents: a case-control study. *Neurotoxicol Teratol*. 2002; 24:711–717 *[ISI][Medline]*.

72. Nevin R. How lead exposure relates to temporal changes in IQ, violent crime, and unwed pregnancy. *Environ Res*. 2000; 83:1–22 *[ISI][Medline]*.

73. Rice DC. Parallels between attention deficit hyperactivity disorder and behavioral deficits produced by neurotoxic exposure in monkeys. *Environ Health Perspect*. 2000; 108(suppl 3):405–408 *[Medline]*.

74. Bellinger DC. Future directions for neurobehavioral studies of environmental neurotoxicants. *Neurotoxicology*. 2001; 22:645–656 *[FISI][Medline]*.

75. Davis JM, Otto DA, Weil DE, Grant LD. The comparative developmental neurotoxicity of lead in humans and animals. *Neurotoxicol Teratol.* 1990; 12:215–229 *[ISI][Medline].*
76. Schwartz J, Otto D. Blood lead, hearing thresholds, and neurobehavioral development in children and youth. *Arch Environ Health.* 1987; 42:153–160 *[ISI][Medline].*
77. Schwartz J, Otto D. Lead and minor hearing impairment. *Arch Environ Health.* 1991; 46:300–305 *[ISI][Medline].*
78. Rice DC. Effects of lifetime lead exposure in monkeys on detection of pure tones. *Fundam Appl Toxicol.* 1997; 36:112–118 *[ISI][Medline].*
79. Lasky RE, Luck ML, Torre P, Laughlin N. The effects of early lead exposure on auditory function in rhesus monkeys. *Neurotoxicol Teratol.* 2001; 23:639–649 *[ISI][Medline].*
80. Counter AS, Vahter M, Laurell G, Buchanan LH, Ortega F, Skerfving S. High lead exposure and auditory sensory-neural function in Andean children. *Environ Health Perspect.* 1997; 105:522–526 *[ISI][Medline].*
81. Buchanan LH, Counter AS, Ortega F, Laurell G. Distortion product oto-acoustic emissions in Andean children and adults with chronic lead intoxication. *Acta Otolaryngol.* 1999; 119:652–658 *[ISI][Medline].*
82. Neisser U, Boodoo H, Bouchard TJ, et al. Intelligence: knowns and unknowns. *Am Psychol.* 1995; 51:77–101 *[ISI].*
83. Mirsky A, Anthony B, Duncan C, Ahearn M, Kellam S. Analysis of the elements of attention: a neuropsychological approach. *Neuropsychol Rev.* 1990; 2:109–145.
84. Paule MG, Chelonis JJ, Buffalo EA, Blake DJ, Casey PH. Operant test battery performance in children: correlation with IQ. *Neurotoxicol Teratol.* 1999; 21:223–230 *[ISI][Medline].*
85. Chelonis JJ, Daniels-Shaw JL, Blake DJ, Paule MG. Developmental aspects of delayed matching-to-sample task performance in children. *Neurotoxicol Teratol.* 2000; 22:683–694 *[ISI][Medline].*
86. Cory-Slechta DA. Exposure duration modifies the effects of low level lead on fixed-interval performance. *Neurotoxicology.* 1990; 11:427–442 *[ISI][Medline].*
87. Paule MG, Adams J. Interspecies comparison of the evaluation of cognitive developmental effects of neurotoxicants in primates. In: Chang LW, ed. Principles of *Neurotoxicology.* New York, NY: Marcel Dekker; 1994:713–731.
88. Bellinger DC. Perspectives on incorporating human neurobehavioral endpoints in risk assessments. *Risk Anal.* 2002; 22:487–498 *[ISI][Medline].*
89. Stiles J. Neural plasticity and cognitive development. *Dev Neuropsychol.* 2000; 18:237–272 *[ISI][Medline].*
90. Gaillard WD, Hertz-Pannier L, Mott SH, Barnett AS, LeBihan D, Theodore WH. Functional anatomy of cognitive development. fMRI of verbal fluency in children and adults. *Neurology.* 2000; 54:180–185 *[Abstract/Free Full Text].*
91. Schlagger BL, Brown TT, Lugar HM, Visscher KM, Miezin FM, Petersen SE. Functional neuroanatomical differences between adults and school-age children in the processing of simple words. *Science.* 2002; 296:1476–1479 *[Abstract/Free Full Text].*
92. Stephenson J. CDC report on environmental toxins: some progress, some concerns. *JAMA.* 2003; 289:1230–1233 *[Free Full Text].*
93. Grosse SD, Matte TD, Schwartz J, Jackson RJ. Economic gains resulting from the reduction in children's exposure to lead in the United States. *Environ Health Perspect.* 2002; 110:563–569 *[ISI][Medline].*
94. Needleman HL, Jackson RJ. Lead toxicity in the 21st century: will we still be treating it? *Pediatrics.* 1992; 89:678–680 *[ISI][Medline].*

STATEMENT OF MIKE KEEGAN, POLICY ANALYST, NATIONAL RURAL WATER ASSOCIATION

THE SOLUTION TO LEAD IN THE WATER IS MORE LOCAL CONTROL

Mr. Chairman, if you relied on the newspapers, television accounts, or commentary from the national environmental groups to understand the situation with lead in the District's drinking water supply, you would only be hearing one perspective. From these sources, you have heard that the only way to solve the lead problem is by removing local authority and transferring it to the Federal Government. This is one solution, however, it is not the only policy that Congress should consider—and by no means, does it have any greater rate of success in solving local problems.

The lead crisis now shrouding the District is a product of a 30-year effort to build a dysfunctional Federal environmental system whose key principal is the antithesis of Home Rule; to take authority away from the local citizens and transfer it to Federal bureaucracies and the interest groups who control them.

The National Rural Water Association has over 23,000 small and rural community members who supply drinking water or wastewater to their communities. These communities are governed by locally-elected officials; they don't make profits and their families drink the water they supply. Therefore, unlike commercial enterprise, there is no incentive nor any reason to supply anything but the safest water possible. The Safe Drinking Water Act, in many instances, directs EPA to override the desired local health policy of these communities—and forces them to pay for EPA's decision. *They all have to comply with the EPA Lead and Copper Rule* just like the District of Columbia (DC) and the Water and Sewer Authority (WASA). We urge you to not let this incident be the predicate for removing more of our local authority to determine local policies to the Federal level.

WHY THE CURRENT LAW IS UNWORKABLE

The country's water protection program (Safe Drinking Water Act) relies on a uniform regulatory compliance program—at the expense of the judgment of locally elected officials—that is too complex and arbitrary to handle local individual problems. This program was guaranteed to fail because (1) it can't possibly manage future local crises that were not dreamed of when it was designed and published in the Federal Register, and (2) it doesn't consider the unintended consequences of its mandates—it operates in a vacuum of reality, and (3) it can't balance competing local priorities. These three flaws appear to have contributed to the current situation in the District. Now when there is a crisis that needs civic leadership—no one is responsible; and the Federal Register isn't talking.

WHAT CAUSED THE PROBLEM IN THE DISTRICT?

Local judgment was overridden by the Federal regulatory system, which was too arbitrary and inflexible to deal with the situation.

One theory, is the EPA Rule to reduce byproducts from disinfection steered the WASA to a new disinfection regiment that caused a change in the water chemistry that resulted in corrosion and increased lead concentrations. The relationship between the correlation rules is not adequately considered when applied in the real world because EPA, in a vacuum, implements them. Additionally, the law prohibits the District from balancing the competing objectives of the two rules.

Another theory, a mechanical action disrupted homeowners' plumbing enough to cause a temporary spike in lead samples. If this were the case, the regulations contain a solution (and public notification protocol) based on the long-term problem including corrosion control and replacing some lead service lines which would not fit if this theory were correct.

The Act and the regulations don't fit in the either case. If the cause is something else, why are we mandating particular solutions for problems we don't understand?

These necessary balancing judgments are beyond the capability of static regulations and beyond the abilities of regulators charged *only* with the enforcement of the specific regulations. Regulators can only regulate the letter of the law, they can't think beyond compliance—which is critical in determining public health policy. It is essential to realize that meeting regulations is not synonymous with public health protection. In the District example, it is likely that WASA and EPA would have looked at the situation differently. EPA was forced to focus on enforcement, regardless of unintended consequences. On the other hand, if WASA retained authority, it would have had the discretion to be more concerned with the overall public health implications and the ability to be more cautious in changing water treatment regimes. It is only elected policy makers with the authority to look at public policy in a holistic manner that can balance public health risks.

Once the lead levels started to increase, WASA and EPA probably knew that the higher lead levels were not as alarming as the environmental community and media would claim because of the conservative nature of the standards, and that this may be a temporary problem which the Federal public notification requirements would not convey. In all of WASA's actions, it appears that after months off the situation, which EPA was aware of reportedly, WASA only violated the public notification elements of the regulations, not any water quality criteria. Just how sacrosanct is the exact application of Federal regulations?

• EPA allowed Columbus, Ohio out of some of the compliance details of the lead rule that was not provided to WASA. EPA's ruling in Columbus[1] shows that they think some higher lead levels in water are not a health risk. They allowed Columbus a safe harbor if they had higher levels. Why was Columbus (allowed this special exemption) and not DC? And more importantly, how much higher than the Federal standards can you go before there is a health risk? This common sense question is one that EPA can't answer.[2]

• Considering the extreme valves in DC's water that exceeded EPA's action level by a factor of over 20, numerous homes tested over the lead action level, all the media uproar over this issue, and the alarm it has fomented in the public, would you ever think the CDC would have said the following just last week: ". . . *although lead in tap water contributed to a small increase in BLLs in DC, no children were identified with BLLs >10 µg/dL, even in homes with the highest water lead levels."*[3] Does this finding by CDC seem consistent with the level of alarm being portrayed to the public?

PUBLIC NOTIFICATION (VIOLATIONS VS. PUBLIC HEALTH RISKS)

The public notification process is another area in the Safe Drinking Water Act that is flawed. Since the relationship between "violation" and public health risk must be evaluated on a case-by-case basis. Mandatory public notice requirements for all violations can be used to mislead the public. Some violations are worse than others and it is the health impact and degree of malice that needs to be conveyed to the public more than the simple fact that there was a violation. The current Federal standard for lead in drinking water is 15 parts per billion (ppb) [based on a percentage of homes tested]. Does that mean that 14.5 ppb is safe and 15.5 ppb is comparably unsafe? It certainly does not. However, this is how the issue is presented to the public. The news reports of the situation commonly report that water with lead levels above 15 ppb is "contaminated" and, inferentially, anything below as not contaminated. Safe, clean, polluted, and contaminated are all characterizations and can be misleading and inflammatory. There is no bright line of concentrations in the parts per billion when lead levels become safe or unsafe. The actual health effects are uncertain and are dependent on the amount of water consumed, age of the person, amount of time exposed and other variables. This is why more public information is better—not just the alarming news. In this instance, the public should have known all along the levels of lead in every test and the balancing that was going on in the water chemistry. This constant public discussion and disclosure would lessen the ability of the media or interest groups to create an appearance of a cover up when there was none.

The public needs to understand that removing all lead from the water supply is technologically impossible and not necessary to protect the public. So the civic policy has dealt with how much lead we can live with and what is the most economical way to get it to acceptable levels. This can be different for separate communities, with unique circumstances, economies, natural environments, demographics, extraordinary local considerations, etc. For example, a community with lead at 16 ppb in their water and numerous public housing units with lead paint should not be forced into the same compliance measures as community with 300 ppb of lead in the water and no houses with lead paint.

THE SOLUTION CONGRESS SHOULD CONSIDER

Ask yourself who cares more about the health of the children in the District (and is more responsive to those families), the local mayors or an EPA regional employee

[1] In the past, the City of Columbus made certain changes to the method it uses to treat drinking water. Inadvertently, the treatment change caused an increase in the level of lead in the drinking water. . . . Through this Agreement, the U.S. EPA would suspend the lead service lines sampling and replacement provisions for up to three years beginning if and when the City exceeds the lead limit . . ." Federal Register: July 27, 2000 (Volume 65, Number 145) [Page 46166–46167].

[2] In a March report, EPA did not find that arsenic concentrations above their standard necessarily present an "unreasonable risk to health." [USEPA, Exemptions & the Arsenic Rule, March 2002, p. 11, #7]. Instead of identifying the levels of arsenic that are "protective of the public" [42USC300g–1(b)(15)(B)l or don't present "an unreasonable risk to health" [42USC300g–5(a)(3)] as named in the Safe Drinking Water Act and that the Agency was requested to name by several Congressmen, EPA creatively chose to identify what these levels are not. "EPA is . . . determining what does not pose an unreasonable risk to health with respect to arsenic, rather than address the much more complex issue of what does constitute an unreasonable risk to health." [USEPA, Exemptions & the Arsenic Rule, March 2002, p. 11, #7].

[3] Blood Lead Levels in Residents of Homes with Elevated Lead in Tap Water—DC, 2004, March 30, 2004 *(http://www.cdc.gov/mmwr)*.

in Philadelphia? If the mayors, or the regional governments of WASA, had the authority over managing the health policy underlying the water supply—we would likely not be in the situation we are in now because they are elected for the exact reason of managing issues that have many variables and impacts. Mayors can manage the balancing of local priorities in a way that regulatory enforcers cannot. Now you are being asked to give more authority to EPA at the expense of the local mayors.

Congress or EPA can expand the regulatory program and require more Federal uniform mandates on locals in response to the District experience. This has been the history of national drinking water legislation. However, this will not solve the problem of drinking water protection because the Federal Government cannot possibly design a program that foresees the infinite challenges that local communities face in providing safe water. The problem with the Safe Drinking Water Act is that improving drinking water in small communities is more of a RESOURCE problem than a REGULATORY problem.

The best way to avoid threats is to have the most educated and responsible local officials overseeing the water supply. We urge you to consider this alternative perspective of local governments and their citizens. The key to finding the best public health policy to tackle the lead issue is for it to be derived and supported by the people that benefit from a safe drinking water and have to pay for the service. If the locals don't like the results, they can elect a new government.

THE ROLE OF NATIONAL ENVIRONMENTAL GROUPS IN LOCAL ISSUES

Why do the environmental groups support a Federal control program? The answer is because they can control it better. Most of their polices would not be accepted at the local level (by the people) if there was an open public health debate. Therefore, the groups have made an expertise at getting national legislation enacted that they can exploit through lawsuits as well as intimidating bureaucrats into publishing over-zealous regulations.

Many interest groups petition this committee to authorize more and more, ever-stringent Federal unfunded mandates on small communities with the intention of improving public health on the community's behalf. Unfortunately, this does not work and things are not that simple. The key to long-term improvement is local support, local education and available resources. We continually ask for the list of the communities that need to improve their drinking water and are not willing to take the steps to do it. Such a list does not exist. We encourage organizations that advocate increasing unfunded mandates on communities to take their case directly to the local community. If they can get the community's support, then we would back any new standard or policy. The problem has been that communities do not support most of these policies at the local level because they waste limited resources on non-priority projects.

A CURRENT EXAMPLE OF THE UNINTENDED CONSEQUENCES OF THE SDWA

It appears that the Stage I rule was the rule that caused WASA to change their treatment to chloramines and resulted in the increase in lead concentrations in the drinking water. The National Rural Water Association is urging EPA to rethink finalizing the Stage 2 Disinfection and Disinfection By-Products (Stage 2) and Long Term 2 Enhanced Surface Water Treatment (LT2) Proposed Rules in light of the recent chloramines study released by the EPA Office of Research and Development. The study concluded that alternatives to drinking water chlorination, such as chloramines, may produce "increased concentrations" of some byproducts.

We are concerned that this rule may result in unintended consequences including exposure to the public of "certain dihalogenated disinfection by-products and iodo-trihalomethanes."

We are particularly concerned by the report's following finding:

> "Important observations included finding the highest levels of iodotrihalomethanes (THMs) at a plant that used chloramination without prechlorination . . . Another important observation involved finding the highest concentration of dichloroacetaldehyde at a plant that used chloramine and ozone disinfection. Therefore, although the use of alternative disinfectants minimized the formation of the four regulated THMs, certain dihalogenated DBPs and iodo-THMs were formed at significantly higher levels than in waters treated with chlorine. Thus, the formation and control of the four regulated THMs is not necessarily an indicator of the formation and control of other halogenated DBPs, and the use of alternative disinfectants does not necessarily control the formation of all halogenated DBPs, and can even result in increased concentra-

tions of some. Moreover, many of these halogenated DBPs—including certain dihalogenated and brominated species—were not studied in the ICR. "

The proposed rules will likely require a significant number of water supplies to switch from their current disinfection process to chloramines which, according to the EPA's recent findings, may have unknown public health risks and may be more harmful than chlorine.

STATEMENT OF RICHARD P. MAAS AND STEVE C. PATCH, UNC, ASHEVILLE ENVIRONMENTAL QUALITY INSTITUTE

UPDATE ON RESEARCH REGULATIONS AND PROPOSITION 65 LITIGATIONS RELATED TO LEAD DISCHARGE FROM BRASS WATER SERVICE PARTS AND METERS

HEALTH EFFECTS RELATED TO LOW LEVEL LEAD EXPOSURE: A CONTINUING ENLIGHTENMENT

- Between 1987 and 1991, Needleman, McMichael and others first discovered that infants and young children even with very low blood lead levels (BLLs) exhibited IQ deficits.
- Infants born with BLLs below 3 μg/dL scored higher on cognitive development Index tests at age 2 than infants born with BLLs of 6–7 μg/dL or 10–12 μg/dL.
- Nine-year-olds with moderately elevated BLLs (10 μg/dL) were found to have higher drop-out rates, behavior problems and criminal behavior at age 19 than 9-years olds with BLLs below 5 μg/dL.
- Numerous recent studies have found that low level lead exposure not only causes IQ reductions but also causes increases in learning disabilities, attention deficit disorder and aggressive behavior.
- Most recently Dr. Bruce Lamphear in a major study (2001) found IQ and learning (esp. reading) deficits in children (ages 6–16) with BLLs as low as 2.5 μg/dL.
- The Centers for Disease Control and Prevention (CDCP) has reviewed and validated this study.
- The USEPA upon review of these studies has officially adopted the position that there is *no threshold dose* below which lead does not cause neurologic damage in infants and young children. Thus, the EPA has set a Maximum Contaminant Level Goal (MCLG) for lead in drinking water of zero.

THE VERY LATEST NEWS ON LEAD POISONING (APRIL 17, 2003: NEW ENGLAND JOURNAL OF MEDICINE)

Researchers from Cornell University, Cincinnati Children's Hospital, and University of Rochester (funded by the National Institute of Environmental Health Sciences NIEHS) were "surprised" to find that the IQ scores of children who had BLLs of 10 μg/dL were about 7 points lower than children with BLLs of 1 μg/dL.

Richard Canfield is a senior research associate at Cornell's Division of Nutritional Sciences (Photo courtesy Cornell)

Most previous studies focused on children with BLLs of 10–30 µg/dL and extrapolated back to lower levels.

It now appears that most of the neurological damage is caused by the first 10 µg/dL.

One in ten North American children (ages 1–5) have BLLs above 5 µg/dL

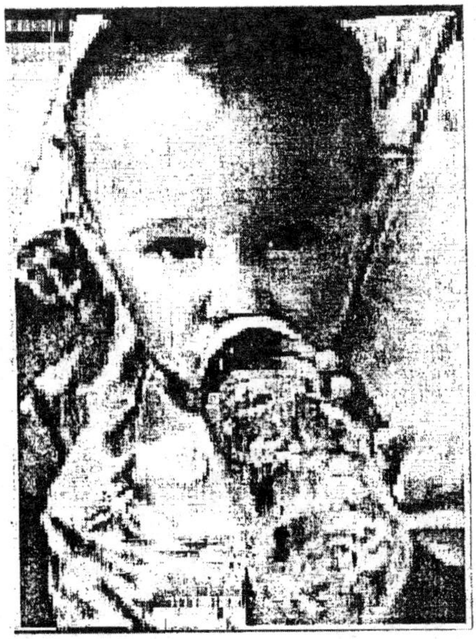

Young children are particularly susceptible to lead poisoning. (Photo courtesy National Center for Healthy Housing)

- For modeling the effects of low level lead exposure in infants and young children, the EPA has calculated a BLL increase of 0.16 μg/dL for each μg/day of lead ingested.
- Various studies have found IQ deficits of 2–6 points for each 10 μg/dL increase in BLLs (mean approx. 6 4 pts). Therefore, a young child drinking 2 liters/day of water with just 10 μg/L of lead (20 μg/day) would experience a BLL increase of approximately 3.2 μg/dL (1.3 IQ point deficit) even if they had no other sources of lead exposure.
- The USEPA estimates that 14–20 percent of total U.S. childhood lead exposure is from drinking water, although nearly all lead exposure could easily come from tapwater in any particular residence.

CHRONOLOGY OF REDUCTION OF LEAD IN DRINKING WATER: CALIFORNIA HAS BEEN THE LLEADER

- 1988: Federal Lead Ban. Elimination of leaded-solder in new buildings (most buildings will continue to have leaded solder for decades to come).
- 1994: Use of leaded brass in submersible well pumps banned nationally by the USEPA.
- June 1998: Kitchen and lavatory faucets. CA Prop 65 settlement agreement requires that residential faucet fixtures meet a very *low* lead discharge std. (Achievable only with no lead or very low lead alloys) Adopted nationally by most faucet companies.
- March 2000: CA Prop 65 settlement agreement to eliminate the use of leaded-brass alloys in residential *water meters*. Specifies Federalloy or Sebiloy (aka EnViroBrass) alloys. (Now available from Schlumberger and others.)
- 2001–present: Ca Prop 65 litigation to require no-lead or very low lead alloys in residential *gate valves, ball valves, backflow preventers*, and *pressure reducing valves*. Reportedly close to final settlement.

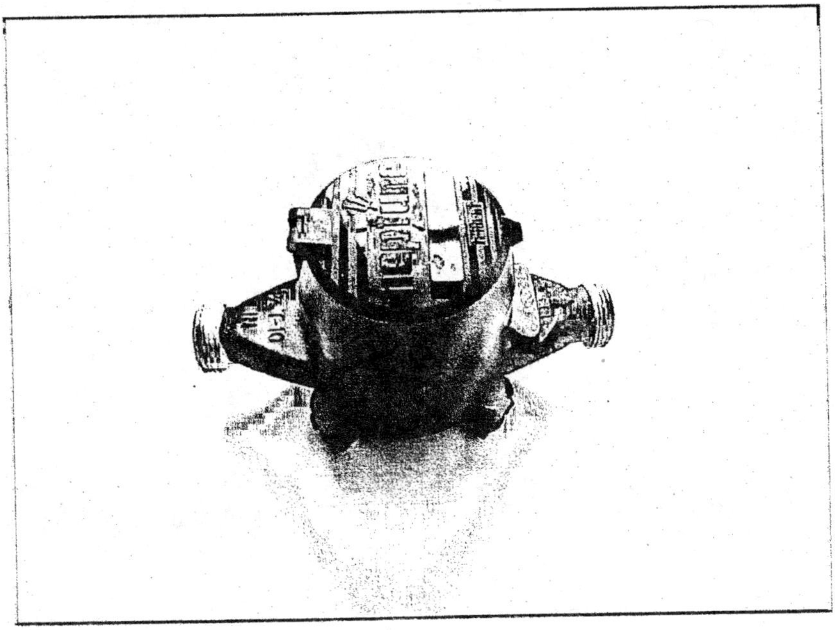

• 2002: Virtually all leaded brass plumbing components have now been banned from use in residential and most other building plumbing systems at least in California. Only leaded-brass water service parts such as curb valves, meter stops, tail pieces, elbows and main (corporation) stops have not been addressed.

• October 10, 2002: A 60-day CA Prop 65 notice was filed with the appropriate CA Public Enforcement Agencies against Mueller, A.Y. McDonald, Ford Meter Box, and James Jones for illegally manufacturing and selling leaded-brass water service components in the State of California.

• January 3, 2003: This lawsuit was expanded to include all distributors of leaded-brass water service parts in California.

ENVIRONMENTAL QUALITY INSTITUTE LABORATORY STUDY OF LEAD DISCHARGE FROM WATER SERVICE PARTS (LEADED-BRASS VS. NO-LEAD BRASS)

METHODS

• Mueller, James Jones, A.Y. McDonald, Ford, *Cambridge Brass* purchased in CA.

• Parts included different types of curb stops, elbows, main stops and compression Ts.

• Extraction water made to simulate average CA public water in terms of lead corrosivity (pH = 8.04, hardness: 100 mg/L, Alk: 82 mg/L, CI residual: 1.0 mg/L.)

• Experiments run for 19 days with samples taken after 16-hour overnight dwell.

• Days 17, 18, 19: shorter dwell time samples of 10 min, 30 min, and 2 hours.

RESULTS

Lead is initially leached quickly from the parts and the rate slows down over time.

- 16-hr dwell = 100 percent

- 2-hr dwell = 58 percent
- 30 min dwell = 31 percent
- 10 min dwell = 19 percent.

Table 5.—Comparison of Lead Discharge (g/L) From "No-Lead" Parts Versus Similar Leaded Brass Parts

No-Lead	Comparable Leaded Brass ID #s	'Q' Stat for Non-lead Part	Mean 'Q' Stat for Leaded Brass Parts	Factor Dif. in Lead Discharge
CB1	13, 23, 33, 42, 44, 45	0.91	20.7	22.8
CB2	13, 23, 33, 42, 44, 45	0.60	20.7	34.5
CB3	18, 24, 28, 38	0.53	8.81	16.6
CB4	13, 23, 33, 42, 44, 45	0.78	20.7	26.5
CB5	17, 22, 40, 43	2.81	17.8	6.35
CB7	15, 20, 26, 30	1.79	5.90	3.30
CB8	14, 19, 25, 29	2.64	16.5	6.25
CB9	31, 35, 36, 39	1.64	47.8	29.1
CB10	34, 44, 45, 47	1.31	32.9	25.1
CB12	16, 21, 27, 32	5.48	56.70	10.3
Mean				18.0

APPROXIMATE CALCULATIONS OF LEAD EXPOSURE FROM WATER SERVICE PARTS

Assumptions:
- 30 water uses/day (1 overnight 4 2-h, 15 30-min, 10 10-min).
- 2 L/day as 8 250 ml ingestions.
- 4.5 liters storage in plumbing system (80 ft of ½″ interior plus 20 ft of ¾″ service line.
- system contains a main stop, elbow, straight coupling, curb stop, also tail pieces and water meter).

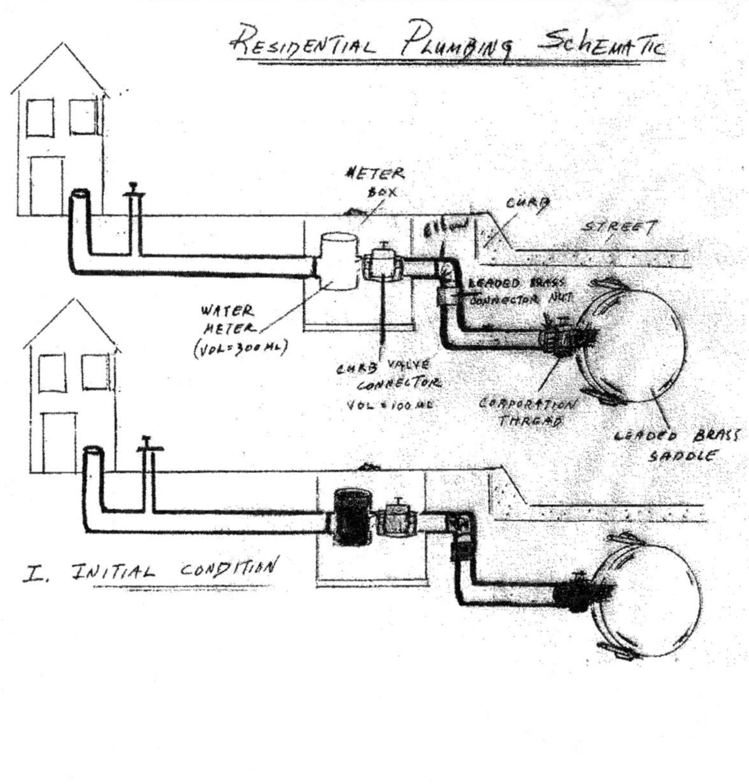

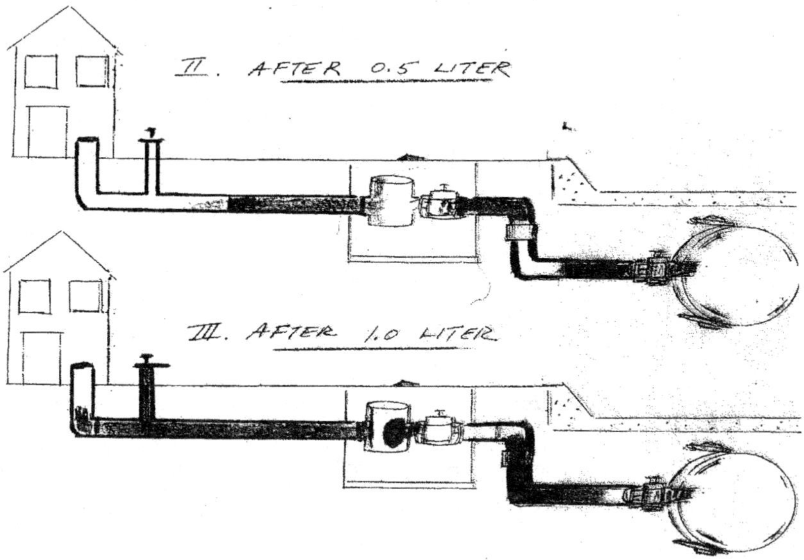

RESULTS

Total Pb discharge (Day 19)

- No lead: No water meter and tail pieces = 28 μg.
- No lead: With lead-free water meter and tailpieces = 50 μg.
- Leaded: No water meter and tail pieces = 205 μg.
- Leaded: With water meter and tail pieces = 332 μg.

Table 7.—Calculated Daily Lead Ingestion for Various Brass Water Service Parts

Lab ID # of Part	Calculated Daily Lead Ingestion	Lab ID # of Part	Calculated Daily Lead Ingestion	Lab ID # of Part	Calculated Daily Lead Ingestion	Lab ID # of Part	Calculated Daily Lead Ingestion
CB1	0.12	CB13	1.19	CB25	1.11	CB37	17.21
CB2	0.10	CB14	2.30	CB26	0.32	CB38	1.82
CB3	0.08	CB15	1.13	CB27	2.16	CB39	8.10
CB4	0.12	CB16	0.95	CB28	1.54	CB40	2.18
CB5	0.24	CB17	1.49	CB29	1.02	CB41	13.96
CB6	0.26	CB18	1.43	CB30	0.16	CB42	1.43
CB7	0.08	CB19	1.79	CB31	2.48	CB43	2.23
CB8	0.21	CB20	1.19	CB32	1.92	CB44	4.63
CB9	0.20	CB21	1.82	CB33	0.92	CB45	1.47
CB10	0.11	CB22	2.15	CB34	2.18	CB46	13.18
CB11	0.40	CB23	0.95	CB35	2.38	CB47	1.58
CB12	0.10	CB24	0.82	CB36	1.82		

Table 8.—Total Calculated Daily Pb Exposures, Childhood Blood Lead Level Increases, and IQ Deficits

Water Delivery System Type	Total Daily Lead Ingestion		BLL μg/dl		IQ Deficit	
	Mean	90th%	Mean	90th%	Mean	90th%
No lead	1.15	1.86	0.18	0.30	0.12	0.21
Conventional leaded brass	5.20	9.50	0.83	1.52	0.58	1.07

Table 8.—Total Calculated Daily Pb Exposures, Childhood Blood Lead Level Increases, and IQ Deficits—Continued

Water Delivery System Type	Total Daily Lead Ingestion		BLL µg/dl		IQ Deficit	
	Mean	90th%	Mean	90th%	Mean	90th%
Conventional leaded-brass in most corrosive 20% of CA system (approx. pop. = 5 million)	12.5	22.8	2.00	3.65	1.40	2.56

SUMMARY AND CONCLUSIONS

We are now aware that even very low lead exposures cause neurologic damage, especially in infants and young children, resulting in IQ reductions, attention deficit disorders, aggressive behavior and reading disabilities.

Leaded-brass water service parts represent a small to moderate additional source of lead exposure to infants and young children. Leading to early measurable BLL increases and IQ deficits of about 0.33 to 1.5 points, along with other lead-related neurological problems.

Some infants and young children, due to unfortunate water consumption habits, will receive lead exposure from drinking water much higher than those estimated from this study.

While the increase in childhood lead exposure from leaded-brass water service parts is usually relatively small, this is a needless extra exposure with the effects additive to other lead exposures.

We have nearly eliminated lead from our drinking water systems, and soon it will be illegal to manufacture and sell leaded-brass water system parts of any type in California.

The city of Los Angeles and many other towns nationwide are already purchasing only no-lead water service components.

WHAT IS THE EXTRA COST OF SWITCHING TO NO-LEAD WATER SERVICE PARTS?

Example: City with Service Area Population of 50,000 people—20,000 residential services.

Approximate Cost of Conventional Leaded and No-Lead Brass Service

	Leaded (USD)	No-Lead (USD)
1. Corporation Stop	$29.50	$36.88
2. Curb Stop	40.50	50.63
3. Tail Pieces (2)	12.00	15.00
4. Water Meter Casing	20.00	25.00
Total	$102.00	$127.50

Price Differential = $25.50 / service.

- Assume City adds new services at 1 percent per year and replaces 1 percent of existing services due to breakage, distribution line upgrades, etc.
- 20,000 services × 2 percent × $25.50 = $10,200/yr.
- Spread over the 20,000 residential services, this will add 51 cents per year to each family's water bill, or about 4.3 cents per month.

BUT MY WATER SERVICE PARTS ARE COMPLIANT WITH NSF–61 SECTION 8!

- NSF–61 Section 8 protocol and standards were developed primarily by the plumbing industry to ensure that most 5 percent and 7 percent brass parts would pass.
- NSF–61 Section 8 is not a health-based standard. It allows a 100mL volume brass part to discharge up to 450 µg/L of lead and still receive verification!
- Children will still receive very substantial doses of lead in drinking water in a home with NSF–61 Section 8 compliant service parts.

SOME FINAL THOUGHTS AND QUESTIONS

1. Given our recent knowledge about health effects of lead, and considering that leaded-brass water service parts installed today will be in service discharging lead

for the next 20–40 years, is it not time for public water suppliers to “do the right thing” by voluntarily and proactively eliminating this last source of lead to our customers?

2. When the next wave of media publicity about the irreversible health effects of low level lead exposure comes to public attention, do we want to have to explain to our customers why we were still installing leaded-brass parts in 2003 when lead-free parts were readily available?

3. Class Action Suits and Personal Injury Suits have gotten completely out of control in the U.S. (78 percent of our Congress are lawyers!). If public water suppliers are shown to have been still installing leaded-brass parts even after all the parts in residences were converted to no-lead brass, how vulnerable could we be to these types of legal actions?

Public Water Suppliers have led the way in reducing lead in drinking water. Let’s finish the job!

Source Volume 10(3). pp 27-28. (CA-NEV AWWA) Journal of the CAL-NEV AWWA

Lead Leaching from In-Service Residential Water Meters: A Laboratory Study

By Richard P. Maas and Steven C. Patch, UNC Asheville Environmental Quality Institute

Lead contamination of tapwater has been an important concern of public water suppliers and consumers alike over the past decade. While leaded solder joints and lead service lines have been recognized as the major sources of lead to residential tapwater, it has become apparent in recent years that leaded-brass plumbing components such as faucet fixtures (1), cut-off valves (2) and water meters (3) associated with the residual water delivery system may also be significant contributors to at-the-tap lead levels. Previous research has found that in 1993-1994 about 7 percent of kitchen faucets in use in California were discharging water with lead concentrations in excess of 15 µg/L after five months when exposed to a test water of about average corrosivity (pH=8.0, alk=50 mg/L, Ca-hardness=100 mg/L, residual chlorine=2.0 mg/L (3)). However, in June 1995 a settlement agreement was reached under Proposition 65 wherein the manufacturers of kitchen faucets agreed to essentially eliminate the use of leaded-brass over the next four years, which is believed to be reducing the lead contribution of this source.

Until very recently, virtually all brass residential water meters sold in the United States were made from an alloy containing about 5 percent lead. Laboratory experiments indicate that these meters discharge water with lead concentrations of about 80-100 µg/L during their first month of usage, decreasing to about 23-30 µg/L after five months when exposed to a test water of about average corrosivity (pH+8.0, alk=50 mg/L, Ca-hardness =100 mg/L, residual chlorine =2.0 mg/L) (3). However, the most important issue in terms of public exposure to lead as well as the ability of public water suppliers to avoid exceeding federal Lead and Copper Rule "Action Levels" relates to how much lead these meters discharge in actual residential usage over time.

Table 1. Public Water Supply Finished Water Characteristics

Public Water Supplier	Avg. pH	Ca-Hardness (As mb/L $CaCO_3$)	Cl Residual (mg/L)	Alk (mg/L as $CaCO_3$)
A	7.9	48	0.30	135
B	7.6	38	0.35	105
C	7.4	55	1.2	62
D	7.8	93	1.5	160
E	8.0	60	0.35	101
F	8.0	163	0.40	235

To address this question we obtained 165 in-service residential meters from six geographically-distributed California public water suppliers who agreed to participate in the study. The meters were removed as part of routine meter change-out activities. The meters were plugged and shipped to the Environmental Quality Institute (EQU) laboratory full of water, and upon receipt we plumbed them to a pressurized lead-free laboratory plumbing system supplied by actual water from the participating water system. Later it was demonstrated that a laboratory-prepared water with the same pH, alkalinity, hardness, and residual chlorine content would produce almost identical lead leaching results, and thereafter these waters were used. The water characteristics of the six participating public suppliers are shown in Table 1.

After being re-plumbed in the laboratory, the meters were conditioned for one week by opening a discharge valve every two hours and flushing water through the meter. After 42 such water changes it was documented that equilibrium had been achieved in terms of lead leaching, and actual test samples were then taken for four consecutive days after a 16-hour overnight dwell period. On the fourth day samples were also taken after shorter dwell times of two hours, 30 minutes, and 10 minutes.

As shown in Table 2, the meters from three of the systems (B, C, D) displayed uniformly low lead concentrations (mean<5 µg/L). Two of the systems (A, E) had meters discharging moderately elevated concentrations (weighted mean=10-15 µg/L) after 16 hours standing time, and one of the systems produced very high meter discharge levels (mean=20-40 µg/L, depending on meter brand). Overall, the two-hour dwell time concentration mens were about 66 percent of the 16-hour dwell concentrations and the 30-minute and 10-minute concentrations averaged 39 percent and 26 percent of the 16-hour dwell concentrations, respectively. These results are consistent with previous studies of water meter leaching which indicate that the dissolution of lead occurs very rapidly at first and proceeds at ever more gradual rates subsequently. These leaching kinetics also indicate that, if an in-serve meter is in fact still leaching lead, it will produce and deliver a significant slug to the residential plumbing system almost any time that water is used in the household. Translating discharge slug concentrations from the meters into actual tapwater exposure is difficult because of the complexity of flow patterns within the residual plumbing system and the difficulty of defining average water use and consumption habits of building residents, as well as the slug dispersion that occurs with travel distance. However, these results strongly suggest that water meters may contribute significantly to US EPA lead "action level" exceedences experienced by thousands of U.S. public water suppliers. Two particularly interesting patterns were noted in the results: 1) There was no significant correlation between age or brand of meter and resulting lead discharge (i.e., for a given water system, on average 40-year-old meters leached the same amounts of lead as five-year-old meters, and 2) Although water supplier was the only significant variable associated with lead discharge, there was no identifiable correlation between water corrosivity and lead concentrations using the water chemistry variables (ex. pH, Alk, hardness, residual CL_2) commonly associated with increased corrosivity for lead. For example, although the meters from public water supplier F consistently produced much higher lead levels than the others, referring to Table 1, it can be seen that, the water from supplier F does not appear to be any more corrosive than suppliers A-E in terms of the parameters commonly believed to determine lead solubility and finished water corrosivity. Clearly, the physical and chemical factors which determine the ultimate dissolution of lead from brass water meters are not simply and accurately predicted from these variables alone.

The discharge of lead from residential water meters has been the subject of a Proposition 65 lawsuit in California since late 1996. A settlement of this suit is imminent as of the time of the submission of this article (July 1999). The settlement will require that the use of leaded brass meters be phased out within two years. One major residential water meter manufacturer is already making available a lead-free brass meter, and it is anticipated that public demand may push both

Lead Leaching: A Laboratory Study

Public Water Supply	Meter Brand	Approximate Age Range (yrs.)	No. of Meters	16-Hour Dwell Mean	2-Hour Dwell Mean	30-Min. Dwell Mean	10-Min. Dwell Mean
A	Brand 1	10-13	20	10.6	8.5	5.3	3.3
A	Brand 2	5	10	15.1	9.0	4.8	2.3
B	Brand 2	20+	5	3.4	1.9	1.3	0.8
B	Brand 1	5-9	2	5.3	2.9	3.1	1.7
		10-50	8	3.0	1.7	1.3	1.1
C	Brand 1	11-12	17	2.0	—	—	—
		17-18	13	1.7	—	—	—
D	Brand 3	20, 38	2	0.0	—	—	—
D	Brand 4	15-19	19	0.6	—	—	—
		28-43	5	1.1	—	—	—
D	Brand 5	6	1	4.8	2.9	2.1	0.9
E	Brand 4	6-12	9	14.3	9.8	5.1	3.2
		14-19	21	7.2	5.1	2.5	1.8
F	Brand 3	32-44	11	25.7	10.5	5.6	3.5
F	Brand 4	< 10	3	29.5	20.0	10.5	7.1
		10-16	3	25.1	18.3	10.7	7.6
		42	1	42.5	30.1	22.3	19.2
F	Brand 6	4-8	5	37.9	26.2	15.5	9.3
		10-12	3	21.3	14.3	8.3	4.7
F	Brand 1	10-11	4	41.6	27.3	12.9	8.8

Table 2. Lead-Leaching Concentrations (in μg/L) From Water Meters at Various Dwell Times

manufacturers and public water suppliers to significantly accelerate the discontinuance of leaded-brass meters. Previous research has shown that virtually all leaded brass water meters leach significant amounts of lead when new. The new research reported here indicates that, while in many water systems, such meters may essentially stop leaching lead after several years of actual service, in other systems substantial lead leaching may persist for decades.

References

1. Maas, R.P., S.C. Patch and J. Pope. 1998. *Lead Leaching from Faucet Fixtures Under Residential Conditions.* JOURNAL OF ENVIRONMENTAL HEALTH. 61(3): 18-21.
2. Maas, R.P., S.C. Patch and A. Parker. 1999. *Lead Leaching from Residential Leaded Brass Valves.* PROCEEDINGS OF THE RESEARCH SYMPOSIUM ON HUMAN EXPOSURES TO ENVIRONMENTAL CONTAMINANTS. Berkeley, CA. April 29, 1999.
3. Maas, R.P., S.C. Patch, D.M. Morgan and H. Kawaguchi. 1997. *Lead Leaching from Brass Water Meters Under Pressurized Flow Conditions.* 1997 AWWA ANNUAL CONFERENCE PROCEEDINGS. Vol. C: 589-602.

Drs. Maas and Patch co-direct the UNCAsheville Environmental Quality Institute, which is widely recognized as the leading U.S. research center for the study of lead in drinking water. The authors wish to thank the California Public Health Trust for their generous support of this research.

Presented at the American Water Works Association Annual Conference
Atlanta, Georgia
June 17, 1997

LEAD LEACHING FROM BRASS WATER METERS UNDER PRESSURIZED FLOW CONDITIONS

by

Richard P. Maas Steven C. Patch

Diane M. Morgan Hitomi Kawaguchi

UNC Asheville Environmental Quality Institute
Asheville, North Carolina

INTRODUCTION

Recent studies indicate that U.S. residences are at significant risk of having tapwater lead contamination regardless of their age, plumbing system type, or whether they are on public or private water supplies (1, 2). In response to these findings, the U.S. Environmental Protection Agency (USEPA) has set an "action level" of 15 $\mu g/L$ for lead and a maximum contaminant level goal (MCLG) of zero $\mu g/L$.

Present federal regulations allow the continued use of plumbing fixtures, water meters and submersible pumps containing up to eight percent lead, even though the federal lead ban of December 1986, restricted the lead content of plumbing solder to less than 0.2 percent. In 1995 a settlement reached under California's Proposition 65 requires that, by the end of 1999, new faucets sold in the state leach less than 5 $\mu g/L$. The National Sanitation Foundation (NSF) has developed a standard protocol for evaluating the lead leaching properties from new faucet fixtures (NSF.61, Section 9)(3). This protocol involves inverting the fixture, filling it with water of a specified composition and, after a 16-hour standing time, dumping the water from the fixture for lead analysis. This "dump and fill" procedure is conducted for 19 days, although only days 3-5, 10-12 and 17-19 are used in the evaluation. Then a Q-statistic, which is "90 percent upper confidence bound of the 75th percentile product dosage," is calculated.

Research based on the NSF protocol has shown that many faucet fixtures containing between two percent and eight percent lead leach high concentrations of lead when new (4, 5, 6). These concentrations decreased rapidly with aging at first, and then more slowly later. High concentrations were also obtained for another dump and fill study using new faucets and a variety of waters representative of possible drinking water compositions (7). A survey of faucets being used in homes found that 17 percent of fixtures four or more years old were still leaching more than 5 $\mu g/L$ after an overnight standing time (8). The amount of lead leached from these leaded-brass alloys appears to be largely related to the manufacturing process employed, with sand-casted fixtures showing greater lead leaching than machined, fabricated, or permanent-mold type units utilizing similar brass alloys (5). Most of these homes were on public water supplies and the composition of the water appeared to have a smaller effect than the manufacturing characteristics of the faucet.

It has been clearly documented that new leaded-brass plumbing fixtures made by the sand-casting process leach significantly more lead than new fixtures using the same alloys which are fabricated or machined (5). There is also some evidence that this trend continues for long time periods under actual usage conditions (8). These findings are not unexpected considering that the sand-casting process creates a much rougher internal surface with much more surface area exposed to water. Being of lower melting point and having lower viscosity at the casting temperatures employed, it is also believed that the lead within the standard leaded-brass alloys will tend to concentrate at the interior surface.

Currently, it appears that nearly all, if not all, commercially-sold brass water meters are sand-casted using leaded brass alloys consisting of approximately 80-85 percent copper, 8-10 percent zinc, 4-7 percent lead and 0-3 percent tin (9). In the only known published study to date, one major brand of U.S. residential water meters was tested for lead leaching in the laboratory using extraction waters of pH 5.0 and pH 10.0 (9). In both cases, very elevated levels of lead were observed after a 24-hour dwell time, with the pH 10 water giving higher concentrations, which were still above 100 parts-per-billion (ppb) after 30 days of testing. A phosphate-based water additive coating initially reduced lead leachate concentrations dramatically, but showed signs of wearing off by the end of the 30-day study.

The current study was intended to determine the lead leaching dynamics of seven brands of residential water meters commonly used throughout the United States, particularly in regard to the effects of aging and water dwell time on resulting water lead concentrations.

METHODS

Testing Protocol. Seven commonly-used brands of residential (three-quarter inch) water meters were tested for their lead leaching characteristics in this study. The specific brands and models studied are listed in Table 1. Each model was tested in triplicate under pressurized plumbing conditions. A plumbing system composed totally of polybutylene was designed and constructed with separate connections and shut off valves to each of the 21 water meters. A small bladder pump connected to the system and piped to a 100 l carboy of laboratory-prepared extraction water provided a constant pressure of 40 psi to the test system.

The extraction water composition, aging procedure, and sampling regime were designed to closely approximate the lead testing procedure adopted by the National Sanitation Foundation (NSF) for residential faucet fixtures (NSF-61, Section 9, November 1994)(3). This protocol, which is used for product certification by NSF, runs for 19 days of testing with sampling of leachate conducted on Days 3, 4, 5, 10, 11, 12, 17, 18, and 19. The extraction water composition as specified by NSF-61, Section 9, is shown in Table 2. Following the initial 19-day test period, the meters continued to be aged by putting the plumbing manifold system on a timer which flushed new water through the meters 30 times per day for the next seven months. This meter aging process was designed to simulate typical residential usage with flushing every 30 minutes for 15 hours each day followed by a nine-hour overnight dwell time.

For the initial 19-day experiments, in addition to the 16-hour overnight dwell samples specified by NSF-61, sampling after several shorter dwell times (10 min., 30 min., 90 min.) was also conducted. The internal volumes of the seven meter types were measured to be between about 250 ml and 350 ml (see Table 1). For rapid sample collection capability, it was necessary to add a short flexible length of 7/16" diameter Tygon™ tubing with a volume of about 50 ml to the outlet of each plumbed meter. Thus, the consistent 400 ml first dwell sample taken from each meter included at least one entire internal meter volume of water. The second 600 ml sample was taken to complete a full one-liter dwell sample. Sampling of each dwell time was immediately followed by a rapid three liter flush (approximately 10 meter volumes) to ensure complete flushing before beginning the subsequent dwell time experiments. The last 100 ml of the three liter flush was collected and analyzed for lead to quantify the extent of any possible carry-over effects.

TABLE 1. Brand, Model, Serial Number, Model Code and Internal Volume of Test Water Meters.

Lab ID #	Model Code	Manufacturer	Model #	Serial #	Internal Volume (ml)
1	1	Western Meter Co.	Performance	12839	257
2		Western Meter Co.	Performance	12840	255
3		Western Meter Co.	Performance	12841	250
4	2	Sensus	Touch Read 3/4" SR II	50716832	298
5		Sensus	Touch Read 3/4" SR II	50716833	297
6		Sensus	Touch Read 3/4" SR II	50716834	297
7	3	Badger Meter Co.	Model 25 Recordall 1195	95841446	355
8		Badger Meter Co.	Model 25 Recordall 1195	95841447	350
9		Badger Meter Co.	Model 25 Recordall 1095	95754157	352
10	4	Sensus	5/8 " SR II	49488016	252
11		Sensus	5/8 " SR II	49488017	253
12		Sensus	5/8 " SR II	49488019	240
13	5	Precision Meters	5/8" PMM Series	95455370	250
14		Precision Meters	5/8" PMM Series	95455372	245
15		Precision Meters	5/8" PMM Series	95455373	245
16	6	Kent Meters	C-700 5/8" MKH4733	94206822	242
17		Kent Meters	C-700 5/8" MKH4733	94206823	233
18		Kent Meters	C-700 5/8" MKH4733	94206826	243
19	7	Neptune	Trident 5/8" T-10 03-96-3026	41548992	250
20		Neptune	Trident 5/8" T-10 03-96-3026	41548993	250
21		Neptune	Trident 5/8" T-10 03-96-3026	41548994	257

TABLE 2. Chemical Composition of Extract Water Used for Leaching Experiments (from NSF-61, Section 9).

Component	Value
pH	8.0 (± 0.5)
Hardness	100 mg/l
Free Chlorine	2 ppm (± 0.5)
Dissolved Inorganic Carbon	122 ppm (± 5)
Alkalinity	50 ppm (± 5)

Extraction water was made up each day in covered 100 l polypropylene carboys. To minimize free chlorine dissipation, free chlorine was added shortly before the initiation of sampling or flushing each morning. All batches were analyzed for lead prior to use. Throughout this study, all batches of extraction water exhibited lead concentrations of less than one part-per-billion (ppb) even after having been pumped through the entire plumbing manifold system, and thus, it was not deemed necessary or appropriate to subtract out extraction water lead concentrations from any of the actual test data.

For the seven months of extended testing, a reduced leachate sampling protocol was employed wherein 16-hour dwell time samples were taken every two weeks and a complete sampling of all four dwell times was conducted every eight weeks.

Sample Analysis and Quality Assurance/Quality Control. All water samples were taken from the water meters in acid-cleaned Nalgene bottles and were preserved with concentrated nitric acid within 24 hours of collection. Following acidification, samples were held for a minimum of 28 hours before analysis. Lead determinations were performed by graphic furnace atomic absorption spectrophotometry (GFAAS) using two Thermo-Jarrel-Ash Model 11 or Model 12 spectrophotometers with TJA Model 755 graphite atomizers. All appropriate laboratory QA/QC procedures regarding standard curve tolerances and standard additions were followed for all samples. The Environmental Quality Institute laboratory maintains certification for drinking water compliance for lead in six states including NY, NC, FL, MD, TN, and KY, with a reportable detection limit of 1.0 ppb for lead.

RESULTS AND DISCUSSION

Descriptive statistics developed for the initial 19-day testing period for each dwell time are displayed in Table 3. From this table, it can be seen that for all models and all standing times, lead concentrations in excess of 5 ppb were generally observed, with mean concentrations for 16-hour dwell times ranging between 32 and 96 ppb.

Results for the seven months of extended testing are shown in Table 4 and in Figures 1a, 1b, 2a, and 2b. As indicated by these data, one-liter adjusted lead concentrations had decreased to about 10-15 ppb for 16-hour dwell times and were typically only about 4-6 ppb for 90-minute dwell times. However, 16-hour first 400 ml draws were still generally in the 20-30 ppb range after seven months. It should be noted that different rates of concentration decrease and different absolute concentrations would be expected for waters of greater or lesser corrosivity than the NSF-61, Section 9 test water used in these experiments. A closer examination of Figures 1a-2b suggests that, after a relatively rapid initial concentration decrease in the first three months, lead levels generally leveled out over the last four months of testing.

TABLE 3. Statistics for the First 19-Day Time Period on the One-Liter Adjusted Lead Concentration in ppb for Water Meter Models Aggregated Over Days and Units. Sample Size is 27 for Each Model-Standing Time Combination.

Model	Standing Time	Mean (Standard Deviation)	Median	Standing Time	Mean (Standard Deviation)	Median
1	10 mins.	7.0 (4.2)	5.4	1.5 hrs.	15.6 (7.5)	12.4
2	10 mins.	23.0 (8.7)	21.4	1.5 hrs.	50.7 (13.4)	47.7
3	10 mins.	9.9 (5.6)	7.5	1.5 hrs.	25.0 (8.9)	22.3
4	10 mins.	26.8 (14.5)	19.9	1.5 hrs.	38.8 (11.9)	35.8
5	10 mins.	15.0 (4.3)	13.5	1.5 hrs.	29.7 (11.5)	26.8
6	10 mins.	8.7 (3.8)	7.4	1.5 hrs.	20.9 (7.9)	18.7
7	10 mins.	15.9 (4.7)	15.6	1.5 hrs.	34.9 (9.4)	33.2
1	30 mins.	8.7 (4.2)	7.3	16 hrs.	32.1 (12.5)	28.0
2	30 mins.	31.6 (5.3)	30.4	16 hrs.	95.8 (16.7)	95.9
3	30 mins.	15.1 (5.7)	13.5	16 hrs.	49.5 (14.1)	43.8
4	30 mins.	29.7 (8.6)	27.6	16 hrs.	69.1 (15.4)	63.6
5	30 mins.	18.0 (7.3)	17.4	16 hrs.	59.4 (15.0)	58.0
6	30 mins.	13.0 (4.8)	11.0	16 hrs.	40.0 (12.1)	36.8
7	30 mins.	25.4 (5.3)	24.7	16 hrs.	63.4 (15.1)	66.7

TABLE 4. Mean 16-Hour Dwell Time Lead Concentrations (first 400 ml draw) for μg/l.

Model #	Month 2	Month 3	Month 4	Month 5	Month 6	Month 7	Month 8
1	98.8	96.6	42.0	38.2	31.0	25.6	23.6
2	47.2	56.5	35.0	44.1	30.5	39.6	26.8
3	51.4	31.8	26.0	26.2	43.0	62.8	28.1
4	95.0	42.6	24.4	23.9	21.5	30.9	29.3
5	48.2	34.7	19.1	16.5	19.0	27.9	20.9
6	30.2	29.3	10.9	8.9	8.3	10.2	7.5
7	86.8	61.6	31.7	41.5	36.5	34.2	33.8

Table 5 compares 16-hour dwell time lead concentrations on the first 400 ml draw for the first (Day 3) and last (Day 19) days of initial testing under the NSF-61, Section 9 protocol. From this table, it can be seen that the initial overnight dwell time leachate samples generally had lead concentrations between 100 and 200 ppb (mean 139.6) with the values dropping by about one-third by the end of the initial 19-day test period. All seven models displayed roughly similar lead levels with Models 2 and 4 giving the highest lead levels over the test period, consistent with the results in Table 3, which combine the first two samples into a one-liter adjusted concentration.

TABLE 5. Comparison of 400 ml 16-Hour Dwell Time lead Concentrations (ppb) Between the Initial and Final Days of Study.

Water Meter Unit #	Water Meter Model	Day 3	Day 19	Day 19 as a Percent of Day 3
1	1	115.5	52.8	45.7%
2	1	91.3	39.8	43.6%
3	1	88.0	36.3	41.3%
4	2	181.5	118.8	65.5%
5	2	178.2	111.1	62.3%
6	2	184.8	129.8	70.2%
7	3	133.1	81.4	61.2%
8	3	161.7	60.5	37.4%
9	3	126.5	64.9	51.3%
10	4	184.8	125.4	67.9%
11	4	162.8	119.9	73.6%
12	4	193.6	124.3	64.2%
13	5	122.1	74.8	61.3%
14	5	129.8	189.2	146.0%
15	5	141.9	136.4	96.1%
16	6	119.9	53.9	45.0%
17	6	88.0	69.3	78.8%
18	6	78.1	55.0	70.4%
19	7	149.6	104.5	69.9%
20	7	156.2	97.9	62.7%
21	7	144.1	101.2	70.2%
Mean		139.6	92.7	66.4%

As shown in Table 6, the large majority of leached lead is expelled from the meters in the first 400 ml sample volume. However, anywhere from about nine percent to 28 percent of the total lead in the first liter is found in the 600 ml of 'rinsate' following the first

400 ml dwell sample. These are lower 'rinsate' percentages than noted previously for similarly plumbed and tested faucet fixtures (6). This is not an unexpected result, however, given that the relatively open cylindrical shape of a water meter provides a significantly lower surface-area-to-volume ratio than the narrow tubing, threaded values, and sharp junctures that comprise the internal shape of most faucet fixtures. Chemical dissolution theory would maintain that the dissolved lead concentrations would quickly become elevated in the aqueous microlayer adjacent to the internal brass surface. As equilibrium solubility maxima are approached in the microlayer, the rates of further dissolution from the leaded brass interface will decrease greatly, and become limited by the rate of molecular diffusion of lead ions from the microlayer to the bulk solution within the faucet fixture or water meter. Flow in this microlayer is reduced by friction to a much greater extent than flow in the bulk of the interior, and thus in a faucet fixture, where much more of the total water exists in close proximity to an interior surface, more of the dissolved lead appears later in the flow expelled from the fixture. From Table 6, it can be assumed that this same principle operates to a lesser extent with flow through water meters. Although turbulent mixing dynamics would also play a role in causing some of the lead to be expelled in the second (600 ml) sample, water meters are expressly designed to minimize such eddy flow dynamics within their cylindrical cavity. A regression analysis on the data from all eight months relating the ratio of the second sample to the first showed that, for three models (2, 3, and 7), this ratio decreased significantly (p-value $< .05$) as the study progressed.

TABLE 6. Statistics on the Percent of the Lead Concentration of the Second Sample (400-1000 ml) to the First Sample (0-400 ml) Aggregaged Over Days (1-19) and Units and Standing Times. Sample Size is 108 for Each Model.

Model	Mean Percent	(Standard Deviation)	Median Percent
1	10.8	(4.6)	9.9
2	39.7	(22.2)	36.1
3	22.2	(6.5)	20.9
4	14.7	(13.6)	12.1
5	9.4	(3.7)	8.2
6	9.9	(3.8)	9.2
7	15.5	(9.8)	14.0

Table 7 examines the relationship between dwell time and lead concentrations. As seen from Table 7, this relationship is far from linear with about 26 percent of the 16-hour dwell concentration achieved within just 10 minutes and slightly more than one-half of the 16-hour dwell levels reached after 1.5 hours. These results are consistent with the surface microlayer dissolution discussion above. Chemical kinetic theory would predict that at room temperature rapid oxidation and dissolution of lead would occur in oxygenated and chlorine-rich water. However, as the surface microlayer becomes saturated with lead ions, and perhaps depleted of oxygen and chlorine, the dissolution rate drops rapidly, and the overall process of further lead leaching becomes rated limited by the rate of molecular diffusion from the surface microlayer to the bulk solution. It is of interest to note that a diffusion controlled system assumes no physical jarring which would disrupt the saturated surface microlayers. If a mechanical perturbance were to occur which caused the saturated microlayer to mix into the bulk, a new time period of rapid surface dissolution would be initiated, and the total lead concentration expected to be observed for a 1.5 hour or 16-hour dwell time would be nearly twice as great. In these plumbed and pressurized laboratory experiments, effects were made to not physically jar the meters in any way. However, in actual suburban and urban usage where meters are often buried directly adjacent to roads and streets, it is possible that frequent jarring sufficient to redistribute the saturated microlayer to the bulk interior meter cavity may occur as a result of heavy traffic, distribution line maintenance, or other vibration-causing activities. Further laboratory study introducing controlled vibration to disturb these microlayers would be necessary to better quantify this effect.

TABLE 7. Shorter Dwell Time Lead Concentrations as a Percentage of 16-Hour Dwell Time Concentrations (Aggregated Over Initial 19 Test Days and Units).

Model	10 Minutes	30 Minutes	90 Minutes
1	21.8%	27.1%	48.6%
2	24.0%	33.0%	52.9%
3	20.0%	30.5%	50.5%
4	38.8%	43.0%	56.2%
5	25.3%	26.9%	50.0%
6	25.0%	32.5%	52.3%
7	25.1%	40.1%	55.0%
Mean	26.0%	34.6%	52.7%

Table 8 shows the relationship between dwell time and lead concentrations for the seven month extended test period. The relationships are approximately equivalent to the ones noted in Table 7 for the initial testing period. The strong implication from this data is that a significant percentage of the overnight dwell-time slug is probably released to the residential plumbing system from relatively new water meters almost anytime water is used in the residence. This hypothesis is further supported by the data presented in Table 9, which indicates that about 1-2 ppb of lead is released by new brass water meters even under continuous flow conditions.

TABLE 8. Mean Percent of 16-Hour Standing Time Lead Concentrations for Entire Study.

Model	Dwell Time		
	10 min.	30 min.	90 min.
1	23.6%	26.1%	45.6%
2	29.9%	25.2%	47.1%
3	17.8%	22.5%	40.7%
4	28.2%	30.2%	45.6%
5	45.4%	37.3%	68.8%
6	17.8%	24.6%	42.5%
7	20.9%	27.9%	43.1%
Mean	26.2%	27.7%	47.6%

TABLE 9. Water Meter Lead Concentrations Produced Under Continuous Flow Conditions, Days 1-19.		
Model	Flushed Sample Concentration (ppb)	Standard Deviation (ppb)
1	0.79	(0.59)
2	2.27	(0.72)
3	0.85	(0.77)
4	2.22	(0.93)
5	1.29	(0.81)
6	0.89	(0.53)
7	1.82	(0.91)

Obviously these concentrations could be higher with more corrosive water and under these conditions could represent the difference between exceeding and not exceeding the action level under the federal lead and copper rule. Although not measured, concentration reductions from other dwell time samples suggest that continuous flow background concentrations after seven months might be about one-third of the values shown in Table 9, although levels could be might higher for more corrosive waters. When broken down by which sampling time the flushed sample followed, the means ranged from 1.28 (following the 90-minute sample) to 1.60 (following the 30-minute sample). The facts that there is relatively little difference in the mean flushed sample concentrations between standard times compared to the first two samples taken at each standing time suggests that almost all of the lead observed in the flushed sample occurs as a result of instantaneous leaching as opposed to residual lead from the previous standing time.

SUMMARY, CONCLUSIONS AND INTERPRETATIONS

These experiments clearly indicate that all seven commonly-used models of leaded brass water meters initially leach lead in concentrations exceeding the USEPA Action Level (15 ppb) and almost certainly would result in human exposures significantly greater than the California Proposition 65 limit of 0.5 μg/day, at least during their initial period of usage. For a 300 ml internal volume water meter connected to standard residential one-half inch plumbing, expulsion of each meter volume (by running 300 ml of water anywhere in the residence) should introduce lead-contaminated water into approximately an eight foot section of the residential plumbing system.

As the data indicate, at least when new, elevated lead concentrations can be expected to be discharged from these meters even after internal dwell times of only 10 minutes. Further study would be needed to determine the rates of lead concentration decreases over longer time periods. The approximate 30-40 percent decrease observed over the initial 19-day study period is substantially less than that observed for other leaded brass products under similar experimental conditions (e.g., 5, 6) for faucet fixtures. Interestingly, even though the rate of initial laboratory concentration decrease was found to be relatively large for faucet fixtures, as noted above, a field study in California found that about 17 percent of faucets which had been in continuous residential usage for more than 10 years were still leaching in excess of 5 ppb lead (8). The percentage was substantially greater for faucets made of sand-casted leaded brass similar to the meters tested in this study. Most likely the chemical corrosivity of the water probably largely determines how long sand-casted leaded brass continues to produce elevated lead levels in drinking water. The NSF-61, Section 9, water is designed to be a relatively non-corrosive water, approximately representative of average U.S. public water supply corrosivity. Many municipalities would be served by much more corrosive water. A non-corrosive water would be expected to form a scale or film, effectively isolating the lead from the water flow after some initial period of time. Similarly, a moderate-corrosivity water would probably initially dissolve away the more physically available lead in the interior brass surface. This would be predicted to be followed by a longer time period where smaller amounts of less physically available lead would slowly be dissolved from micropores and micropitting on the inside brass surface. In marked contrast, however, it is reasonable to assume that more corrosive waters will continue to dissolve away all of the metals (copper, zinc, tin, lead) on the interior of the brass surface, continually exposing new lead-containing surface throughout the life of the plumbing product.

REFERENCES

1. Consumer Union. 1993. *Is There Lead in Your Water?* **Consumer Reports**. February 1993:73-78.

2. Maas, R.P., S.C. Patch, B.T. Peek and R.L. Winfield. 1993. *Controlling Lead at the Tap: A Statistical Investigation of Key Parameters.* **UNC Asheville Environmental Quality Institute Technical Report #93-006.**

3. ANSI/NSF International, Standard 61, Section 9. 1994. *Drinking Water System Components-Health Effects.*

4. Lee, R.G., W.C. Becker and D.W. Collins. 1989. *Lead at the Tap: Sources and Control.* **JAWWA**. 81(7):52-62.

5. Maas, R.P., G.M. Brown, S.C. Patch, D.M. Morgan and E.R. Rainey. 1993. *The Rate and Extent of Lead Leaching From Faucet and Hot Water Dispensers Under Laboratory Conditions.* **UNC Asheville Environmental Quality Institute Technical Report #93-009.** 20 p.

6. Maas, R.P., S.C. Patch, C. Moore and D. Morgan. 1996. *A Study of the Lead Leaching Characteristics of New Water Faucets Under Pressurized Conditions.* **UNC Asheville Environmental Quality Institute Technical Report #96-030.** 24 p.

7. Samuels, E.R. and J.C. Meranger. 1984. *Preliminary Studies on the Leaching of Some Trace Metals From Kitchen Faucets.* **Water Res.** 18(1):75-80.

8. Maas, R.P., S.C. Patch, J.P. Pope and D.M. Morgan. 1993. *Contribution of Faucet Fixtures to Residential Tapwater Contamination in California.* **UNC Asheville Environmental Quality Institute Technical Report #93-007.** 21 p.

9. Maas, R.P., S.C. Patch, J.P. Pope and E.R. Rainey. *The Effects of Acidity, Pre-Cleaning and Surface Coating on Lead Leaching From Standard Leaded-Brass Water Meters.* **UNC Asheville Environmental Quality Institute Technical Report.** 20 p.

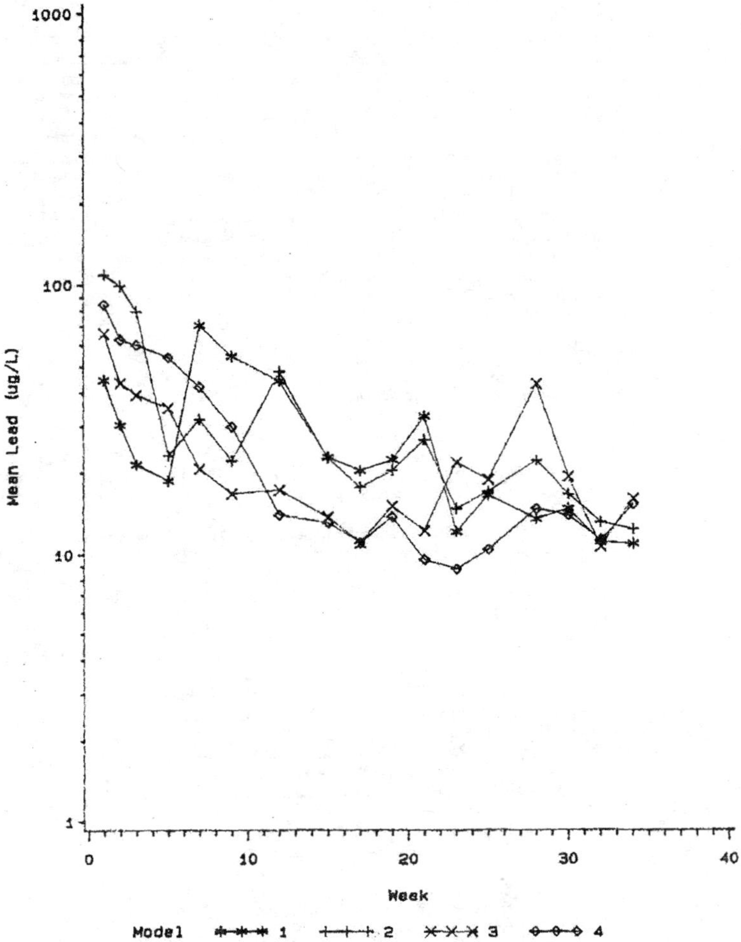

Figure 1a. Mean Lead Concentrations (ug/L) for Calculated One-Liter Draw of 16-Hour Standing Time Versus Week by Model Number for Models 1-4

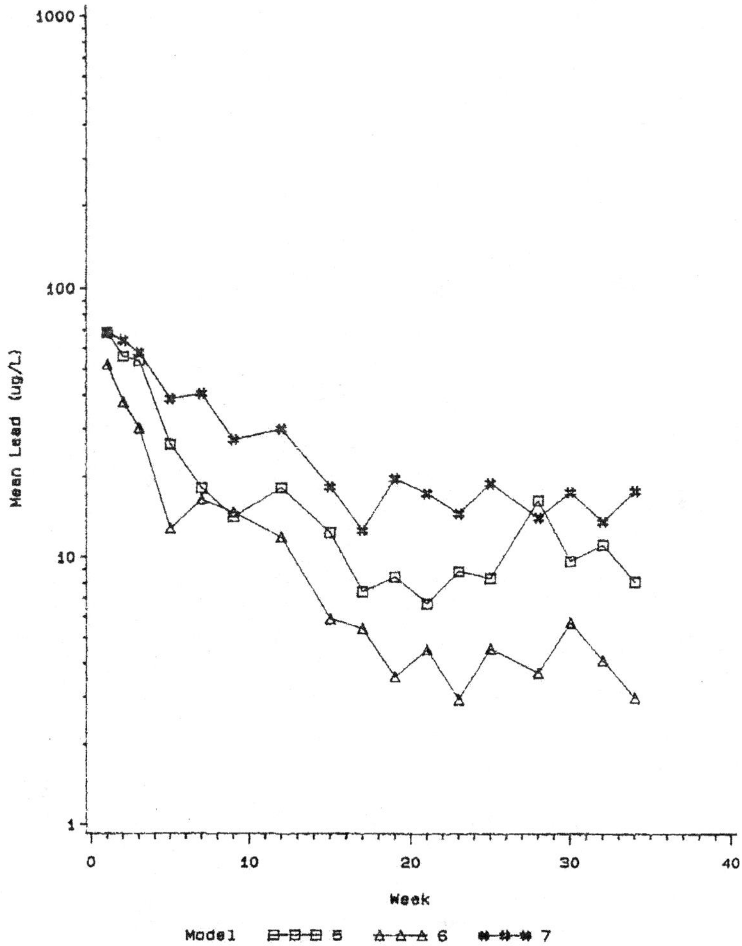

Figure 1b. Mean Lead Concentrations (ug/L) for Calculated One-Liter Draw of 16-Hour Standing Time Versus Week by Model Number for Models 5–7

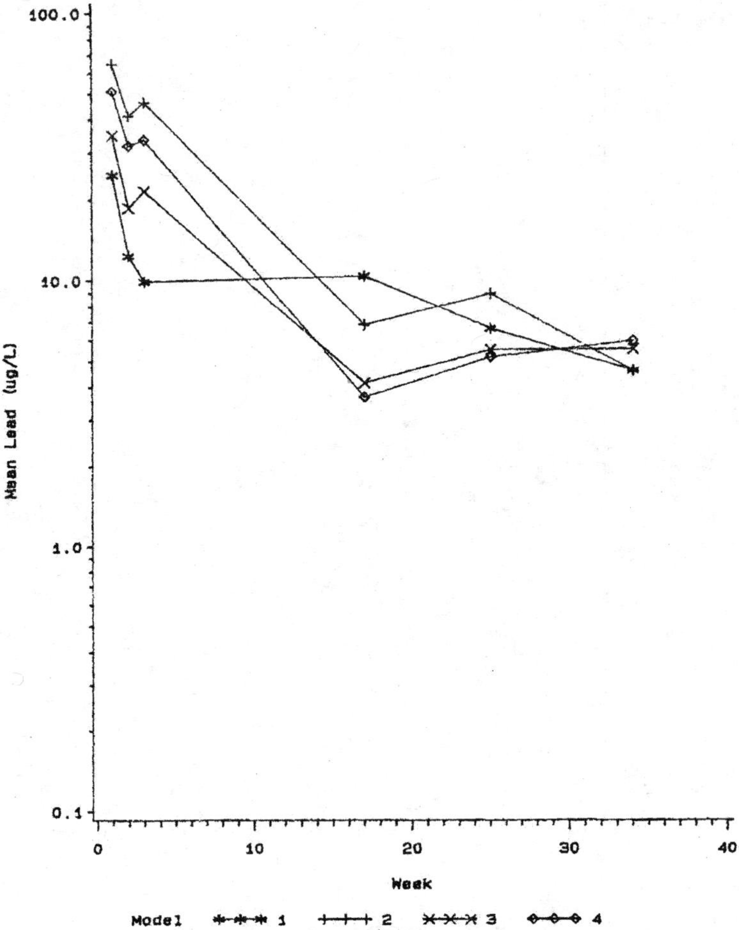

Figure 2a. Mean Lead Concentrations (ug/L) for Calculated One-Liter Draw of 90-Minute Standing Time Versus Week by Model Number for Models 1-4

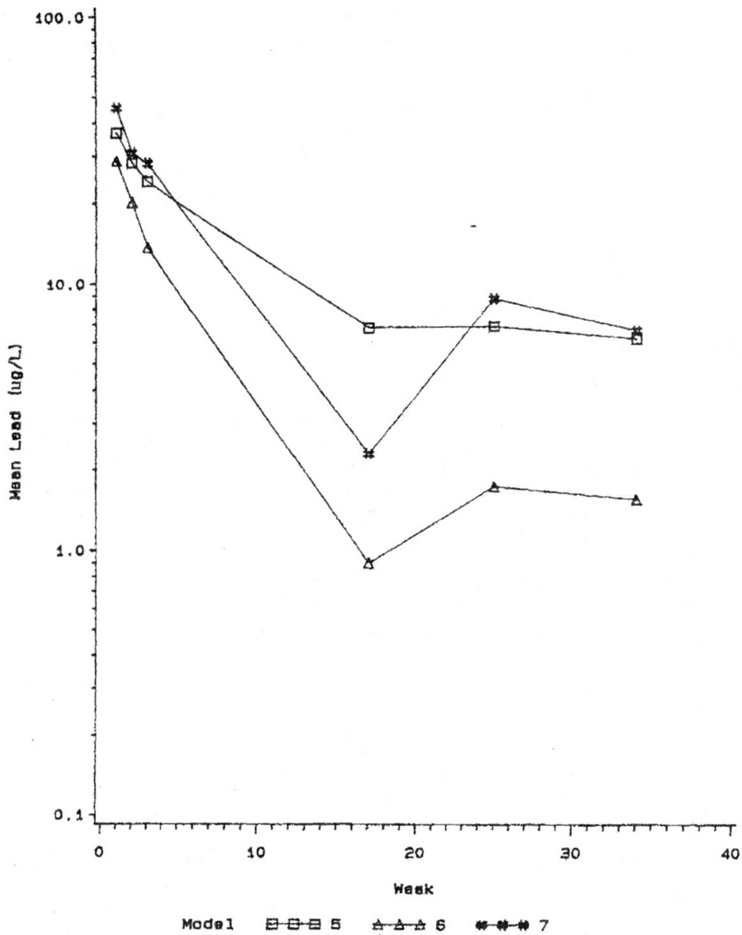

Figure 2b. Mean Lead Concentrations (ug/L) for Calculated One-Liter Draw of 90-Minute Standing Time Versus Week by Model Number for Models 5–7

GOVERNMENT OF THE DISTRICT OF COLUMBIA

February 27, 2004

The Honorable James M. Inhofe
Chairman, Committee on Environment and Public Works
United States Senate
Washington, D.C. 20510

Dear Senator Inhofe:

We write today concerning the problem with unacceptable levels of lead in some District residents' drinking water. On Wednesday, February 11, we established the Interagency Task Force on Lead in Drinking Water. It has been meeting weekly to look into ways the District of Columbia Water and Sewer Authority (WASA) can reduce lead in water, ensure that high-risk populations receive priority lead pipe replacement, identify funding sources to help pay for replacements and make certain WASA and the DC Department of Health communicate critical information to citizens promptly and clearly.

Because there are two essential components to this matter, we feel it is important that each component be investigated by the appropriate entity. On the local side, WASA is a quasi-independent District government agency with direct oversight by the city. On the federal side, there is the Washington Aqueduct, which is responsible for collecting, purifying and pumping the potable water utilized by WASA and is a division of the U.S. Army Corps of Engineers, and the Environmental Protection Agency (EPA).

Although the Council's Committee on Public Works and the Environment has had the Washington Aqueduct present and testifying at its hearings, and it is a member of our Task Force, we think it is important to have an additional forum where officials from federal entities, such as the Washington Aqueduct and EPA, could be questioned by a body with appropriate legislative oversight authority. The House of Representatives, partly in response to a letter of request from us, has scheduled a hearing for March 5, 2004. We feel that Senate involvement would also complement and expedite that work, as well as the work of the Council's Committee and the Task Force we co-chair. Therefore, we are respectfully requesting that you, as Chairman of the United States Senate Committee on Environment and Public Works, hold hearings on this topic.

We look forward to your response to our request.

Sincerely,

Anthony A. Williams
Mayor
District of Columbia

Carol Schwartz
Councilmember, At-Large
Council of the District of Columbia
Chair, Committee on Public Works and the Environment

○

[From the Washington Post (1877–1954), Sept. 15, 1895]

Potomac Water and Lead Pipe

SOURCE OF DANGER WHERE SUCH A PIPE HAS BEEN USED A LONG TIME

A. W. Dow, inspector of asphalt and cements, yesterday made his report to the Engineer Commissioner. In it he says considerable change has been made in the past year in asphalt pavement by the addition of a fine sand to a sand similar to that formerly used. Under the present circumstances this is the best that can be done. The only fine sand now available is that dredged off the foot of Seventeenth street.

The inspector deals also with the public wells analyzed. There were found to be 96 good ones, 41 suspicious, and 57 condemned.

The most interesting part, of the report deals with the investigation of the action of Potomac water on lead pipe, to determine if enough lead is dissolved by the water to be injurious to public health. In order to have all conditions corresponding as near as possible with those of actual service, the inspector had one new 40-foot lead service pipe in Anacostia and 50 feet of new lead pipe attached to the high service main at the U street pumphouse. From the investigation the inspector concludes that the only great source of danger is where the coating becomes detached by a rapid flow of water after the pipe had remained unused for some time. He will continue the investigation.

[From the Washington Post (1877–1954), June 9, 1893]

Lead Pipes Unsatisfactory

LOOKING FOR A GOOD SANITARY PIPE FOR SUPPLYING WATER

Capt. Powell, the Engineer Commissioner, has determined that a substitute must be found for lead pipes which, according to the present plumbing regulations, must be used in providing a water service for residences. The general fear that such pipes might cause lead poisoning under certain conditions makes their general adoption in the District a menace to the health of the people.

It has been shown that the chemical character of Potomac water causes such pipes to become coated on the inside with an insulation of carbonate of lime, soda, and clay, held in solution in the water. This coating, it has been argued, is a sure protection from danger of lead poisoning, but the engineer department has decided that it is too **???** safeguard. It is probable that the city's supply of water will be filtered at some future day, as sand filtration of drinking water has been adopted in many large cities abroad and is rapidly becoming popular.

Just what effect the filtered water may have on the coating of lead pipes has not been determined. The fact that iron pipes become thickly rusted on the inside, which causes a material loss of water pressure, makes their use unsatisfactory. Yesterday, Capt. Derby, in charge of the division of water and sewers, examined the first substitute for lead pipe that has been presented since the investigation began. It was what is known as the improved BoWar-Barff process, being a steel pipe coated inside and out with black oxide of iron. Capt. Derby reported it as "worth experimenting with," and tests of the pipe will be commenced at once. Several other styles of pipe are to be examined.

[From the Washington Post, March 28, 2004]

District Residents Applaud Planned Inquiry by Senate

(By David Nakamura, Staff Writer)

A U.S. Senate subcommittee has scheduled an oversight hearing for April 7 to investigate the Federal role in the lead contamination of the District's drinking water, residents were told at a special meeting on the contamination yesterday. The Fisheries, Wildlife and Water Subcommittee, chaired by Senator Michael D. Crapo (R-Idaho), will hear from the U.S. Environmental Protection Agency, the Army Corps of Engineers and the DC Water and Sewer Authority, a staff member confirmed later. The Senate's involvement comes after the House Government Reform Committee held a hearing March 5, at which members blasted the two Federal agencies and WASA for failing to provide a safe water supply and to inform the public of the health risks. So far, tests have shown that at least 5,000 DC homes have water with lead levels that exceed the Federal limit. The problem is caused by lead leaching off pipes and plumbing fixtures.

At the meeting of environmentalists and residents yesterday at Van Ness Elementary School in Southeast Washington, some parents cheered the news of the Senate's action. Government leaders "are not moving fast enough," said Liz Pelcyger, who lives on Capitol Hill and has year-old twins. Valerie Jablow, a Capitol Hill resident who has a 2-year-old son, said that she had met with staff members of Sen. James Jeffords (I-Vt.), the Ranking Minority Member on the committee. "They urged us to be present at the hearing," Jablow told about 40 residents at the meeting yesterday. "They need to hear from ordinary citizens on this issue." Members of the coalition that organized yesterday's meeting said they want to force District and Federal leaders to include residents in decision-making as they deal with the lead contamination problems. Damu Smith, executive director of the National Black Environmental Justice Network, part of the coalition, told the audience that it is unfair that DC leaders had assembled an interagency lead task force that includes no ordinary citizens. "We need to drive this process," Smith said. "This is not an issue the politicians or agency should be leading. They are responsible for the crisis in the first place. We'll meet with them and work with them." The only District leader at the meeting was DC Council member Jim Graham (D-Ward 1), whose staff handed out free water test kits.

Graham criticized his colleague, DC Council member Carol Schwartz (R-At Large), who co-chairs the task force with Mayor Anthony A. Williams (D), for not including residents and for holding meetings behind closed doors. Schwartz has argued that the task force can move faster to combat the lead problems if members are not distracted by reporters or residents. Last week, Schwartz allowed Smith and other environmentalists to meet with the task force for an hour. Not everyone was critical of the way city leaders are handling the crisis. Robert Brannum, who lives in the Bloomingdale neighborhood, cautioned residents to "be careful before we cast the blame. We can talk about getting the lead out of the water, or we can cast blame and be political." But most other residents voiced less patience. Michael Smith, a firefighter from Northeast, said, "I do not have any confidence in WASA's ability to manage this." Ethel Meachum of Southwest said she was outraged that the agency, which first knew of lead problems during the 2001–2002 testing period, had "waited 3 years to tell me about this." Another woman complained that she has "gone all over the world and the first thing people tell me is, 'Be careful of the water.' Now I find that in DC the water is just as bad."

U.S. ENVIRONMENTAL PROTECTION AGENCY,
Philadelphia, PA.

Jerry N. Johnson,
General Manager,
District of Columbia Water and Sewer Authority,
5000 Overlook Ave., SW,
Washington, DC.

DEAR MR. JOHNSON: As you may be aware, over the past several weeks, the United States Environmental Protection Agency Region III ("EPA") has been conducting an audit of the District of Columbia Water and Sewer Authority's ("DCWASA") compliance with the Lead and Copper Rule, specifically focusing on 40 C.F.R. Sections 141.84, 141.85 and 141.90. That compliance audit remains ongoing, and EPA is continuing to evaluate additional information as it becomes available.

Based on the information reviewed to date, EPA believes that DCWASA failed to comply with the provisions listed below. As EPA's continues to review DCWASA's compliance with the Lead and Copper Rule, EPA may identify other areas of noncompliance.

1. On information and belief, DCWASA failed to comply with the lead service line replacement sampling requirements of 40 C.F.R. § 141.84(d)(1), by failing to conduct follow-up sampling within 72 hours after the completion of the partial replacement of a lead service line during the compliance period ending September 30, 2003.

2. On information and belief, DCWASA failed to comply with the Public Education requirements of 40 C.F.R. § 141.85(b) by failing to use the required language for public service announcements submitted to television and radio stations for broadcasting during the 6-month compliance periods ending October 2002, April 2003, and October 2003.

3. On information and belief, DCWASA failed to comply with the Public Education requirements of 40 C.F.R. § 141.85(c)(2)(i) by failing to use the required language in notices inserted in each customer's water utility bill during August 2003.

4. On information and belief, DCWASA failed to conduct public service announcements every 6 months as required of 40 C.F.R. § 141.85(c)(3) during the compliance period beginning April 2003.

5. On information and belief, Respondent failed to submit tap water monitoring for lead and copper within the first 10 days following the end of the monitoring period ending June 30, 2002, as required of 40 C.F.R. § 141.90(a).

6. On information and belief, Respondent failed to comply with the Public Education reporting requirements of 40 C.F.R. § 141.90(f) by failing to send written documentation to EPA within 10 days after the end of each period in which the system is required to perform public education during the compliance period ending October 31, 2002.

If DCWASA believes it has not violated the provisions set forth above, or if DCWASA has any information relevant to its compliance with the provisions set forth above that it believes EPA should consider, please provide any relevant information to EPA within twenty-one (21) days of receipt of this letter. If this information has been provided in your response to the Information Request dated March 31, 2004, please note which response provides documentation of compliance. The information should be sent to: Karen D. Johnson (3WP32), Chief, Safe Drinking Water Act Branch, United States Environmental Protection Agency, Region III, 1650 Arch Street, Philadelphia, PA 19103–2029.

We appreciate your cooperation, and the cooperation of your staff, in connection with EPA's compliance audit. Please be aware that neither this letter nor EPA's decision to conduct a compliance audit limits EPA's ability to take an enforcement action against any person, including, but not limited to DCWASA. If you have any questions, please contact Karen Johnson at (215) 814–5445. Thank you for your cooperation in this matter.

Sincerely,

JON M. CAPACASA, *Director,*
Water Protection Division.

INFORMATION REQUEST

This information is requested pursuant to Section 1445(a) of the Safe Drinking Water Act, 42 U.S.C. § 300j-4(a). The Instructions and Definitions for responding to this Information Request are as follows:

A. INSTRUCTIONS & DEFINITIONS

1. A separate narrative response must be made for each question set forth below, and for each subpart of each question.

2. Precede each answer with the corresponding number of the question and subpart to which it responds.

3. Provide all documents in your possession which relate to the responses given. With respect to each document, identify the date, author, addressee, current location, and custodian and identify the question or subpart to which it relates.

4. Provide documents in both hard copy and electronic form, where available. The term "document" refers to "writings," "recordings" and "photographs" as those terms are defined in Rule 1001 of the Federal Rules of Evidence. Documents should be produced as they are kept in the usual course of business.

5. If any question cannot be answered in full, answer to the extent possible along with an explanation of why the question cannot be answered in full. If your responses are qualified in any manner, please explain.

6. If information or documents not known or not available to you as of the date of submission of your response to this request should later become known or available to you, you must supplement your response to EPA. Moreover, should you find at any time after the submission of your response that any portion of the submitted information is false or misrepresents the truth, you must notify EPA of this fact as soon as possible and provide a corrected response.

7. The term "you" or "your" refers to the District of Columbia Water and Sewer Authority ("DCWASA").

8. The term "LCR" refers to EPA's Lead and Copper Rule, 40 C.F.R. Sections 141.80–.90.

9. The term "lead service line" means "a service line made of lead which connects the water main to the building inlet and any lead pigtail, gooseneck or other fitting which is connected to such lead line." *See* 40 C.F.R. § 141.2.

10. To the extent you provide information in electronic format, contact Karen D. Johnson at (215) 814–5445 prior to providing the information in order to verify compatibility with EPA's equipment.

11. The following certification must accompany each submission pursuant to this request and must be signed by a management representative of DCWASA authorized to respond on behalf of DCWASA:

> "I certify that the information contained in or accompanying this submission is true, accurate, and complete. As to the identified portion(s) of this submission for which I cannot personally verify its truth and accuracy, I certify as the company official having supervisory responsibility for the person(s) who, acting under my direct instructions, made the verification, that this information is true, accurate, and complete. I am aware that there are significant penalties for submitting false information, including the possibility of fines and imprisonment."

12. All information shall be submitted within twenty-one (21) days of receipt of this Request for Information to: Karen D. Johnson (MC 3WP32), Chief, Safe Drinking Water Act Branch, United States Environmental Protection Agency, Region III, 1650 Arch Street, Philadelphia, PA 19103–2029.

B. REQUEST FOR INFORMATION

DCWASA is hereby required, pursuant to Section 1445(a) of the Safe Drinking Water Act, 42 U.S.C. § 300j-4(a), to submit the following information pursuant to the Instructions set forth above:

1. With regard to the person providing answers to these questions, State your name, business address, business telephone number and position with DCWASA.

2. With regard to any person who participated in or contributed to DCWASA's response to this Request for Information, provide that person's name, business address, business telephone number, and position with DCWASA, including whether the person is an employee or a contractor.

3. (a) Provide all results from lead sampling or monitoring performed on drinking water supplied by DCWASA in the District of Columbia since January 1, 1994. This request includes all results in the possession or control of DCWASA or its agents or representatives (including contractors) regardless of whether the sampling and/or analysis was performed by DCWASA, a representative or agent of DCWASA (including a contractor), a homeowner or building owner, a representative of the Washington Aqueduct, a representative of any other Federal or District of Columbia agency, or by any other person. This request seeks all results from lead sampling or monitoring in the possession or control of DCWASA or its agents or representatives (including contractors) regardless of whether the sample was taken from a building served by a known or suspected lead service line or not. This request seeks all results from lead sampling or monitoring in the possession or control of DCWASA or its agents or representatives (including contractors) regardless of whether the samples were required by EPA's Lead and Copper Rule ("LCR"), 40 C.F.R. § 141.80, *et seq.*, or whether the samples were used to calculate the 90th percentile pursuant to the LCR. This request seeks all results from lead sampling or monitoring in the possession or control of DCWASA or its agents or representatives (including contractors) regardless of whether or not those samples were invalidated.

(b) To the extent any samples taken to determine the concentration of lead in drinking water provided by DCWASA were invalidated, DCWASA shall identify the invalidated samples, the lead concentration of each sample, and provide an explanation as to why the samples were invalidated. Such explanation shall include the procedures followed for such invalidation, including but not limited to identifying who in DCWASA made the decsion to invalidate the sample, and who if anyone at EPA approved the invalidation. Provide copies of all approvals by EPA for any lead sample invalidation.

4. Identify all lead service lines within DCWASA's service area that were physically replaced from 1996 to the present for any reason, including lead service lines physically replaced in connection with normal maintenance and/or other road work.

(a) Provide all work orders, daily construction reports, or any other documents reflecting physical replacement of lead service lines from 1996 to the present.

(b) With respect to all lead service lines physically replaced from 1996 to the present, state what portions of the lead service lines were physically replaced and what portions were not physically replaced (for example, those portions downstream of the property line). When the lead service line was replaced only up to the property line, identify what, if any, portion was replaced by the homeowner and/or building owner.

(c) Provide the location of each lead service line that has been replaced. Identify all lead samples taken following full or partial replacement of lead service lines from 1996 to the present, including but not limited to those taken to comply with the requirements of 40 C.F.R. § 141.84(d)(1).

(d) With respect to each lead service line that has been replaced from 1996 to the present, State the reason the line was replaced (i.e., exceedance of EPA action level for lead of 0.015 mg/L, routine maintenance, etc.)

5. Provide the locations by address of all known or suspected lead service lines. This may be provided in electronic or written format.

6. (a) Provide the location of all lead service lines that have been tested for lead since 2000. This may be provided in electronic or written format.

(b) Identify each lead service line that has tested below the EPA action level for lead of 0.015 mg/L and been counted by DCWASA toward fulfilling the requirements of 40 C.F.R. § 141.84. This may be provided in electronic or written format.

(c) Provide the location of all lead service lines that have exceeded the EPA action level for lead of 0.015 mg/L. This may be provided in electronic or written format.

7. Identify the type and composition of any pipe, collar or shut off valve used for service line replacement since January 1, 1994.

8. Provide copies of all instructions provided by you to residents from December 1999 to the present for the purpose of obtaining samples for compliance with the lead action level under the LCR, sampling in connection with DCWASA's lead service line replacement program, sampling after physical replacement of a lead service line, or any other purpose related to sampling for lead in drinking water. Provide all versions of these instructions. To the extent the instructions changed over time, provide all versions and identify the timeframes in which each version of the instructions was used.

9. Provide all lead service line replacement sampling results from 2000 to the present, including the date the lead service line was replaced, when the sample was taken, when the sampling results were received from the laboratory, and the date the results were sent to the homeowner and/or residents served by the lead service line. Provide representative samples of all notification provided from 2000 to the present to homeowners and/or residents served by a service line that exceeded 0.015 mg/L of lead. Provide the addresses that received the notice and the dates of such notices.

10. Provide representative samples of all transmittals of lead sampling results to residents sent by you since December 1999, including the cover letter(s) and any attachment(s). To the extent different versions of the cover letter were used to transmit lead sampling results to different populations (i.e., residences sampled for compliance with the LCR, lead service line sampling, post-replacement sampling, or any other purpose), provide samples of each version. To the extent the wording of the transmittals changed over time, provide all versions and identify the timeframe(s) in which each version was used. State whether sample results were transmitted to all residences that were sampled or only to a subset (such as residences that tested over the LCR action level).

11. (a) Provide copies of all documents (including bill inserts) produced or distributed by you since December 2000 in any language for the purpose of educating the public about lead in drinking water.

(b) Produce all information regarding lead in drinking water used as a reference by persons answering telephone help lines for DCWASA in any language other than English since December 2000.

12. Provide copies of all newspaper advertisements you have purchased since December 2000 for the purpose of educating the public about lead in drinking water. Identify the newspaper(s) in which each advertisement was published and the date(s) of publication.

13. (a) Provide copies of all transmittal documents, letters or other documents since December 2001 that accompanied any document or public service announcement regarding lead in drinking water distributed by DCWASA to television stations, radio stations, newspapers, the Department of Health of the District of Columbia, libraries, hospitals, clinics, City Council or any other person or agency pursuant to 40 C.F.R. § 141.85(c).

(b) Provide copies of all public service announcements regarding the lead content of drinking water distributed by DCWASA since December 2001 to television stations, radio stations, newspapers, the Department of Health of the District of Columbia, libraries, hospitals, clinics, City Council or any other person or agency pursuant to 40 C.F.R. § 141.85(c).

(c) Identify the date each public service announcement was distributed and provide any document demonstrating the date each public service announcement was distributed.

14. Provide copies of all policies and/or procedures that DCWASA has for lead testing, lead service line replacement and public notification/education regarding the presence of lead in drinking water.

15. Provide copies of all preliminary, draft and final reports for all tap water monitoring for lead and copper submitted by DCWASA to EPA pursuant to 40 C.F.R. § 141.90(a) since December 2001.

16. Provide copies of all written documentation submitted by DCWASA to EPA pursuant to 40 C.F.R. § 141.90(f) since December 2001.

○